上海高等学校一流本科课程教材

INTRODUCTION TO OPTOELECTRONICS

光电子技术导论

第 2 版

主　编　郑继红　冯吉军

副主编　刘　一　袁　帅　刘学静

编　委（按姓氏笔画排序）

冯吉军　刘　一　刘学静

金钻明　郑继红　袁　帅

梁青青　游冠军

中国科学技术大学出版社

内容简介

光电子技术是光学和电子技术相结合而产生的新技术，涉及光电信息技术的方方面面，是未来信息产业发展的核心，在现代社会中起着举足轻重的作用。本书主要介绍了光学基本知识，半导体基础知识，发光二极管、半导体激光器等光电信息转换器件，光纤与光波导等波导器件，光敏二极管、电荷耦合器件等常用光电子器件，光电控制器件，以及相应的应用实例。此外，为拓宽学生知识面，本书还介绍了本领域内最新的研究进展和相关产业的发展等内容。本书可供相关专业本科生和研究生作为教材使用，也可作为高校教师的教学参考用书。

图书在版编目(CIP)数据

光电子技术导论：英文/郑继红，冯吉军主编. —2 版. —合肥：中国科学技术大学出版社，2023. 5

ISBN 978-7-312-05603-1

Ⅰ. 光…　Ⅱ. ①郑…②冯…　Ⅲ. 光电子技术—英文　Ⅳ. TN2

中国国家版本馆 CIP 数据核字(2023)第 027962 号

光电子技术导论

GUANGDIANZI JISHU DAOLUN

出版　中国科学技术大学出版社
安徽省合肥市金寨路 96 号，230026
http://press. ustc. edu. cn
https://zgkxjsdxcbs. tmall. com

印刷　安徽国文彩印有限公司

发行　中国科学技术大学出版社

开本　710 mm×1000 mm　1/16

印张　20. 75

字数　571 千

版次　2015 年 2 月第 1 版　2023 年 5 月第 2 版

印次　2023 年 5 月第 3 次印刷

定价　68. 00 元

Preface
前　言

随着社会经济全球化、教育国际化的不断深入,我国迫切需要培养一大批掌握光电信息工程专业知识、具有创新精神和实践能力、具备国际化视野的高素质工程专业人才。"光电子学"是光电信息科学与工程专业的专业基础课,开设双语"光电子学"课程对于光电专业学生的培养无疑是非常有意义的。光电子学是一门严谨、科学且应用广泛的学科,涉及光学、电子学、半导体物理等多个学科门类。光电子技术发展非常迅速,现有教材很难将这些新技术包含进去,这些技术(如大功率半导体激光器、集成光子芯片、太阳能电池等)的应用价值巨大,但多还处于研究阶段,很多理论和技术还不是十分成熟。因此,迫切需要一本均衡处理光电前沿技术与基本理论知识的教材,这也是本书编写的出发点。本书详细介绍了相关光电基础知识,并结合光电子技术的前沿研究领域,如半导体光源、光纤与光波导、光探测与显示技术等,进行了重点介绍。

本书的内容体系主要沿袭了上海理工大学光电信息与计算机工程学院为光电信息专业的学生教学所编撰的《光电信息技术》,并在《光电子技术导论》第1版的基础上,从国外相关书籍和互联网上搜集并更新了很多有益的英文资料,新增了一些当前光电领域热点技术的介绍,同时进一步丰富了本学院教师在科研工作中设计的光电信息探测系统的教学实例。为了便于学生自学,本书中还特意增加了中英文专业词汇生词对照表。

本书以"光电子学"课堂教学为着力点,重点解决"国际化"水平不高,"专业"和"思政"融合不足的问题,拓展教学时间与空间,将课堂内外、学校内外、线上线下相结合,力争实现全过程"课程思政"的教学改革。通过双语教学,使学生在掌握基础知识的同时提高专业英语能力,培养学生勇于探索与实践的科研素养,将光电专业知识传授、国际化能力培养与培养合格建设者和接班人的价值引领相结合。

本书为上海理工大学一流本科教材,其编写得到了上海理工大学研究生创新教材建设的大力支持，由上海理工大学郑继红、冯吉军主编。上海理工大学光电信息与计算机工程学院的部分研究生也参与了资料的搜集和整理工作，为书稿的最终完成做了很多细致的工作，在此一并向他们表示感谢。

由于时间仓促，书中难免存在疏漏之处，恳请广大读者批评指正，以便进一步修订改善。

编　者

2022年10月

Contents

目　录

Chapter 1
Physical Basis of Optoelectronic Information Technology

The development of the transistor and the ***integrated circuit*** (***IC***) has led to many remarkable capabilities. The IC permeates almost every aspect of our daily lives, including such things as the compact disk player, the fax machine, laser scanners at the grocery store, and the cellular telephone. The semiconductor electronics field continues to be a fast-changing one. Moreover, optics is a field of science which is particularly lucid, logical, challenging and beautiful. This chapter introduces basics of semiconductor, like energy bands, doping and so on. Fundamentals of optics, like reflection and refraction and basics of circuits are also introduced.

Before we start introducing the content of this chapter, let's first introduce Mozi, who made significant contributions to ancient physics in China.

Mozi (see Fig. 1-1), also known as Zhai, was a native of the State of Song in the late Spring and Autumn Period and the early Warring States Period. He was the only peasant philosopher in Chinese history who founded Mohism. Mohism had a great influence in the pre-Qin period and was called "eminent school" with Confucianism.

Fig. 1-1 Mozi

With universal love as the core, saving and advocating excellence as the fulcrum, he founded Mozi theory. He also established a set of scientific theories with geometry, physics and optics as outstanding achievements. The outstanding contribution of Mohism in science and technology is mostly recorded in the book *Mohism*. Mohism's thought of science and technology is manifested in the scientific and technological view of the unification of Taoism and technology.

The law of straight-line propagation of light, as it is interpreted today, means that light travels in a straight line in a uniform medium. Shadows are created because light cannot penetrate opaque material. The linear propagation of light has been widely used in ancient Chinese astronomical calendar. Our ancestors made sundials to measure the length and orientation of the sun's shadow to determine the time, the winter solstice, and the summer solstice. A speculum is mounted on an astronomical instrument to observe the sky and measure the position of the stars.

1.1 Introduction of Theoretical Semiconductor Basis

This section gives a brief introduction to semiconductors. Semiconductors play a vital role in optics both as sources and as detectors of light. The ***light-emitting diode*** (***LED***) and ***laser diode*** (***LD***) are widely used as the various forms of ***photodiode*** detector. ***Electrons*** and ***holes*** are carriers of electrical current in semiconductors and they are separated by an ***energy gap***. ***Photons*** are the smallest energy packets of light waves and their interaction with electrons is the key physical mechanism in optoelectronic devices.

1.1.1 Energy Bands and Electrical Conduction

A semiconductor is a material which has electrical conductivity to a degree between that of a metal (such as copper) and that of an ***insulator*** (such as glass). Semiconductors are the foundation of modern electronics, including ***transistors***, ***solar cells***, light-emitting diodes, ***quantum dots*** and digital and analog integrated circuits.

A semiconductor may have a number of unique properties, one of which is the ability to change conductivity by the addition of impurities (***doping***) or by interaction with another phenomenon, such as an electric field or light; this ability makes a semiconductor very useful for constructing a device that can amplify, switch, or convert an energy input. The modern understanding of the properties of a semiconductor relies on quantum physics to explain the movement of electrons inside a lattice of atoms.

Semiconductors are defined by their unique electric conductive behavior, somewhere between that of a metal and an insulator. The differences between these materials can be understood in terms of the quantum states for electrons, each of which may contain zero or one electron (by the ***Pauli exclusion principle***). These states are associated with the electronic band structure of the material. Electrical conductivity arises due to the presence of electrons in states that are delocalized (extending through the material). However, in order to transport electrons, a state must be partially filled, and it contains an electron only part of the time. If the state is always occupied with an electron, then it is inert, blocking the passage of other electrons via that state. The energies of these quantum states are critical, since a state is partially filled only if its energy is near to the ***Fermi level***.

High conductivity in a material comes from it, having many partially filled states and much state delocalization. Metals are good electrical conductors and have many partially filled states with energies near their Fermi level. Insulators, by contrast, have few partially filled states, their Fermi levels sit within band gaps with few energy states to occupy. Importantly, an insulator can be made to conduct by increasing its temperature: heating provides energy to promote some electrons across the band gap, inducing partially filled states in both the band of states beneath the band gap (***valence band***) and the band of states above the band gap (***conduction band***). An (intrinsic) ***semiconductor*** has a band gap which is smaller than that of an insulator and at room temperature significant numbers of electrons can be excited to cross the band gap.

A ***intrinsic*** (***pure***) ***semiconductor*** is not very useful, as it is neither a very good insulator nor a very good conductor. However, one important feature of semiconductors (and some insulators, known as semi-insulators) is that their conductivity can be increased and controlled by doping with impurities and gating with electric fields. By doping and gating, either the conduction or valence band are moved much closer to the Fermi level, and the number of partially filled states are greatly increase.

Some semiconductor materials which have wider-band gap are sometimes referred to as semi-insulators. When undoped, these have electrical conductivity nearer to that of electrical insulators; when doped, they are useful as semiconductors. Semi-insulators find niche applications in micro-electronics, such as substrates for high electron mobility transistor (HEMT). An example of a common semi-insulator is gallium arsenide. Some materials, such as titanium dioxide, can even be used as insulating materials for some applications, while being treated as wide-gap semiconductors for other applications.

1.1.2 Charge Carriers (Electrons and Holes)

The partial filling of the states at the bottom of the conduction band can be understood as adding electrons to that band. The electrons do not stay indefinitely (due to the natural thermal recombination) but they can move around for some time. The actual concentration of electrons is typically very dilute, and so (unlike in metals) it is possible to think of the electrons in the conduction band of a semiconductor as a sort of classical ideal gas, where the electrons fly around freely without being subject to the Pauli exclusion principle. In most semiconductors, the conduction bands have a parabolic dispersion relation, and so these electrons respond to forces (electric field, magnetic field, etc.) much like they would in a vacuum, though with a different effective mass. Because the electrons behave like an ideal gas, one may also think about conduction in very simplistic terms such as the Drude model, and introduce concepts such as electron mobility.

For partial filling at the top of the valence band, it is helpful to introduce the concept of an electron hole. Although the electrons in the valence band are always moving around, a completely full valence band is inert, not conducting any current. If an electron is taken out of the valence band, then the trajectory that the electron would normally have taken is now missing its charge. For the purposes of electric current, this combination of the full valence band, minus the electron, can be converted into a picture of a completely empty band which contains a positively charged particle that moves in the same way as the electron.

Combined with the negative effective mass of the electrons at the top of the valence band, we arrive at a picture of a positively charged particle that responds to electric and magnetic fields just as a normal positively charged particle would do in vacuum, again with some positive effective mass. This particle is called a hole, and the collection of holes in the valence band can again be understood in simple classical terms (as with the electrons in the conduction band).

1.1.3 Carrier Generation and Recombination

When ionizing radiation strikes a semiconductor, it may excite an electron out of its ***energy level*** and consequently leave a hole. This process is known as ***electron-hole pair*** generation. Electron-hole pairs are constantly generated from thermal energy as well, in the absence of any external energy source.

Electron-hole pairs are also apt to recombine. Conservation of energy demands that these recombination events, in which an electron loses an amount of energy larger than the band gap, be accompanied by the emission of thermal energy (in the form of phonons) or radiation (in the form of photons).

In some states, the generation and recombination of electron-hole pairs are in equipoise. The number of electron-hole pairs in the steady state at a given temperature is determined by quantum statistical mechanics. The precise quantum mechanical mechanisms of generation and recombination are governed by conservation of energy and conservation of momentum.

As the probability that electrons and holes meet together is proportional to the product of their amounts, the product is in steady state nearly constant at a given temperature, providing that there is no significant electric field (which might "flush" carriers of both types, or move them from neighbour regions containing more of them to meet together) or externally driven pair generation. The product is a function of the temperature, as the probability of getting enough thermal energy to produce a pair increases with temperature, being approximately $\exp(-E_G/kT)$, where k is Boltzmann's constant, T is absolute temperature and E_G is band gap.

The probability of meeting is increased by carrier traps — impurities or dislocations which can trap an electron or hole and hold it until a pair is completed. Such carrier traps are sometimes purposely added to reduce the time needed to reach the steady state.

1.1.4 Doping[①]

By introducing small amounts of impurities into an otherwise-pure crystal, it is possible to obtain a semiconductor in which the concentration of carriers of one polarity is much in excess of the other type. Such semiconductors are referred to as extrinsic semiconductors vis-à-vis the intrinsic case of a pure and perfect crystal. For example, by adding pentavalent impurities, such are arsenic, which have a valence one more than Si, we can obtain a semiconductor in which the electron concentration is much larger than the hole concentration. In this case, we will have an ***n-type semiconductor***. If we add trivalent impurities, such as boron, which have a valence of one less than four, we then have an excess of holes over electrons — a ***p-type semiconductor***.

An arsenic (As) atom has five valence electrons whereas Si has four. When the Si crystal is doped with small amounts of As, each As atom substitutes for one Si atom and is surrounded by four Si atoms. When an As atom bonds with four Si atoms, it has one electron left unbounded. This fifth electron cannot find a bond to go into so it is left orbiting around the As atom, which looks like an As^+, as illustrated in Fig. 1-2(a). The As^+ ionic center, with an electron e^- orbiting it, resembles a hydrogen atom in a silicon environment. We can easily calculate how much energy is required to free this electron away from the As site, thereby ionizing the As ***impurity*** by using our knowledge on the ionization of a hydrogen atom (removing the electron from the H-atom). This energy turns out to be a few hundredths of an electron volt, that is, ~ 0.05 eV, which is comparable to the thermal energy at room temperature ($\sim k_B T = 0.025$ eV). Thus, the fifth valence electron can be readily freed by thermal vibrations of the Si lattice. The electron will then be "free" in the semiconductor, or in other words, it will be in the CB. The energy required to excite the electron to the CB is therefore ~ 0.05 eV. The addition of As atoms introduces localized electronic states at the As sites because the fifth electron has a localized wave function, of

① Kasap S O. Optoelectronics and Photonics: Principles & Practices[M]. 2nd ed. New Jersey: Prentice Hall, Inc., 2012: 203-205.

the hydrogenic type, around As^+. The energy of these states, E_d, is ~0.05 eV below E_c because this is how much energy is required to take the electron away into the CB. Thermal excitation by lattice vibrations at room temperature is sufficient to ionize the As atom, that is, excite the electron from E_d into CB. This process creates free electrons; however, the As^+ ions remain immobile as shown in the energy band diagram of an n-type semiconductor in Fig. 1-2(b).

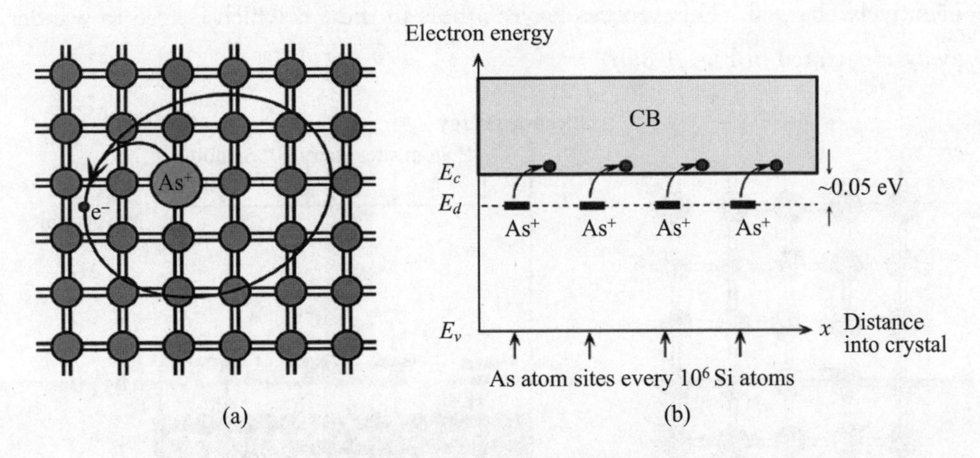

Fig. 1-2 Si crystal is doped with small amounts of As

(a) The four valence electrons of As allow it to bond just like Si but the fifth electron is left orbiting the As site. The energy required to release to free fifth-electron into the CB is very small. (b) Energy band diagram for an n-type Si doped with 1 ppm As. There are donor energy levels just below E_c around As^+ sites.

We should, by similar arguments to the above, anticipate that doping a Si crystal with a trivalent atom (valence of 3) such as B (boron) will result in a p-type Si that has an excess of holes in the crystal. Consider doping Si with small amounts of B as shown in Fig. 1-3(a). Because B has only three valence electrons, when it shares them with four neighboring Si atoms one of the bonds has a missing electron which is of course a "hole". A nearby electron can tunnel into this hole and displace the hole further away from the B atom. As the hole moves away it gets attracted by the negative charge left behind on the B atom. The binding energy of this hole to the B^- ion (a B atom that has accepted an electron) can be calculated using the hydrogenic atom analogy just like in the n-type Si case. This binding energy also turns out to by very small, ~0.05eV, so that at room temperature the thermal vibrations of the lattice can free the hole away from the B^- site, A free hole, we recall, exists in the VB. The escape of the hole from the B^- site involves the B atom accepting an electron from a neighboring Si-Si bond (from the VB) which effectively results in the hole being displaced away, and its eventual escape to freedom in the VB. The B atom introduced into the Si crystal therefore acts as an electron acceptor impurity. The electron accepted by the B atom comes from a nearby bond. On the energy band diagram, an electron leaves the VB and gets accepted by a B atom which becomes negatively charged. This process leaves a hole in the VB which is free to wander away illustrated in Fig. 1-3(b).

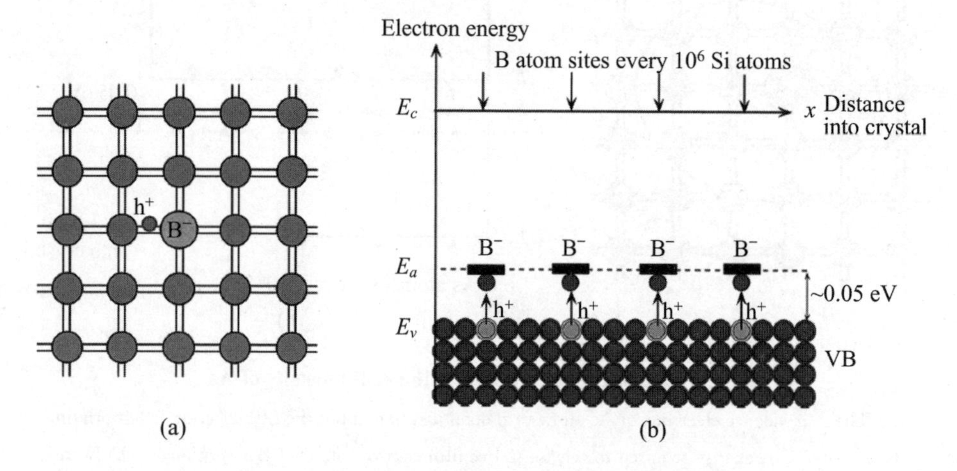

Fig. 1-3 Si crystal is doped with small amounts of B

(a) Boron doped Si crystal. B has only three valence electrons. When it substitutes for a Si atom one of its bonds has an electron missing and therefore a hole. (b) Energy band diagram for a p-type Si doped with 1 ppm B. There are acceptor energy levels just above E_v around B^- sites. These acceptor levels accept electrons from the VB and therefore create holes in the VB.

1.1.5 P-N Junction

A ***p-n junction*** (see Fig. 1-4) is a boundary or interface between two types of semiconductor material, p-type and n-type, inside a single crystal of semiconductor. It is created by doping, for example by ion implantation, diffusion of ***dopants***, or by epitaxy (growing a layer of crystal doped with one type of dopant on top of a layer of crystal doped with another type of dopant). If two separate pieces of material were used, this would introduce a grain boundary between the semiconductors that severely inhibits its utility by scattering the electrons and holes.

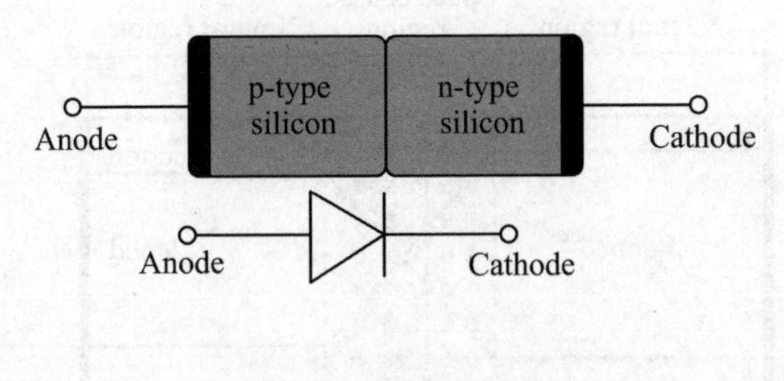

Fig. 1-4 A p-n junction

The circuit symbol is shown: the triangle corresponds to the p side.

P-n junctions are elementary "building blocks" of most semiconductor electronic devices such as diodes, transistors, solar cells, LEDs, and integrated circuits. They are the active sites where the electronic action of the device takes place. For example, a common type of transistor, the bipolar junction transistor, consists of two p-n junctions in series, in the form n-p-n or p-n-p.

The discovery of the p-n junction is usually attributed to American physicist Russell Ohl of Bell Laboratories. A Schottky junction is a special case of a p-n junction, where metal serves the role of the p-type semiconductor.

In a p-n junction, without an external applied voltage, an equilibrium condition is reached in which a potential difference is formed across the junction. This potential difference is called built-in potential V_{bi}.

After joining p-type and n-type semiconductors, electrons from the n region near the p-n interface tend to diffuse into the p region. As electrons diffuse, they leave positively charged ions (donors) in the n region. Likewise, holes from the p-type region near the p-n interface begin to diffuse into the n-type region, leaving fixed ions (acceptors) with negative charge. The regions nearby the p-n interfaces lose their neutrality and become charged, forming the ***space charge region*** (***depletion layer***) (see Fig. 1-5).

The electric field created by the space charge region opposes the diffusion process for both electrons and holes. There are two concurrent phenomena: the diffusion process that tends to generate more space charge, and the electric field generated by the space charge that tends to counteract the diffusion. The carrier concentration profile at equilibrium is shown in Fig. 1-5. Also shown are the two counterbalancing phenomena that establish equilibrium.

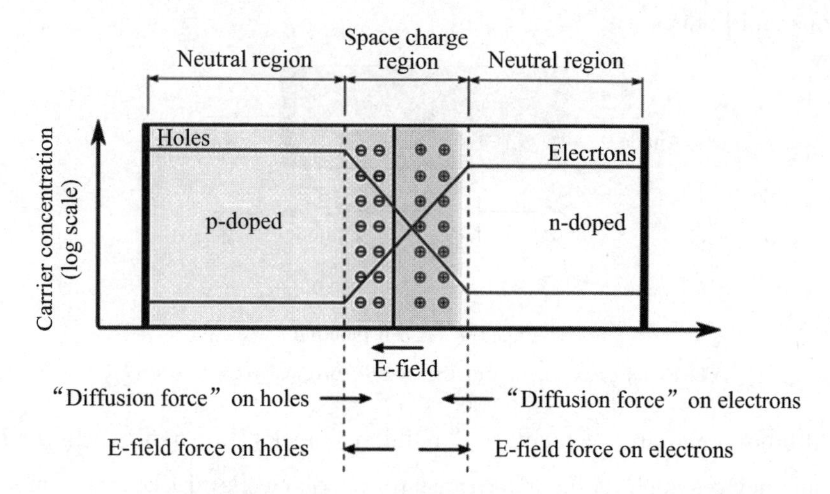

Fig. 1-5 A p-n junction in thermal equilibrium with zero-bias voltage applied

Electron and hole concentration are reported with blue and red lines, respectively. Gray regions are charge-neutral. Light-red zone is positively charged. Light-blue zone is negatively charged. The electric field is shown on the bottom, the electrostatic force on electrons and holes and the direction in which the diffusion tends to move electrons and holes.

The space charge region is a zone with a net charge provided by the fixed ions (donors or acceptors) that have been left uncovered by majority carrier diffusion. When equilibrium is reached, the charge density is approximated by the displayed step function. In fact, the region is completely depleted of majority carriers (leaving a charge density equal to the net doping level), and the edge between the space charge region and the neutral region is quite sharp [see Fig. 1-6, $Q(x)$ graph]. The space charge region has the same magnitude of charge on both sides of the p-n interfaces, thus it extends farther on the less doped side (the n side in Fig. 1-5 and Fig. 1-6).

In forward bias, as shown in Fig. 1-7, the p-type is connected with the positive terminal and the n-type is connected with the negative terminal. With a battery connected this way, the holes in the p-type region and the electrons in the n-type region are pushed toward the junction. This reduces the width of the depletion zone. The positive charge applied to the p-type material repels the holes, while the negative charge applied to the n-type material repels the electrons. As electrons and holes are pushed toward the junction, the distance between them decreases. This lowers the barrier in potential. With increasing

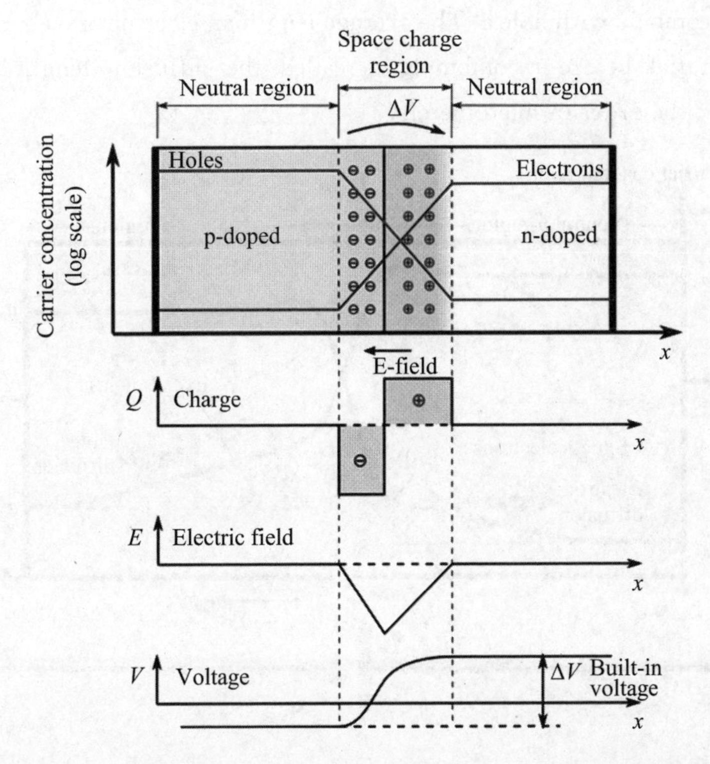

Fig. 1-6 A p-n junction in thermal equilibrium with zero-bias voltage applied

Under the junction, plots for the charge density, the electric field, and the voltage are reported.

forward-bias voltage, the depletion zone eventually becomes thin enough that the zone's electric field cannot counteract charge carrier motion across the p-n junction, as a consequence reducing electrical resistance. The electrons that cross the p-n junction into the p-type material (or holes that cross into the n-type material) will diffuse in the near-neutral region. Therefore, the amount of minority diffusion in the near-neutral zones determines the amount of current that may flow through the diode.

Only majority carriers (electrons in n-type material or holes in p-type) can flow through a semiconductor for a macroscopic length. With this in mind, consider the flow of electrons across the junction. The forward bias causes a force on the electrons pushing them from the n side toward the p side. With forward bias, the depletion region is narrow enough that electrons can cross the junction and inject into the p-type material. However, they do not continue to flow through the p-type material indefinitely, because it is energetically favorable for them to recombine with holes. The average length an electron travels through the p-type material before recombining is called the diffusion length, and it is typically on the order of micrometers.

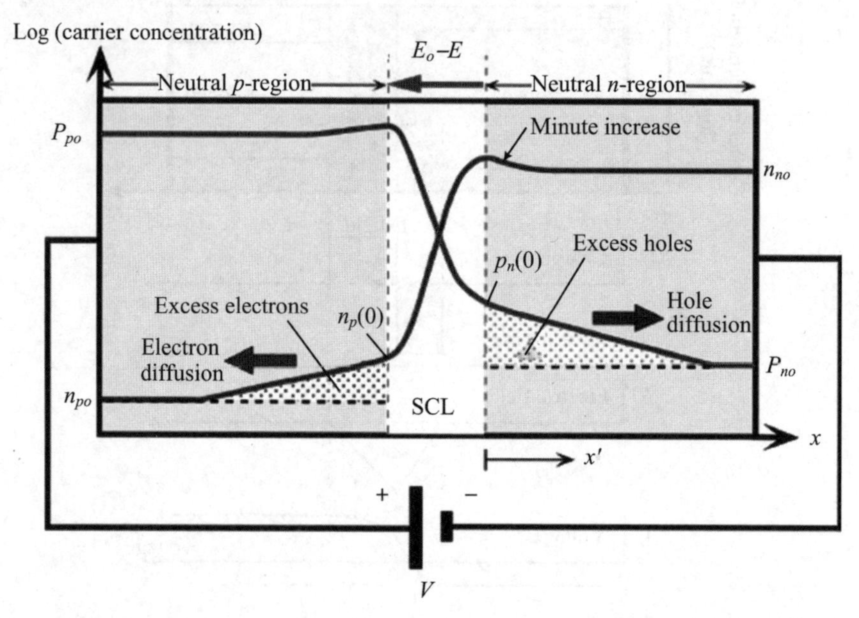

Fig. 1-7 Forward biased p-n junction and the injection of minority carriers①

Although the electrons penetrate only a short distance into the p-type material, the electric current continues uninterrupted, because holes (the majority carriers) begin to flow in the opposite direction. The total current (the sum of the electron and hole currents) is constant in space, because any variation would cause charge buildup over time (this is Kirchhoff's current law). The flow of holes from the p-type region into the n-type region is exactly analogous to the flow of electrons from n to p (electrons and holes swap roles and the signs of all currents and voltages are reversed).

Therefore, the macroscopic picture of the current flow through the diode involves electrons flowing through the n-type region toward the junction, holes flowing through the p-type region in the opposite direction toward the junction, and the two species of carriers constantly recombining in the vicinity of the junction. The electrons and holes travel in opposite directions, but they also have

① Kasap S O. Optoelectronics and Photonics: Principles & Practices[M]. 2nd ed. New Jersey: Prentice Hall, Inc., 2012:21.

opposite charges, so the overall current is in the same direction on both sides of the diode, as required.

1.1.6 Energy Level[①]

All matter ultimately consists of atoms. Each atom, in turn, consists of a nucleus surrounded by electrons. For the sake of simplicity, you can use a solar model of an atom, where electrons rotate on different orbits around the nucleus. The fact that the nucleus is approximately 10^5 times smaller than the whole atom might help you visualize the model. Bear in mind that this model, developed by Niels Bohr at the dawn of the quantum physics era (1913), does not describe atomic properties as we understand them today, but it does facilitate a presentation of the basic ideas.

Bohr's model assumes that electrons rotate on stationary orbits and therefore possess a stationary value of energy. Bohr's breakthrough was the assumption that rotating electrons do not radiate. That is, they do not change their energy value during rotation, as classic electromagnetic theory suggests. Any change in energy occurs only discretely, such as when electrons jump from one orbit to another. This implies that an entire atom possesses discrete values of energy; in other words, an atoms's energy is quantized.

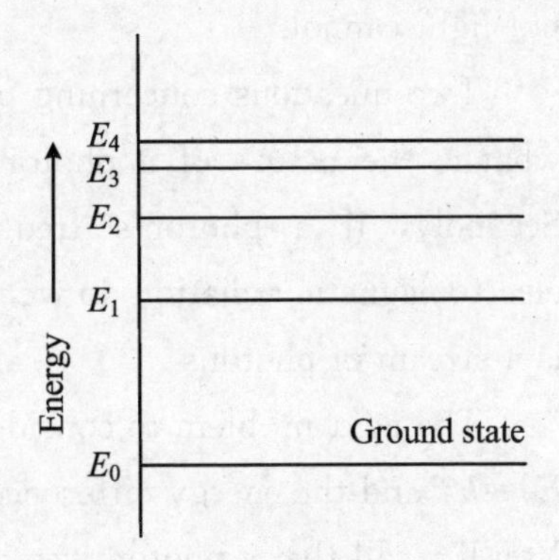

Fig. 1-8 Energy-level diagram

Fig. 1-8 introduces all energy-level diagram. Observe that there is only a vertical axis in the diagram and this axis shows an energy value. While it is normal for the position of the horizontal axis to represent an energy level, its length means nothing.

Possible discrete energy values (that is, those that are allowed by the law of quantum physics) are always separated by energy gaps. The lowest energy level is called the ***ground state***. An atom can be at any of these levels, or states, and it can change its energy only by jumping from one level to another. In other words,

① Kasap S O. Optoelectronics and Photonics: Principles & Practices[M]. New Jersey: Prentice Hall, Inc. , 2001: 195-200.

it can change its energy level only discretely. It should be emphasized that there can be no smooth transition between these states. An atom can have as its energy value, say, E_2 or E_3, but nothing between E_2 and E_3.

What happens if an atom jumps from an ***upper level*** to a ***lower level***, say, from level E_3 to level E_2? There is an energy gap between these two levels, $\Delta E = E_3 - E_2$, and this difference will be released as a quantum of energy, which is called a photon. You can think of a photon as a particle — not a mechanical particle like a speck of dust, for instance, but as an elementary particle that carries a quantum of energy, E_P, and that travels with the speed of light, c. A photon's energy, E_P, is defined as follows:

$$E_P = hf \tag{1-1}$$

where h is ***Planck's constant*** and f is the photon's frequency. Equation (1-1) introduces one of the fundamental concepts of modem physics: The energy of a photon, which is an elementary particle, depends on its frequency, which we always associate with waves. Also, remember that the higher the photon's frequency, the more energy it carries. That's why X-rays can penetrate our body but light cannot.

Two questions concerning photons must be considered at this point: First, what is the nature of a photon? Simply put, it is ***electromagnetic radiation***. Secondly, if a photon's frequency, f, is about 10^{14} Hz, what sort of electromagnetic radiation do we mean? Simply put again, it is light. Hence, light is a stream of photons.

The next problem to consider is the relationship between a photon's energy, $E_P = hf$ and the energy difference, $\Delta E = E_3 - E_2$, of the energy levels E_3 and E_2. You'll recall that a photon was created when an atom jumped from E_3 to E_2 and released energy $(E_3 - E_2)$. Therefore,

$$E_P = \Delta E = E_3 - E_2 \tag{1-2}$$

Meanwhile, $E_P = hf$; hence, $hf = E_3 - E_2$ and $f = (E_3 - E_2)/h$. On the other hand, $\lambda = c/f$, Therefore,

$$\lambda = ch/(E_3 - E_2) \tag{1-3}$$

However, the product ch is the constant, and the only variable in equation (1-2) is the energy gap $\Delta E = E_3 - E_2$. Since a stream of photons makes light, we arrive at this very important conclusion: The wavelength (the color) of radiated light is determined by the energy levels of the radiating material.

You may wonder whether we can change the energy levels of a given material to obtain a desired color of radiated light. The answer is no. These energy levels are given by nature and we cannot control them. But we can choose another material to achieve different colors of radiated light.

Atoms want to exist at the lowest possible energy levels, that's the law of nature. To raise them to higher levels, which is necessary for atoms to be able to jump down to produce light radiation, we must energize them from an external source. When atoms absorb external energy, they jump to the higher energy levels and then drop to the lower levels, radiating photons — that is, light. The process of making atoms jump to the higher levels by feeding them external energy is called ***pumping***.

Fig. 1-9 demonstrates these processes. Keep in mind that this illustration is no more than a convenient model and is only a representation of the real pumping and radiation processes. To visualize these processes in everyday life, look at a lamp in your room. You know, of course, that no light radiates when the switch is off. But did you ever wonder why? When the switch is on, electrical energy is delivered to the light-source material (gas in luminescent lamps and a filament in light bulbs), atoms absorb this energy, jump to the upper energy levels, and then drop down to the lower levels, radiating light.

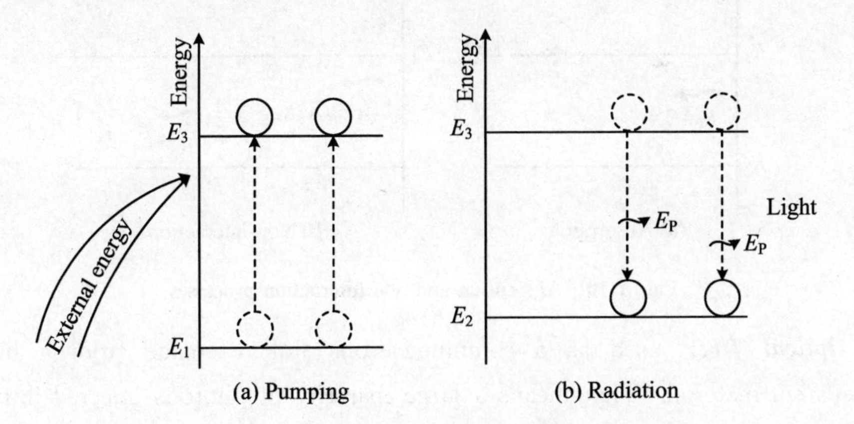

Fig. 1-9 Pumping and radiation processes

Why atoms jump to the higher energy levels when they absorb energy from an external source is the next obvious question. Suppose an atom is at level E_1 as shown in Fig. 1-9 (a), that means the atom possesses an energy value of E_1. Now it absorbs external energy in the amount of $\Delta E = E_2 - E_1$. Its new energy becomes equal to $E_1 + \Delta E = E_1 + (E_2 - E_1) = E_2$. In reality, after absorbing external energy, atoms have a new energy value, which we can demonstrate by placing them at different energy levels in our energy-level diagram. In other words, an energy-level diagram is a convenient model that helps us understand the pumping and radiation processes by visualizing them.

What happens if an external photon (light) strikes a medium? If its energy, $E_P = hf$, is equal to the energy gap, ΔE, the photon will be absorbed by an atom and the atom will jump to the appropriate higher level. If E_P is not equal to ΔE, the photon will pass by the material without interaction. Fig. 1-10 demonstrates both absorption and non-interaction processes.

You know, of course, that sunlight is not as bright inside a room as it is outside. But why? It's because the light is partially absorbed by the window and completely absorbed by the wall. But what happens to the energy that the sun's photons transmit to the window and the wall? This light energy is absorbed by these objects, which become warmer as a result. Analyze your own everyday experience in terms of the absorption and non-interaction processes shown in Fig. 1-10.

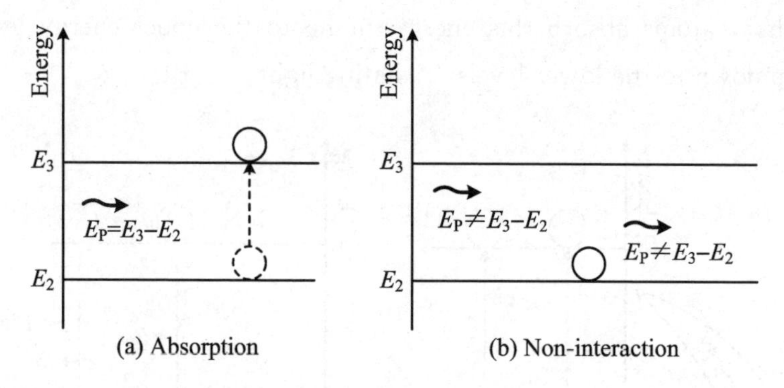

Fig. 1-10 Absorption and non-interaction processes

Optical fiber used as a communications link is made from a highly transparent material. That means a large majority of photons injected into the fiber by an LED or LD will travel through it without interacting with the fiber material. But some impurities have energy gaps close to the energy level of the photons, and that poses a problem.

Another important point to note about absorption and non-interaction processes is conveyed in Fig. 1-10(a), which shows that we can use photons (light, remember) to pump atoms to upper energy levels.

Example 1.1.1

Problem

Suppose you use an LED whose energy gap equals 2.5 eV. What color will it radiate?

Solution

When we calculate radiating wavelength, we need to know only energy gap ΔE, that is, the energy difference between the upper and the lower energy levels involved in the process. Atomic energy and energy gaps are measured in electron-volts (eV). 1 eV$=1.602\times10^{-19}$ J (note how small 1 eV is). Note also that ΔE (eV)$=\Delta E$ (joules/e), where e, the electron (unit) charge, equals 1.602×10^{-19} coulomb.

Typical energy gaps of semiconductor materials used in fiber-optic LEDs, LDs, and PDs are in units of eV, which means on the order of 10^{-19} J. This is why eV is a common energy unit when dealing with atomic processes.

Now, back to the problem. Since $\Delta E=2.5$ eV and $E_P=\Delta E$, then $E_P=2.5$ eV. On the other hand, $E_P=hf=hc/\lambda$; therefore,

$$\lambda = h/E_P$$

where E_P is measured in joules.

$$ch = 3\times10^8\ \text{m/s}\times6.6261\times10^{-34}\ \text{J}\cdot\text{s}\approx20\times10^{-26}\ \text{m}\cdot\text{J}$$

Converting E_P from eV to J:

$$E_P = 2.5\ \text{eV}\times1.602\times10^{-19}\approx4\times10^{-19}\ \text{J}$$

Substituting numbers in the above formula, we get:

$$\lambda = hc/E_P = 20\times10^{-26}\ \text{m}\cdot\text{J}/4\times10^{-19}\ \text{J} = 5\times10^{-7}\ \text{m} = 500\ \text{nm}$$

We can say that the color is green according to the electromagnetic spectrum.

This type of calculation is made even easier if we simply use three constants: h, c, and the conversion coefficient from eV to J. Now let's redo this calculation but more quickly:

$$\lambda(\text{m}) = hc/1.602\times10^{-19}E_P(\text{eV})$$
$$= 20\times10^{-26}\ \text{m}\cdot\text{J}/1602\times10^{-19}E_P(\text{eV}) = 12.48\times10^{-7}/E_P(\text{eV})$$

Hence, you can use the formula:

$$\lambda(\text{nm}) = 1248/E_{\text{P}}(\text{eV})$$

where E_{P} is in eV and λ will be in nm.

New Words and Expressions

[1] **integrated circuit (IC)**：集成电路。它是一种微型电子器件或部件，通过采用一定的工艺，把一个电路中所需的晶体管、二极管、电阻、电容和电感等元件及布线互连在一起，制作在一小块或几小块半导体晶片或介质基片上，然后封装在一个管壳内，成为具有所需电路功能的微型结构。

[2] **semiconductor**：半导体。在常温下，其导电性能介于导体(conductor)与绝缘体(insulator)之间。

[3] **light-emitting diode (LED)**：发光二极管。它是由镓(Ga)、砷(As)、磷(P)、氮(N)、铟(In)的化合物制成的二极管，当电子与空穴复合时能辐射出可见光，因而可以用来制成发光二极管。在电路及仪器中作为指示灯，或者组成文字或数字显示。

[4] **laser diode (LD)**：半导体激光器。它是以一定的半导体材料作为工作物质而产生受激发射作用的器件。

[5] **photodiode**：光电二极管。它是由一个 pn 结组成的半导体器件，具有单方向导电特性。但在电路中它不是整流元件，而是把光信号转换成电信号的光电传感器件。

[6] **electrons**：电子。

[7] **holes**：空穴，在固体物理学中指共价键上流失一个电子，最后在共价键上留下空位的现象。

[8] **energy gap**：能隙。在固态物理学中泛指半导体或绝缘体的价带顶端至传导带底端的能量差距。

[9] **photon**：光子。

[10] **insulator**：绝缘体。

[11] **transistor**：晶体管。它是一种固体半导体器件，可以用于检波、整流、放大、开关、稳压、信号调制等。

[12] **solar cell**：太阳能电池。它是通过光电效应或者光化学效应直接把光能转化为电能的装置。

[13] **quantum dot**：量子点。它是准零维的纳米材料，由少量原子构成。

[14] **doping**：掺杂。

[15] **Pauli exclusion principle**：泡利不相容原理。

[16] **Fermi level**：费米能级。

[17] **valence band**：价带。通常是指在半导体或绝缘体中，在绝对零度下能被电子占满的最高能带。

[18] **conduction band**：导带。它是由自由电子形成的能量空间，即固体结构内自由运动的电子所具有的能量范围。

[19] **optical fiber**：光纤。它是一种利用光在玻璃或塑料制成的纤维中的全反射原理而制成的光传导工具。

[20] **energy level**：能级。

[21] **electron-hole pair**：电子-空穴对。

[22] **impurity**：杂质。

[23] **dopant**：掺杂剂。

[24] **intrinsic (pure) semiconductor**：纯半导体。

[25] **p-type semiconductor**：p 型半导体，也称为空穴型半导体，是指空穴浓度远大于自由电子浓度的杂质半导体。

[26] **n-type semiconductor**：n 型半导体，也称为电子型半导体，是指自由电子浓度远大于空穴浓度的杂质半导体。

[27] **p-n junction**：pn 结。采用不同的掺杂工艺，通过扩散作用，将 p 型半导体与 n 型半导体制作在同一块半导体（通常是硅或锗）基片上，它们的交界面就形成了空间电荷区，称为 pn 结。pn 结具有单向导电性，是电子技术中许多元件，如半导体二极管、双极性晶体管的物质基础。

[28] **space charge region (depletion layer)**：空间电荷区（耗尽层）。

[29] **forward-bias voltage**：正向偏置电压。

[30] **ground state**：基态。

[31] **upper level**：上能级。

[32] **lower level**：下能级。

[33] **Planck's constant**：普朗克常数。它是一个物理常数，用以描述量子的大小，约为 6.62×10^{-34} J · s。

[34] **electromagnetic radiation**：电磁辐射。

[35] **pumping**：泵浦。在激光器中，外部能量通常会以光或电流的形式输入产生激光的媒质中，把处于基态的电子激励到较高的能级和高能态，物理学家将这种状态称为激发态。

1.2 Optical Basis[①]

Light is an ***electromagnetic wave***: Light is emitted and absorbed as a stream of discrete photons, carrying packets of energy and momentum. How can these two statements be reconciled? Similarly, while light is a wave, it nevertheless travels along straight lines or rays, allowing us to analyze lenses and mirrors in terms of geometric optics. Can we use these descriptions of waves, rays and photons interchangeably, and how should we choose between them? These problems, and their solutions, recur throughout this book, and it is useful to start by recalling how they have been approached as the theory of light has evolved over the last three centuries.

1.2.1 The Nature of Light

In his famous book *Opticks*, published in 1704, Isaac Newton described light as a stream of particles or corpuscles. This satisfactorily explained rectilinear propagation, and allowed him to develop theories of ***reflection*** and ***refraction***, including his experimental demonstration of the splitting of sunlight into a spectrum of colours by using a prism. The particles in rays of different colours were supposed to have different qualities, possibly of mass, size or velocity. White light was made up of a compound of coloured rays, and the colours of transparent materials were due to selective absorption. It was, however, more difficult for him to explain the coloured ***interference*** patterns in thin films, which we called Newton's rings. For this, and for the partial reflection of light at a glass surface, he suggested a kind of periodic motion induced by his corpuscles, which reacted on the particles to give "fits of easy reflection and transmission". Newton also realized that double refraction in a calcite crystal (Iceland spar) was best explained by attributing a rectangular cross-section (or "sides") to light

① Smith F G, King T A, Wilkins D. Optics and Photonics: An Introduction[M]. 2nd ed. New York: John Wiley & Sons, Inc., 2007:1-3.

rays, which we would now describe as ***polarization***. He nevertheless argued vehemently against an actual wave theory, on the grounds that waves would spread in angle rather than travel as rays, and that there was no medium to carry light waves from distant celestial bodies.

The idea that light was propagated as some sort of wave was published by René Descartes in *La Dioptrique* (1637); he thought of it as a pressure wave in an elastic medium. Christiaan Huygens, a Dutch contemporary of Newton, developed the wave theory; his explanation of rectilinear propagation is now known as "Huygens' construction". He correctly explained refraction in terms of a lower velocity in a denser medium. Huygens' construction is still a useful concept, and we use it later in this chapter.

It was not, however, until 100 years after Newton's *Opticks* that the wave theory was firmly established and the wavelength of light was found to be small enough to explain rectilinear propagation. In ***Thomas Young's double slit experiment***, monochromatic light from a small source passed through two separate slits in an opaque screen, creating interference fringes where the two beams overlapped; this effect could only be explained in terms of waves. Augustin Fresnel, in 1821, then showed that the wave must be a transverse oscillation, as contrasted with the longitudinal oscillation of a sound wave; following Newton's ideas of rays with "sides", this was required by the observed polarization of light as in double refraction. Fresnel also developed the theories of partial reflection and transmission, and of ***diffraction*** at shadow edges. The final vindication of the wave theory came with James Clerk Maxwell, who synthesized the basic physics of electricity and magnetism into the four Maxwell equations, and deduced that an electromagnetic wave would propagate at a speed which equaled that of light.

The end of the nineteenth century therefore saw the wave theory on an apparently unassailable foundation. Difficulties only remained with understanding the interaction of light with matter, and in particular the "blackbody spectrum" of thermal radiation. This was, however, the point at which the corpuscular theory came back to life. In 1900, Max Planck showed that the form of the blackbody spectrum could be explained by postulating that the walls of the body which contains the radiation consisted of harmonic oscillators with a range of frequencies, and that the energies of those with frequency ν were restricted to integral multiples of the quantity $h\nu$. Each oscillator therefore had a fundamental energy quantum:

$$E = h\nu \tag{1-4}$$

where h became known as Planck's constant.

In 1905 Albert Einstein explained the ***photoelectric effect*** by postulating that electromagnetic radiation was itself quantized, so that electrons are emitted from a metal surface when radiation is absorbed in discrete quanta. It seemed that Newton was right after all! Light was again to be understood as a stream of particles, later to become known as photons. What had actually been shown, however, was that light energy and the momentum carried by a light wave existed in discrete units, or quanta; photons should be thought of as events at which these quanta are emitted or absorbcd.

If light is a wave that has properties usually associated with particles, could material particles correspondingly have wave-like properties? This was proposed by Louis de Broglie in 1924, and confirmed experimentally three years later in two classical experiments by George Thomson and by Clinton Davisson and Lester Germer. Both showed that a beam of particles, like a light ray encountering an obstacle, could be diffracted, behaving as a wave rather than a geometric ray. The diffraction pattern formed by the spreading of an electron beam passing through a hole in a metal sheet, for example, was the same as the diffraction pattern in light. Furthermore, the ***wavelength*** λ involved was simply related to the momentum p of the electrons by

$$\lambda = \frac{h}{p} \tag{1-5}$$

The constant h was again Planck's constant, as in the theory of quanta in electromagnetic radiation; for material waves λ is the de Broglie wavelength. A general wave theory of the behavior of matter, wave mechanics, was developed in 1926 by Erwin Schrödinger following de Broglie's ideas. Wave mechanics revolutionized our understanding of how microscopic particles were described and placed limitations on the extent of information one could have about such systems — the famous Heisenberg uncertainty relationship. The behavior of both matter and light evidently has dual aspects: They are in some sense both particles and waves, which aspect best describes their behavior depends on the circumstances; light propagates, diffracts and interferes as waves, but are emitted and absorbed discontinuously as photons, which are discrete packets of energy and momentum. Photons do not have a continuous existence, as does for example an electron in the beam of an accelerator machine; in contrast with a material particle, it is not possible to say where an individual photon is located within a light beam. In some contexts, we nevertheless think of the light within some experimental apparatus, such as a cavity or a laser, as consisting of photons, and we must then beware of following Newton and being misled by thinking of photons as particles with properties like those of material particles.

Although photons and electrons have very similar wave-like characteristics, there are several fundamental differences in their behavior. Photons have zero mass; the momentum p of a photon in equation (1-5) is related to its kinetic energy E by $E=pc$, as compared with $E=p^2/2m$ for particles moving well below light speed. Unlike electrons, photons are not conserved and can be created or destroyed in encounters with material particles. Again, their statistical behavior is different in situations where many photons or electrons can interact, as for example the photons in a laser or electrons in a metal. No two electrons in such a system can be in exactly the same state, while there is no such restriction for photons: This is the difference between Fermi-Dirac and Bose-Einstein statistics respectively for electrons and for photons.

1.2.2 The Electromagnetic Spectrum①

The wavelength range of ***visible light*** covers about one octave of the ***electromagnetic spectrum***, approximately from 400 nm to 800 nm (1 nm$=10^{-9}$ m). The electromagnetic spectrum covers a vast range, stretching many decades through infrared light to radio waves and many more decades through ***ultraviolet light*** and ***X-rays*** to ***gamma rays*** (γ-rays) (see Fig. 1-11). The differences in behavior across the electromagnetic spectrum are very large. Frequencies (ν) and wavelengths (λ) are related to the velocity of light (c) by $\lambda\nu=c$. The frequencies vary from 10^4 Hz for long radio waves (one hertz equals one cycle per second), to more than 10^{21} Hz for commonly encountered gamma rays; the highest energy

① Smith F G, King T A, Wilkins D. Optics and Photonics: An Introduction[M]. 2nd ed. New York: John Wiley & Sons, Inc., 2007:10-11.

cosmic gamma rays so far detected reach to 10^{35} Hz (4×10^{20} eV). It is unusual to encounter a quantum process in the radio frequency spectrum, and even more unusual to hear a physicist refer to the frequency of a gamma ray, instead of the energy and the momentum carried by a gamma ray photon.

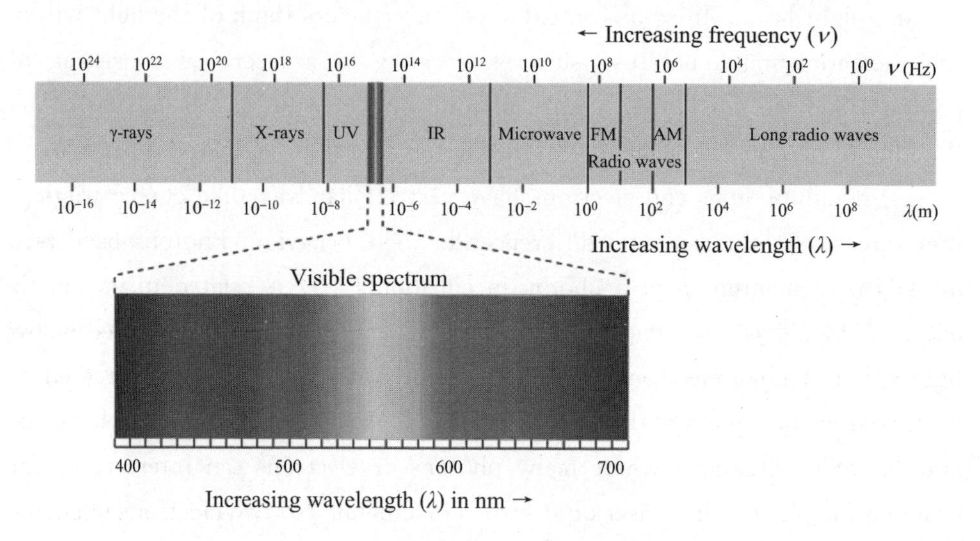

Fig. 1-11 The electromagnetic spectrum

Although wave aspects dominate the behavior of the longest wavelengths, and photon aspects dominate the behavior of short-wavelength X-rays and gamma rays, the whole range is governed by the same basic laws. It is in the optical range (waves in or near the visible range) that we most usually encounter the "wave particle duality" which requires a familiarity with both concepts.

The propagation of light is determined by its wave nature, and its interaction with matter is determined by quantum physics. The relation of the energy of the photon to common levels of energy in matter determines the relative importance of the quantum at different parts of the spectrum: Cosmic gamma rays, with a high photon energy and a high photon momentum, can act on matter explosively or like a high-velocity billiard ball, while long infrared or radio waves, with low photon energies, usually only interact with matter through classical electric and magnetic induction.

General absorption is characterized by low absorption, the absorption of visible light by quartz (almost transparent). Selective absorption is large and varies dramatically with wavelength. For example, quartz pairs strongly in infrared light 3. 5～5 μm.

Beer-Lambert law is also known as Beer law, it is the basic law of light absorption (see Fig. 1-12), applicable to all electromagnetic radiation and all light-absorbing substances, including gases, solids, liquids, molecules, atoms and ions. Beer-Lambert law is the quantitative basis of absorbance, colorimetry and photoelectric colorimetry.

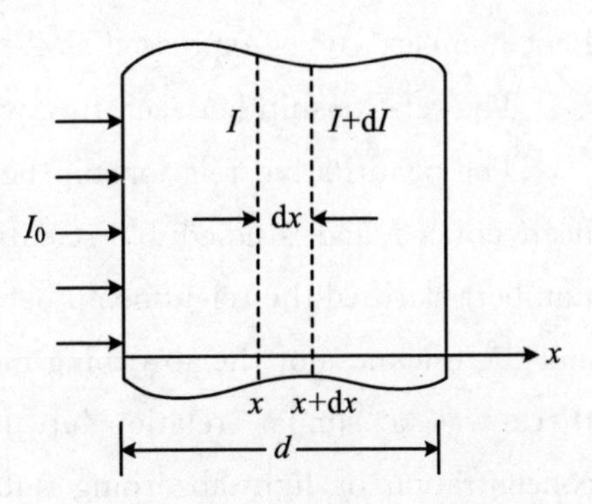

Fig. 1-12 Light absorption medium

Beer-Lambert law is a beam of monochromatic light shines on the surface of an absorbing medium. After passing through the medium of a certain thickness, the intensity of the transmitted light will be weakened because the medium absorbs part of the light energy. The greater the concentration of the absorbing medium, the greater the thickness of the medium, and the more significant reduction of light intensity, the relationship is as follows:

$$A = -\lg \frac{I_t}{I_0} = \lg \frac{1}{T} = Klc \tag{1-6}$$

where A is absorbance. I_0 is incident light intensity. I_t is transmitted light intensity. T is transmittance or light transmittance. K is absorption coefficient or molar absorption coefficient. l is the thickness of the absorption medium, generally in cm unit. c is the concentration of a light-absorbing substance, which can be in grams per liter or moles per liter.

The physical meaning of Beer-Lambert law is that when a beam of parallel monochromatic light passes vertically through a uniform unscattered light-absorbing material, its absorbance A is proportional to the concentration of the light-absorbing material c and the thickness of the absorption layer l.

Additivity of absorbance: When medium contains multiple components, the total absorbance of the medium at a certain wavelength is the addition of the absorbance of each component at that wavelength, this law is called the additivity of absorbance.

Coefficient K: When the thickness of the medium l is in unit of cm and the concentration of the absorbant c is in unit of g/L, K is denoted by a, which is called the absorption coefficient in the unit of $L \cdot g^{-1} \cdot cm^{-1}$. In this case, the

Beer-Lambert law is expressed as $A=alc$.

The relationship between the two absorption coefficients is $\kappa=aM_m$.

The quantitative relationship between matter and light absorption has long been noticed and studied by scientists. Pierre Bouguer and Johann Heinrich Lambert clarified the relationship between the amount of light absorbed by matter and the thickness of the absorbing medium. In 1852, August Beer proposed that there was a similar relationship between the absorption of light and the concentration of light-absorbing substances, and the combination of the two resulted in the basic law of light absorption — Bouguer-Lambert-Beer law, or Beer-Lambert law for short.

Suppose a beam of parallel monochromatic light (incident light) with the intensity of I_0 is irradiated vertically on the surface of an isotropic uniform absorption medium. After passing through an absorption layer of thickness (optical path) l, due to the absorption of light by particles in the absorption layer, the intensity of the incident light of the beam is reduced to I_1, which is called transmitted light intensity. The capacity of a substance to absorb light is proportional to the cross-sectional area of all light-absorbing particles.

Assumed that the thickness l of the absorption layer can be divided into several small thin layers of infinitesimal thickness $\mathrm{d}l$ in the direction perpendicular to the incident light, with a cross-sectional area of S, and each thin layer contains $\mathrm{d}n$ light-absorbing particles, with a cross-sectional area of a. Thus, the total cross-sectional area of all light-absorbing particles in the thin layer is $\mathrm{d}S=a\mathrm{d}n$.

Suppose that the incident light with intensity I is reduced to $\mathrm{d}I$ through the thin layer. $\mathrm{d}I$ is proportional to the total cross-sectional area $\mathrm{d}S$ of the light-absorbing particles and incident light intensity I, i. e.,

$$-\mathrm{d}I = k_1 I \mathrm{d}S = k_1 I a \mathrm{d}n \tag{1-7}$$

The minus sign indicates that the light intensity reduction by absorption, and k_1 is the proportionality coefficient.

Assuming that the concentration of the light-absorbing substance is c, the number of light-absorbing particles in the above thin layer is

$$\mathrm{d}n = 6.02 \times 10^{23} cS\mathrm{d}l$$

Substitute into the above equation, combine the constant term and set $k_2=6.02\times10^{23}k_1aS$, we get:

$$-\frac{\mathrm{d}I}{I} = k_2 c \mathrm{d}l \tag{1-8}$$

If we take the definite integral of the above, we get:

$$-\int_{I_0}^{I_1} \frac{\mathrm{d}I}{I} = \int_0^l k_2 c \mathrm{d}l \tag{1-9}$$

$$-\ln \frac{I_t}{I_0} = k_2 cl \tag{1-10}$$

$$\lg \frac{I_0}{I_t} = 0.434 k_2 cl = Klc \tag{1-11}$$

In the above equation, $\lg \frac{I_0}{I_t}$ is called absorbance (A). The ratio between the intensity of transmitted light and the intensity of incident light $\frac{I_t}{I_0}$ is called transmittance, or transmittance (T), and its relationship is

$$A = \lg \frac{I_0}{I_t} = \lg \frac{1}{T} = Klc \tag{1-12}$$

The Beer-Lambert law is established on the conditions that:

(1) Incident light is parallel monochromatic light and vertical irradiation.

(2) The light-absorbing material is a uniform non-scattering system.

(3) There is no interaction between the light-absorbing particles.

(4) The interaction between radiation and matter is limited to the process of light absorption, and no fluorescence and photochemistry occur.

1.2.3 Refractive Index and Dispersion

In optics the ***refractive index*** or index of refraction n of a substance (optical medium) is a dimensionless number that describes how light, or any other radiation, propagates through that medium. It is defined as:

$$n = \frac{c}{v} \tag{1-13}$$

where c is the speed of light in vacuum and v is the speed of light in the substance. For example, the refractive index of water is 1.33, meaning that light travels 1.33 times slower in water than it does in vacuum. The refractive index determines how much light is bent, or refracted, when entering a material. The refractive indices also determine the amount of light that is reflected when reaching the interface, as well as the critical angle for total internal reflection and

Brewster's angle.

Example 1.2.1

Problem

What is the light velocity within glass?

Solution

When a light ray from the air strikes and penetrates glass, as shown in Fig. 1-13, its rate of movement slows. Taking the refractive index of glass, $n=1.5$, the light velocity within glass will be found as:

$$v = c/n = 3\times 10^{8}\,(\mathrm{m/s})/1.5 = 2\times 10^{8}\,(\mathrm{m/s}) \tag{1-14}$$

Fig. 1-13 A ray of light being refracted in glass

As you can see, the calculations ignore the slight difference in the refractive indexes of vacuum and air. Obviously, the higher the refractive index, the denser the material from an optical standpoint.

This simple example demonstrates one of the basic ideas of optics: All characteristics of light in free space are changed inside the material with the refractive index n. Velocity becomes c/n, wavelength becomes λ/n, and so forth.

The refractive index can be seen as the factor by which the speed and the wavelength of the radiation are reduced with respect to their vacuum values: The speed of light in a medium is $v=c/n$, and similarly the wavelength in that medium is $\lambda=\lambda_0/n$, where λ_0 is the wavelength of that light in vacuum. This implies that vacuum has a refractive index of 1, and that the frequency ($f=v/\lambda$) of the wave is not affected by the refractive index.

The refractive index varies with the wavelength of light. This is called ***dispersion*** and causes the splitting of white light into its constituent colors in prisms and rainbows, and ***chromatic aberration*** in lenses (see Fig. 1-14 and Fig. 1-15). In optics, dispersion is the phenomenon in which the phase velocity of a wave depends on its frequency, or alternatively when the group velocity depends on the frequency. Media having such a property are termed dispersive media. Dispersion is sometimes called chromatic dispersion to emphasize its wavelength-dependent nature, or group-velocity dispersion (GVD) to emphasize the role of the group velocity. Dispersion is most often described for light waves, but it may occur for any kind of wave that interacts with a medium or passes through an inhomogeneous geometry (e.g., a waveguide), such as sound waves. A material's dispersion is measured by its ***Abbe number***, V, with low Abbe numbers corresponding to strong dispersion.

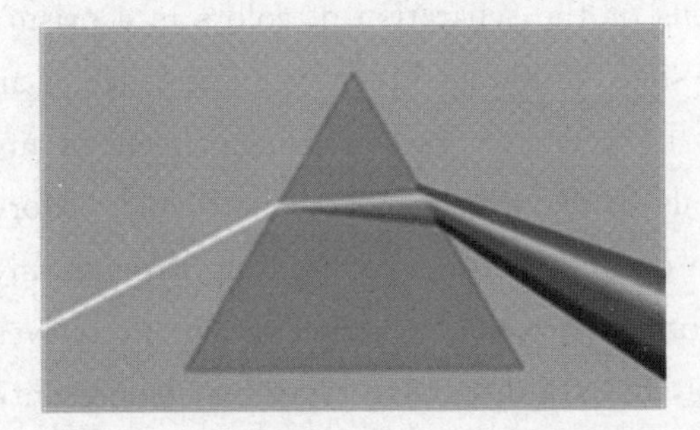

Fig. 1-14 Dispersion phenomenon in a prism

In a prism dispersion causes different colors to refract at different angles, splitting white light into a rainbow of colors.

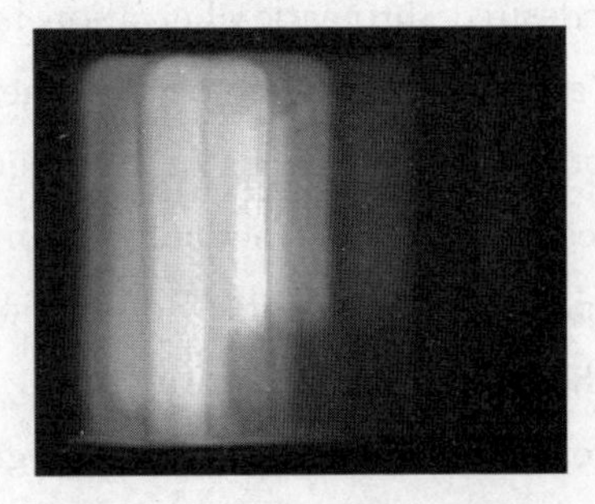

Fig. 1-15 A compact fluorescent lamp seen through an Amici prism

The most familiar example of dispersion is probably a rainbow, as shown in Fig. 1-16, in which dispersion causes the spatial separation of a white light into components of different wavelengths (different colors). However, dispersion also has an effect in many other circumstances: For example, GVD causes pulses to spread in optical fibers, degrading signals over long distances; also, a cancellation between group-velocity dispersion and nonlinear effects leads to soliton waves.

There are generally two sources of dispersion: material dispersion and waveguide dispersion. Material dispersion comes from a frequency-dependent response of a material to waves. For example, material dispersion leads to

Fig. 1-16 Colorful rainbow in the sky

Light of different colors have slightly different refractive indices in water and therefore show up at different positions in the rainbow.

undesired chromatic aberration in a lens or the separation of colors in a prism. Waveguide dispersion occurs when the speed of a wave in a waveguide (such as an optical fiber) depends on its frequency for geometric reasons, independent of any frequency dependence of the materials from which it is constructed. More generally, "waveguide" dispersion can occur for waves propagating through any inhomogeneous structure (e. g. , a photonic crystal), whether or not the waves are confined to some region. In general, both of dispersion types may be present, although they are not strictly additive. Their combination leads to signal degradation in optical fibers for telecommunications, because the varying delay in arrival time among different components of a signal "smears out" the signal in time.

Material dispersion can be a desirable or undesirable effect in optical applications. The dispersion of light by glass prisms is used to construct spectrometers and spectroradiometers. Holographic gratings are also used, as they allow more accurate discrimination of wavelengths. However, in lenses, dispersion causes chromatic aberration, an undesired effect that may degrade images in microscopes, telescopes and photographic objectives.

Optical fibers, which are used in telecommunications, are among the most abundant types of waveguides. Dispersion in these fibers is one of the limiting factors that determine how much data can be transported on a single fiber.

In photographic and microscopic lenses, dispersion causes chromatic aberration, which causes the different colors in the image not to overlap properly. Various techniques have been developed to counteract this, such as the use of ***achromats***, multielement lenses with glasses of different dispersion. They are constructed in such a way that the chromatic aberrations of the different parts cancel out.

1.2.4 Reflection and Refraction

Reflection is the change in direction of a ***wavefront*** at an interface between two different media so that the wavefront returns into the medium from which it originated. Reflection of light is either specular (mirror-like) or diffuse (retaining the energy, but losing the image) depending on the nature of the interface. A mirror provides the most common model for ***specular*** light ***reflection***, and typically consists of a glass sheet with a metallic coating where the reflection actually occurs. Reflection is enhanced in metals by suppression of wave propagation beyond their skin depths. Reflection also occurs at the surface of transparent media, such as water or glass.

In Fig. 1-17, a light ray *PO* strikes a vertical mirror at point *O*, and the reflected ray is *OQ*. By projecting an imaginary line through point *O* perpendicular to the mirror, known as the normal, we can measure the angle of incidence, θ_i and the angle of reflection, θ_r. The law of reflection states that $\theta_i = \theta_r$, or in other words, the angle of incidence equals the angle of reflection.

If the reflecting surface is very smooth, the reflection of light that occurs is called specular or regular reflection. The laws of reflection are as follows:

(1) The incident ray, the reflected ray and the normal to the reflection surface at the point of the incidence lie in the same plane.

(2) The angle which the incident ray makes with the normal is equal to the angle which the reflected ray makes to the same normal.

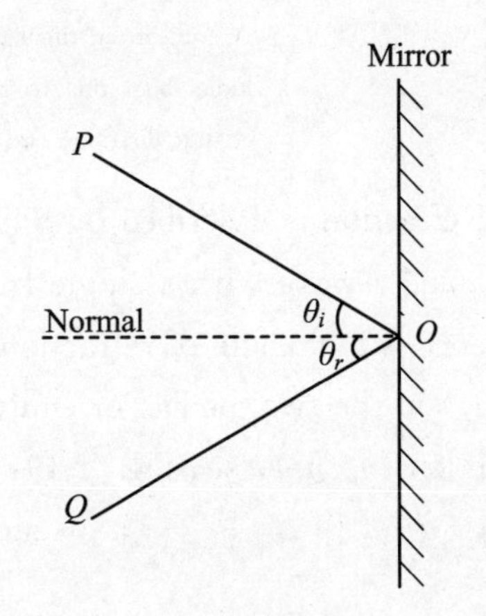

Fig. 1-17 Diagram of specular reflection

(3) The reflected ray and the incident ray are on the opposite sides of the normal.

In fact, reflection of light may occur whenever light travels from a medium of a given refractive index into a medium with a different refractive index. In the most general case, a certain fraction of the light is reflected from the interface, and the remainder is refracted.

In optics, refraction is a phenomenon that often occurs when waves travel from a medium with a given refractive index to a medium with another at an oblique angle as shown in Fig. 1-18. At the boundary between the media, the wave's phase velocity is altered, usually causing a change in direction. Its wavelength increases or decreases but its frequency remains constant. For example, a light ray will refract as it enters and leaves glass, assuming there is a change in refractive index. A ray traveling along the normal (perpendicular to the boundary) will change speed, but not direction. Refraction still occurs in this case.

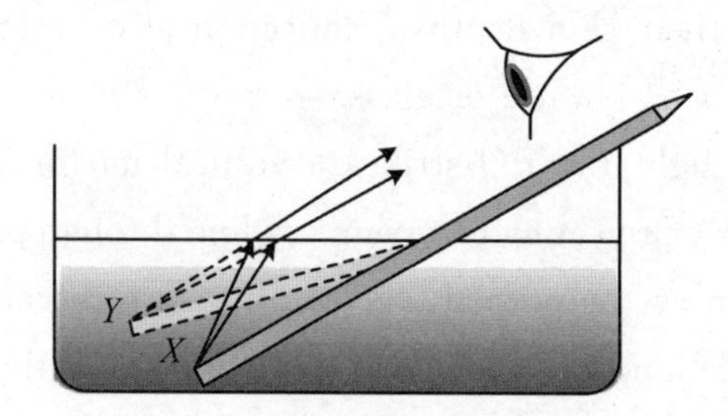

Fig. 1-18 A pencil looks bent in the water

An object (in this case, a pencil) part immersed in water looks bent due to refraction: The light waves from X change direction and so seem to originate at Y.

Refraction is described by ***Snell's law***, which states that for a given pair of media and a wave with a single frequency, the ratio of the sines of the angle of incidence θ_1 and angle of refraction θ_2 is equivalent to the ratio of phase velocities (v_1/v_2) in the two media, or equivalently, to the opposite ratio of the indices of refraction (n_2/n_1), see Fig. 1-19:

$$\frac{\sin\theta_1}{\sin\theta_2}=\frac{v_1}{v_2}=\frac{n_2}{n_1} \tag{1-15}$$

In general, the incident wave is partially refracted and partially reflected; the details of this behavior are described by the ***Fresnel's equations***.

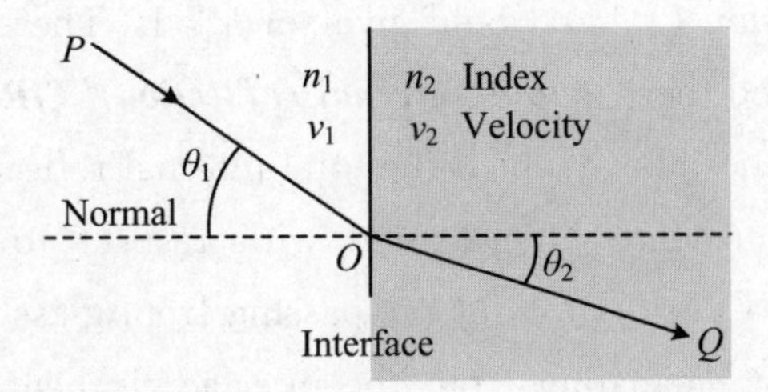

Fig. 1-19 Refraction of light at the interface between two media of different refractive indices, with $n_2 > n_1$

Example 1.2.2

Problem

Let $n_1 = 1$, $\theta_1 = 30°$, and $n_2 = 1.5$ (see Fig. 1-20). What are θ_3 and θ_2?

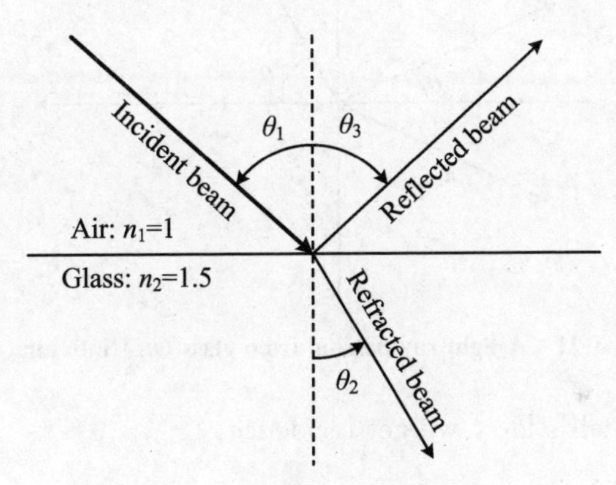

Fig. 1-20 Light travels from air to glass

Solution

$\theta_3 = 30°$ and θ_2 is calculated as follows:

$\sin\theta_1 = 30° = 0.5$; hence, $\sin\theta_2 = 0.5/1.5 = 0.333$, and $\theta_2 = \sin^{-1}(0.333) = 19.5°$.

1.2.5 Total Internal Reflection

Referring again to Fig. 1-19, and noting that the geometry is the same if the ray direction is reversed, we consider what happens if a ray inside the refracting medium meets the surface at a large angle of incidence θ_2, so that $\sin\theta_2$ is greater than n_1/n_2 and equation (1-15) would give $\sin\theta_1 > 1$. There can then be no ray above the surface, and there is ***total internal reflection*** (***TIR***). The critical angle is the angle of incidence above which the total internal reflection occurs.

The angle of incidence is measured with respect to the normal at the refractive boundary. Consider a light ray passing from glass into air as shown in Fig. 1-21. The light emanating from the interface is bent towards the glass. When the incident angle is increased sufficiently, the transmitted angle (in air) reaches 90°. It is at this point no light is transmitted into air. The critical angle θ_c is given by Snell's law:

$$n_1 \sin\theta_i = n_2 \sin\theta_t \tag{1-16}$$

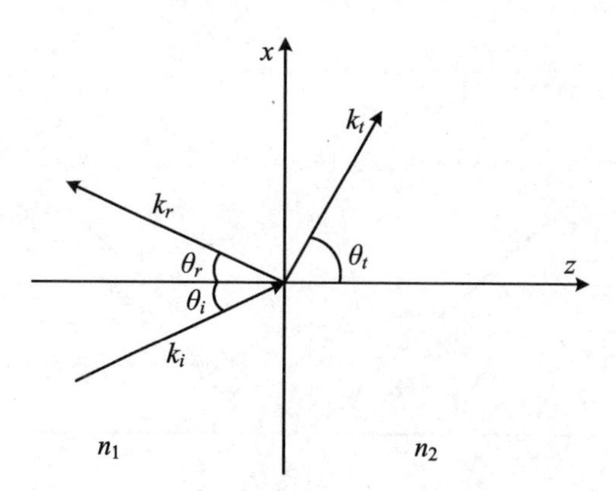

Fig. 1-21 A light ray passing from glass (n_1) into air (n_2)

Rearranging Snell's law, we get incidence:

$$\sin\theta_i = \frac{n_2}{n_1}\sin\theta_t \tag{1-17}$$

To find the critical angle, we find the value for θ_i when $\theta_t = 90°$ and thus $\sin\theta_t = 1$. The resulting value of θ_i is equal to the critical angle θ_c.

Now, we can solve for θ_i, and we get the equation for the critical angle:

$$\theta_c = \theta_i = \arcsin\left(\frac{n_2}{n_1}\right) \tag{1-18}$$

If the incident ray is precisely at the critical angle, the refracted ray is tangent to the boundary at the point of incidence. If for example, visible light was traveling through acrylic glass (with an index of refraction of approximately 1.5) into air (with an index of refraction of 1), the calculation would give the critical angle for light from acrylic into air, which is

$$\theta_c = \arcsin\left(\frac{1}{1.5}\right) = 41.8^\circ \tag{1-19}$$

Light incident on the border with an angle less than 41.8° would be partially transmitted, while light incident on the border at larger angles with respect to normal would be totally internally reflected.

The phenomenon of total internal reflection is put to good use in the reflecting prism as shown in Fig. 1-22 (a), in which light entering the cant surface of a reflecting prism is reflected directly and emerges at the horizontal surface. The total internal reflection principle is applicable to the transmission of light down thin optical fibers, as shown in Fig. 1-22(b), but here the relation of the wavelength of light to the fiber diameter must be taken into account.

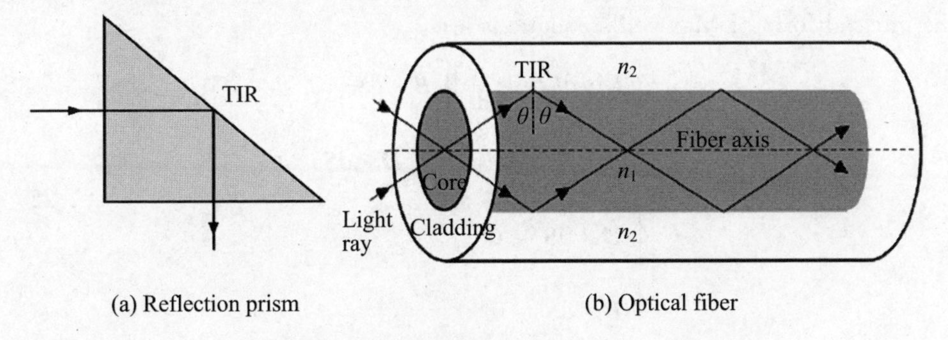

(a) Reflection prism (b) Optical fiber

Fig. 1-22 The application of total internal reflection

1. Maxwell's Equations and Matter Equations

Differential equation of ***Maxwell's equations*** is

$$\begin{cases} \nabla \times \boldsymbol{E} = -\dfrac{\partial \boldsymbol{B}}{\partial t} \\ \nabla \times \boldsymbol{H} = \boldsymbol{J} + \dfrac{\partial \boldsymbol{D}}{\partial t} \\ \nabla \cdot \boldsymbol{D} = \rho \\ \nabla \cdot \boldsymbol{B} = 0 \end{cases} \tag{1-20}$$

$\boldsymbol{J}$ is the ***conduction current*** density with the unit of $\mathrm{A/m^2}$, free ***charge density*** is ρ with the unit of $\mathrm{C/m^2}$. At the same time, there is a relationship between the ***electromagnetic field*** on the material medium, namely the ***matter equation*** (or constitutive equation) is

$$\begin{cases} \boldsymbol{D} = \varepsilon \boldsymbol{E} = \varepsilon_0 \boldsymbol{E} + \boldsymbol{P} \\ \boldsymbol{B} = \mu \boldsymbol{H} = \mu_0 (\boldsymbol{H} + \boldsymbol{M}) \\ \boldsymbol{J} = \sigma \boldsymbol{E} \end{cases} \tag{1-21}$$

Maxwell's equations and matter equations describe the distribution and variation of electromagnetic field in space and time. Therefore, all problems related to the generation and propagation of electromagnetic waves can be attributed to the problem of solving Maxwell's equations under given initial conditions and boundary conditions, which is also the key and core of solving the propagation problem of light waves in various media and ***boundary conditions***.

2. Integral Form and Boundary Conditions

In some cases, the vectors ***E***, ***D***, ***B***, ***H*** jumps on the interface between two media, so the derivatives of these things tend to be discontinuous. In this case, Maxwell's equations in differential form cannot be directly applied on the interface, but the boundary conditions must be derived from its integral form. The integral form of Maxwell's equations are

$$\begin{cases} \oint_l \boldsymbol{E} \cdot \mathrm{d}\boldsymbol{l} = -\dfrac{\mathrm{d}}{\mathrm{d}t}\iint_S \boldsymbol{B} \cdot \mathrm{d}\boldsymbol{S} \\ \oint_l \boldsymbol{H} \cdot \mathrm{d}\boldsymbol{l} = \boldsymbol{I} + \dfrac{\mathrm{d}}{\mathrm{d}t}\iint_S \boldsymbol{D} \cdot \mathrm{d}\boldsymbol{S} \\ \oint_S \boldsymbol{D} \cdot \mathrm{d}\boldsymbol{S} = Q \\ \oint_S \boldsymbol{B} \cdot \mathrm{d}\boldsymbol{S} = 0 \end{cases} \tag{1-22}$$

The boundary conditions between two medium surfaces are

$$\begin{cases} \boldsymbol{n} \times (\boldsymbol{E}_2 - \boldsymbol{E}_1) = 0 \\ \boldsymbol{n} \times (\boldsymbol{H}_2 - \boldsymbol{H}_1) = \boldsymbol{\alpha} \\ \boldsymbol{n} \cdot (\boldsymbol{D}_2 - \boldsymbol{D}_1) = \sigma \\ \boldsymbol{n} \cdot (\boldsymbol{B}_2 - \boldsymbol{B}_1) = 0 \end{cases} \tag{1-23}$$

From equation (1-20) to equation (1-23), specific explanations are as follows:

(1) The tangential component of the electric field intensity vector ***E*** is continuous and ***n*** is the normal component of the interface.

(2) $\boldsymbol{\alpha}$ is the linear density of surface conduction current at the interface. When there is no conduction current at the interface, $\boldsymbol{\alpha}=0$, and the tangential component of ***H*** is continuous. For example, there is no free charge or conduction current on the surface of an insulating medium.

(3) σ is the surface density of free charge at the interface.

(4) The normal component of the magnetic induction intensity vector ***B*** is

continuous at the interface.

3. Laws of Reflection and Refraction

Laws of reflection and refraction see Fig. 1-23.

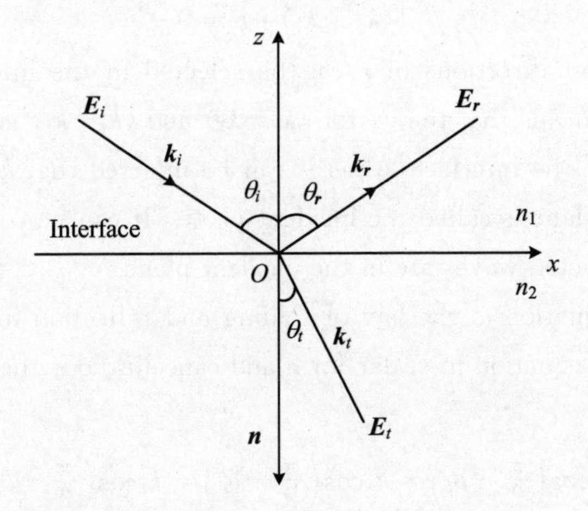

Fig. 1-23 Laws of reflection and refraction

When light is incident from one medium to another, reflection and refraction occur at the interface. Now assume that the two media are homogeneous, transparent and isotropic media, the interface is an infinite plane, the incoming, reflected and refracted light are all plane light waves, and the electric field expression is

Incident wave: $\boldsymbol{E}_i = \boldsymbol{E}_{0i}\exp[\mathrm{i}(\omega_i t - \boldsymbol{k}_i \cdot \boldsymbol{r})]$

Reflective wave: $\boldsymbol{E}_r = \boldsymbol{E}_{0r}\exp[\mathrm{i}(\omega_r t - \boldsymbol{k}_r \cdot \boldsymbol{r})]$

Refractive wave: $\boldsymbol{E}_t = \boldsymbol{E}_{0t}\exp[\mathrm{i}(\omega_t t - \boldsymbol{k}_t \cdot \boldsymbol{r})]$

The total electric field on both sides of the interface is

$$\begin{cases} \boldsymbol{E}_1 = \boldsymbol{E}_i + \boldsymbol{E}_r = \boldsymbol{E}_{0i}\exp[\mathrm{i}(\omega_i t - \boldsymbol{k}_i \cdot \boldsymbol{r})] + \boldsymbol{E}_{0r}\exp[\mathrm{i}(\omega_r t - \boldsymbol{k}_r \cdot \boldsymbol{r})] \\ \boldsymbol{E}_2 = \boldsymbol{E}_t = \boldsymbol{E}_{0t}\exp[\mathrm{i}(\omega_t t - \boldsymbol{k}_t \cdot \boldsymbol{r})] \end{cases} \tag{1-24}$$

According to the boundary conditions of the electric field, $\boldsymbol{n}\times(\boldsymbol{E}_2 - \boldsymbol{E}_1) = 0$, there is

$$\boldsymbol{n}\times\boldsymbol{E}_{0i}\exp[\mathrm{i}(\omega_i t - \boldsymbol{k}_i \cdot \boldsymbol{r})] + \boldsymbol{n}\times\boldsymbol{E}_{0r}\exp[\mathrm{i}(\omega_r t - \boldsymbol{k}_r \cdot \boldsymbol{r})]$$
$$= \boldsymbol{n}\times\boldsymbol{E}_{0t}\exp[\mathrm{i}(\omega_t t - \boldsymbol{k}_t \cdot \boldsymbol{r})] \tag{1-25}$$

If the above equation is true for any time t and the interface $\boldsymbol{r}$, it must be

$$\omega_i = \omega_r = \omega_t = \omega \tag{1-26}$$

$$\boldsymbol{k}_i \cdot \boldsymbol{r} = \boldsymbol{k}_r \cdot \boldsymbol{r} = \boldsymbol{k}_t \cdot \boldsymbol{r} \tag{1-27}$$

It can be seen that the time frequency ω is the inherent characteristic of incident electromagnetic wave or light wave, which is not different from the medium, nor will it change because of refraction or reflection.

$$\begin{cases}(\boldsymbol{k}_r - \boldsymbol{k}_i) \cdot \boldsymbol{r} = 0 \\ (\boldsymbol{k}_t - \boldsymbol{k}_i) \cdot \boldsymbol{r} = 0\end{cases} \tag{1-28}$$

Since different directions of $\boldsymbol{r}$ can be selected in the interface, the above formula actually means that the vector $(\boldsymbol{k}_r - \boldsymbol{k}_i)$ and $(\boldsymbol{k}_t - \boldsymbol{k}_i)$ are both parallel to the normal line of the interface, thus it can be inferred that $\boldsymbol{k}_i$、$\boldsymbol{k}_r$、$\boldsymbol{k}_t$ and $\boldsymbol{n}$ are coplanar, which plane is called the incident plane. It can be concluded that both reflected and refracted waves are in the incident plane.

The above equation is the law of folding and reflection in vector form. By writing the above equation in scalar form and canceling out the common position quantities, we get:

$$\boldsymbol{k}_i \cos\left(\frac{\pi}{2} - \theta_i\right) = \boldsymbol{k}_r \cos\left(\frac{\pi}{2} - \theta_r\right) = \boldsymbol{k}_t \cos\left(\frac{\pi}{2} - \theta_t\right) \tag{1-29}$$

and according to $\boldsymbol{k}_i = n_1\omega/c$, $\boldsymbol{k}_r = n_1\omega/c$, $\boldsymbol{k}_t = n_2\omega/c$, we get:

$$\begin{cases}\theta_i = \theta_r \quad \text{(Incident angle equals refraction angle)} \\ n_1 \sin\theta_i = n_2 \sin\theta_t \quad \text{(Refraction law)}\end{cases} \tag{1-30}$$

4. Fresnel's Equation

The law of refraction and reflection gives the relationship between the propagation directions of reflected, refracted and incident waves. The quantitative relationship between the amplitude and phase of reflected wave, refracted wave and incident wave is described by Fresnel's equation.

The electric field $\boldsymbol{E}$ is a vector, which can be divided into a pair of orthogonal electric field components, one vibrating perpendicular to the incident plane, called the S component, and the other vibrating in (or parallel to) the incident plane, called the P component.

First, two special cases of incident wave containing only S component and only P component are studied. When two components exist at the same time, the refraction and reflection electric fields of a single component are calculated respectively. The result is then obtained by adding the vectors according to the principle of vector superposition.

(1) The S Component Alone

S and P components of reflection and refraction see Fig. 1-24.

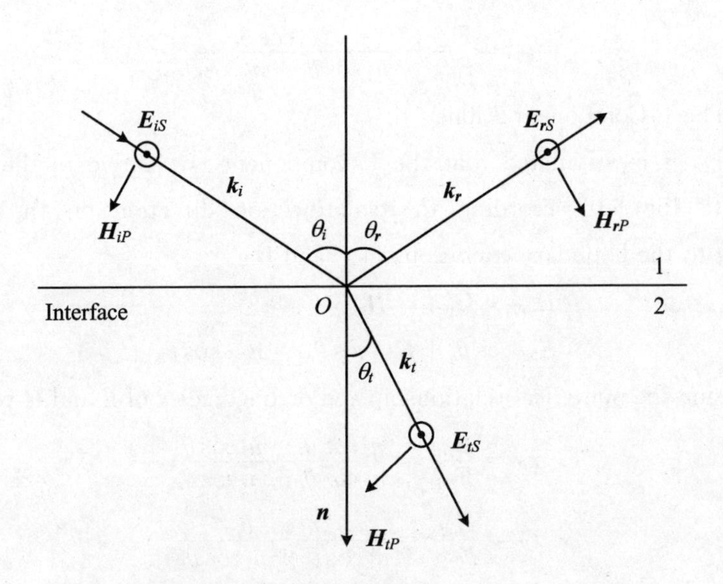

Fig. 1-24 *S* and *P* components of reflection and refraction

Firstly, it is stipulated that the S component of the electric and magnetic fields is perpendicular to the surface of the paper. Outward is positive and inward is negative. The tangential component of the electric field is continuous at the interface:

$$\boldsymbol{n} \times (\boldsymbol{E}_2 - \boldsymbol{E}_1) = 0$$

In addition, according to equations (1-24) and (1-25), it can be obtained:

$$\boldsymbol{E}_{0iS} + \boldsymbol{E}_{0rS} = \boldsymbol{E}_{0tS} \tag{1-31}$$

The tangential component of the magnetic field at the interface is continuous:

$$\boldsymbol{n} \times (\boldsymbol{H}_2 - \boldsymbol{H}_1) = 0$$

Notice $\boldsymbol{H} = \frac{1}{\mu\omega}\boldsymbol{k} \times \boldsymbol{E}$, as shown in the Fig. 1-24. So similarly:

$$-\boldsymbol{H}_{0iP}\cos\theta_i + \boldsymbol{H}_{0rP}\cos\theta_r = -\boldsymbol{H}_{0tP}\cos\theta_t \tag{1-32}$$

The relationship between $\boldsymbol{E}$ and $\boldsymbol{H}$ in non-magnetic isotropic media:

$$\begin{cases} \boldsymbol{H} = \dfrac{\boldsymbol{B}}{\mu_0} = \dfrac{\boldsymbol{n}}{\mu_0 c}\boldsymbol{E} \\ \boldsymbol{E} \perp \boldsymbol{H} \end{cases} \tag{1-33}$$

Then the equation (1-32) is rearranged as:

$$-n_1\boldsymbol{E}_{0iS}\cos\theta_i + n_1\boldsymbol{E}_{0rS}\cos\theta_r = -n_2\boldsymbol{E}_{0tS}\cos\theta_t \tag{1-34}$$

Solve simultaneous equations (1-31) and (1-32), we get:

$$\boldsymbol{r}_S = \frac{\boldsymbol{E}_{0rS}}{\boldsymbol{E}_{0iS}} = \frac{n_1\cos\theta_i - n_2\cos\theta_t}{n_1\cos\theta_i + n_2\cos\theta_t} \tag{1-35}$$

$$t_S = \frac{\boldsymbol{E}_{0tS}}{\boldsymbol{E}_{0iS}} = \frac{2n_1 \cos\theta_i}{n_1 \cos\theta_i + n_2 \cos\theta_t} \tag{1-36}$$

(2) The P Component Alone

Firstly, it is stipulated that the P component is positive to the right and negative to the left according to its projection direction on the interface. According to the boundary conditions of $\boldsymbol{E}$ and $\boldsymbol{H}$:

$$\boldsymbol{H}_{0iS} + \boldsymbol{H}_{0rS} = \boldsymbol{H}_{0tS} \tag{1-37}$$

$$\boldsymbol{E}_{0iP} \cos\theta_i + \boldsymbol{E}_{0rP} \cos\theta_r = \boldsymbol{E}_{0tP} \cos\theta_t \tag{1-38}$$

By using the numerical relationship and orthogonality of $\boldsymbol{E}$ and $\boldsymbol{H}$ we can get:

$$\boldsymbol{r}_P = \frac{\boldsymbol{E}_{0rP}}{\boldsymbol{E}_{0iP}} = \frac{n_2 \cos\theta_i - n_1 \cos\theta_t}{n_2 \cos\theta_i + n_1 \cos\theta_t} \tag{1-39}$$

$$t_P = \frac{\boldsymbol{E}_{0tP}}{\boldsymbol{E}_{0iP}} = \frac{2n_1 \cos\theta_i}{n_2 \cos\theta_i + n_1 \cos\theta_t} \tag{1-40}$$

To sum up, the reflection coefficient and transmission coefficient of S and P waves can be expressed as:

$$\begin{cases} \boldsymbol{r}_S = \dfrac{\boldsymbol{E}_{0rS}}{\boldsymbol{E}_{0iS}} = \dfrac{n_1 \cos\theta_i - n_2 \cos\theta_t}{n_1 \cos\theta_i + n_2 \cos\theta_t} \\ \boldsymbol{r}_P = \dfrac{\boldsymbol{E}_{0rP}}{\boldsymbol{E}_{0iP}} = \dfrac{n_2 \cos\theta_i - n_1 \cos\theta_t}{n_2 \cos\theta_i + n_1 \cos\theta_t} \\ t_S = \dfrac{\boldsymbol{E}_{0tS}}{\boldsymbol{E}_{0iS}} = \dfrac{2n_1 \cos\theta_i}{n_1 \cos\theta_i + n_2 \cos\theta_t} \\ t_P = \dfrac{\boldsymbol{E}_{0tP}}{\boldsymbol{E}_{0iP}} = \dfrac{2n_1 \cos\theta_i}{n_2 \cos\theta_i + n_1 \cos\theta_t} \end{cases} \tag{1-41}$$

$$\begin{cases} \boldsymbol{r}_S = -\dfrac{\sin(\theta_i - \theta_t)}{\sin(\theta_i + \theta_t)} \\ \boldsymbol{r}_P = \dfrac{\tan(\theta_i - \theta_t)}{\tan(\theta_i + \theta_t)} \\ t_S = \dfrac{2\cos\theta_i \sin\theta_t}{\sin(\theta_i + \theta_t)} \\ t_P = \dfrac{2\cos\theta_i \sin\theta_t}{\sin(\theta_i + \theta_t)\cos(\theta_i - \theta_t)} \end{cases} \tag{1-42}$$

The one on the top left is the famous Fresnel's equation. Using the law of refraction, Fresnel's equation can also be written in the right-hand form.

5. Properties of Reflected and Transmitted Waves

(1) $n_1 < n_2$

① Reflection coefficient and transmission coefficient (see Fig. 1-25).

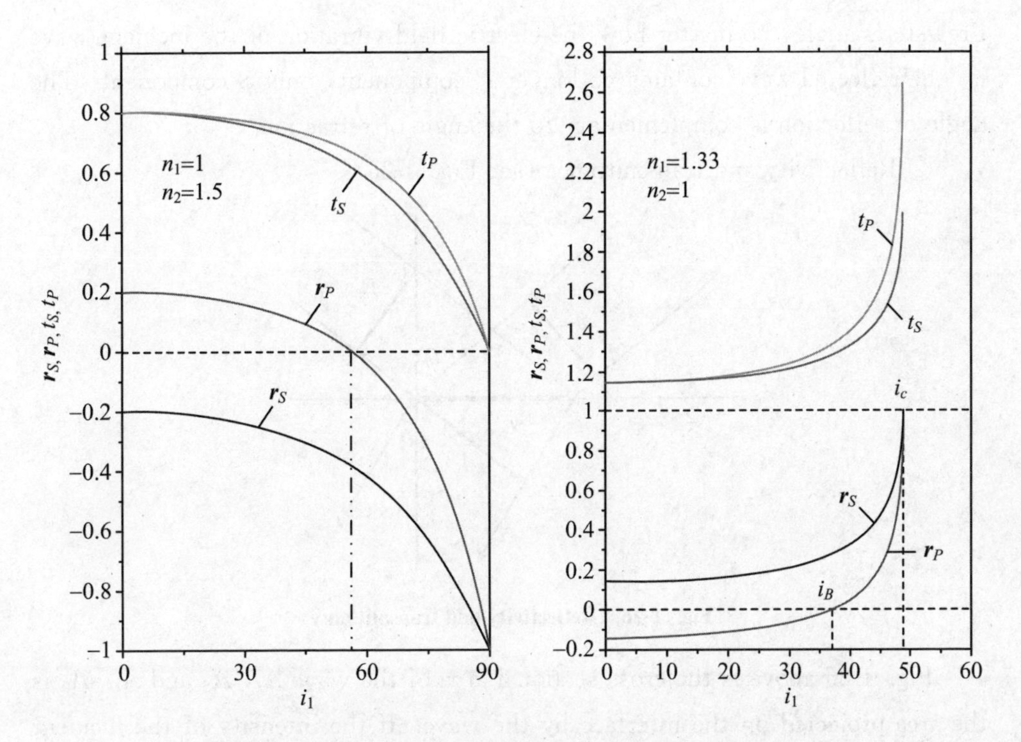

Fig. 1-25 Reflection coefficient and transmission coefficient

The two transmission coefficients t_S and t_P decrease monotonically with the increase of incident angle θ_i, that is, the more inclined the incident wave is, the weaker the transmitted wave is. In addition, under the definition of positive direction, both t_S and t_P are greater than zero, that is, no phase change occurs in the refracted light.

$\boldsymbol{r}_S$ is always negative, and its absolute value increases monotonically with the incident angle. According to the positive direction, the vibration direction of the S component of the reflected wave electric field on the interface is always opposite to that of the incident wave S component, and there is a phase change of π (also called half wave loss).

For $\boldsymbol{r}_P$, its substitution decreases monotonically with the incident angle θ_i, but it undergoes a positive to negative change. According to the formula $\boldsymbol{r}_P = \frac{\tan(\theta_i - \theta_t)}{\tan(\theta_i + \theta_t)}$. When $\boldsymbol{r}_P = 0$, $\theta_i + \theta_t = 90°$, namely $\sin\theta_i = \cos\theta_t$, by the law of refraction $n_1 \sin\theta_i = n_2 \sin\theta_t$. In combination, the incident angle is Brewster's angle $\theta_B = \tan^{-1}\frac{n_2}{n_1}$. Content of Brewster's law: If the plane wave is incident at

Brewster's angle, no matter how the electric field vibration of the incident wave is, the reflected wave contains no longer P component, only S component. The angle of reflection is complementary to the angle of refraction.

② Reflectivity and transmittance (see Fig. 1-26).

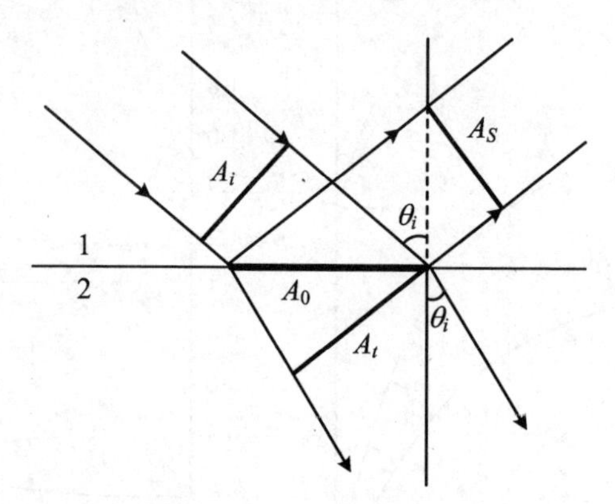

Fig. 1-26 Reflectivity and transmittance

Fig. 1-26 above is the cross-sectional area of the wave A_i, A_S and A_t. A_0 is the area projected on the interface by the wave. If the intensity of the incident light wave is I_{iS}, the energy incident on the interface area A_0 per second is

$$W_{iS} = I_{iS}A_i = I_{iS}A_0 \cos\theta_i$$

And by the intensity of light expression $I = \dfrac{\boldsymbol{n}}{2\mu_0 c}|\boldsymbol{E}_0|^2$, the above equation can be written as:

$$W_{iS} = \frac{n_1}{2\mu_0 c}|\boldsymbol{E}_{0iS}|^2 A_0 \cos\theta_i \tag{1-43}$$

Similarly, the energy of reflected and refracted light can be expressed as:

$$W_{rS} = \frac{n_1}{2\mu_0 c}|\boldsymbol{E}_{0rS}|^2 A_0 \cos\theta_i \tag{1-44}$$

$$W_{tS} = \frac{n_2}{2\mu_0 c}|\boldsymbol{E}_{0tS}|^2 A_0 \cos\theta_t \tag{1-45}$$

so the reflectance and the refractive index are

$$\begin{cases} R_S = \dfrac{W_{rS}}{W_{iS}} = \dfrac{I_{rS}}{I_{iS}} = |\boldsymbol{r}_S|^2 \\ T_S = \dfrac{W_{tS}}{W_{iS}} = \dfrac{\cos\theta_t}{\cos\theta_i} \cdot \dfrac{I_{tS}}{I_{iS}} = \dfrac{n_2 \cos\theta_t}{n_1 \cos\theta_i} \cdot |t_S|^2 \end{cases} \tag{1-46}$$

Similarly, when the incident wave contains only the P component, the reflectance R_P and transmittance T_P of the P component can be calculated:

$$\begin{cases} R_P = \dfrac{W_{rP}}{W_{iP}} = \dfrac{I_{rP}}{I_{iP}} = |\, \boldsymbol{r}_P \,|^2 \\ T_P = \dfrac{W_{tP}}{W_{iP}} = \dfrac{\cos\theta_t}{\cos\theta_i} \cdot \dfrac{I_{tP}}{I_{iP}} = \dfrac{n_2 \cos\theta_t}{n_1 \cos\theta_i} \cdot |\, t_P \,|^2 \end{cases} \tag{1-47}$$

There is a complementary relationship between R_S and T_S, R_P and T_P, namely:

$$\begin{cases} R_S + T_S = 1 \\ R_P + T_P = 1 \end{cases} \tag{1-48}$$

This indicates that at the interface, the energy of incident wave is completely converted into the energy of reflected wave and refracted wave (condition: there is no energy loss such as ***scattering*** and absorption at the interface).

When the incident wave contains both S component and P component, since the directions of the two components are perpendicular to each other, there is

$$|\, \boldsymbol{E}_i \,|^2 = |\, \boldsymbol{E}_{iS} \,|^2 + |\, \boldsymbol{E}_{iP} \,|^2 \tag{1-49}$$

which is

$$I_i = I_{iS} + I_{iP} \Rightarrow W_i = W_{iS} + W_{iP} \tag{1-50}$$

similarly:

$$W_r = W_{rS} + W_{rP}, \quad W_t = W_{tS} + W_{tP} \tag{1-51}$$

Reflectance R and transmittance T can be defined as:

$$R = \frac{W_r}{W_i}, \quad T = \frac{W_t}{W_i} \tag{1-52}$$

Note: The S component (P component) of the incident light wave only contributes to the refractivity and reflectivity of S component (P component).

If the intensity ratio of the S and P components in the incident wave is α, $W_i = \alpha W_{iS} + W_{iP}$, then:

$$R = \frac{1}{1+\alpha}(\alpha R_S + R_P) \tag{1-53}$$

$$T = \frac{1}{1+\alpha}(\alpha T_S + T_P) \tag{1-54}$$

That is, R and T are the weighted average of R_S, R_P, T_S and T_P respectively. But there are still:

$$R + T = 1$$

At normal incidence, the difference between the S and P components disappears. If R_0 and T_0 are used to represent the reflectance and transmittance at this time, then:

$$R_0 = r_0^2 = \left(\frac{n_1 - n_2}{n_1 + n_2}\right)^2 \tag{1-55}$$

$$T_0 = \frac{n_2}{n_1} t_0^2 = \frac{4n_1^2 n_2^2}{(n_1 + n_2)^2} \tag{1-56}$$

These two equations can be used to estimate the reflectance and transmittance of non-normal incidence but with a small incident angle ($\theta_i < 30°$).

(2) $n_1 > n_2$

This is the case where an optically dense medium is incident into an optically thinner medium.

According to the law of refraction, $\theta_i < \theta_t$.

The corresponding incident angle $\theta_t = 90°$ is called the critical angle of total reflection, denoted by θ_c. Namely, $\sin\theta_c = \frac{n_2}{n_1}$.

So we'll talk about it in two ways of $\theta_i \leqslant \theta_c$ and $\theta_i > \theta_c$.

① $\theta_i \leqslant \theta_c$.

In the case of $\theta_t \leqslant 90°$, the Fresnel's equation can be directly used to discuss the properties of reflected and refracted waves, and the analysis method is exactly the same as that of $n_1 < n_2$. The phase shift of refraction and reflection see Fig. 1-27.

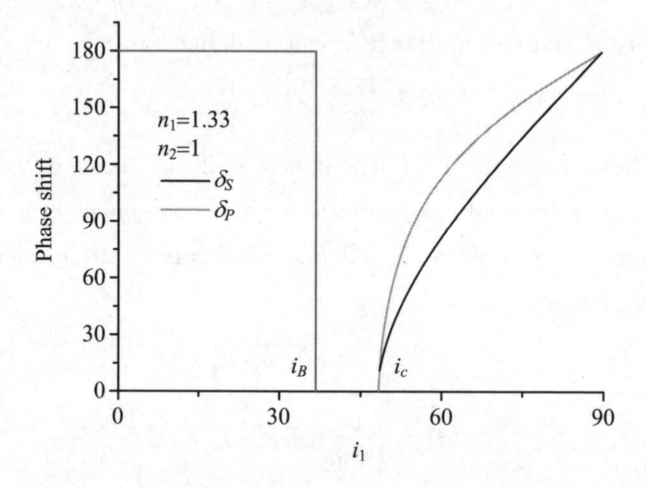

Fig. 1-27 The phase shift of refraction and reflection

For the S component, when $\theta_i < \theta_c$, $r_S > 0$, indicates that there is no half-wave loss, as shown by the black line in the Fig. 1-27. For the P component, in the range $\theta_i < \theta_B$, $r_P < 0$, means that there is half wave loss, while in the range, $\theta_B < \theta_i < \theta_c$, means that there is no half wave loss.

Notice $\sin\theta_c = \tan\theta_B = \frac{n_2}{n_1}$, so it has to be $\theta_B < \theta_c$, which means Brewster's law still holds, and it also means that Brewster's law holds whether $n_1 > n_2$ or $n_1 < n_2$.

Both t_S and t_P are greater than one and increase with the increase of θ_i, but this does not mean that transmission T is greater than one and T must increase with the increase of θ_i.

$$T_S = \frac{n_2 \cos\theta_t}{n_1 \cos\theta_i} \cdot |t_S|^2 \tag{1-57}$$

$$T_P = \frac{n_2 \cos\theta_t}{n_1 \cos\theta_i} \cdot |t_P|^2 \tag{1-58}$$

② $\theta_i > \theta_c$.

Because the critical angle of total reflection satisfies $\sin\theta_c = \frac{n_2}{n_1}$. From the formula, it can be seen that, when $\theta_i > \theta_c$, the phenomenon $\sin\theta_i > \frac{n_2}{n_1}$ will appear, which is obviously unreasonable. The law of refraction no longer holds. But in order to be able to apply Fresnel's equation to the total reflection case, we still have to formally use the relationship $\sin\theta_t = \frac{n_1}{n_2}\sin\theta_i$.

Since θ_t does not exist in the real number range, we can extend the relevant parameters to the complex number field, and is always a real parameter, for which it should be written in the following imaginary form:

$$\cos\theta_t = \sqrt{1-\sin^2\theta_t} = \mathrm{i}\sqrt{\sin^2\theta_t - 1} = \mathrm{i}\sqrt{\left(\frac{n_1}{n_2}\sin\theta_i\right)^2 - 1} \tag{1-59}$$

The physical meaning of imaginary number of $\cos\theta_2$ and the reason for the positive sign will be explained later. Substitute the above equation into Fresnel's equation formula to obtain the complex reflection coefficient:

$$\tilde{\boldsymbol{r}}_S = \frac{\cos\theta_i - \mathrm{i}\sqrt{\sin^2\theta_i - \boldsymbol{n}^2}}{\cos\theta_i + \mathrm{i}\sqrt{\sin^2\theta_i - \boldsymbol{n}^2}} = |\tilde{\boldsymbol{r}}_S| \exp(\mathrm{i}\varphi_{rS}) \tag{1-60}$$

$$\tilde{\boldsymbol{r}}_S = \frac{\boldsymbol{n}^2\cos\theta_i - \mathrm{i}\sqrt{\sin^2\theta_i - \boldsymbol{n}^2}}{\boldsymbol{n}^2\cos\theta_i + \mathrm{i}\sqrt{\sin^2\theta_i - \boldsymbol{n}^2}} = |\tilde{\boldsymbol{r}}_P| \exp(\mathrm{i}\varphi_{rP}) \tag{1-61}$$

and we get:

$$|\tilde{\boldsymbol{r}}_S| = |\tilde{\boldsymbol{r}}_P| = 1 \tag{1-62}$$

$$\tan\frac{\varphi_{rS}}{2} = \boldsymbol{n}^2 \tan\frac{\varphi_{rP}}{2} = -\frac{\sqrt{\sin^2\theta_i - \boldsymbol{n}^2}}{\cos\theta_i} \tag{1-63}$$

where, $\boldsymbol{n}=n_2/n_1$ is the relative refractive index of the two media. $|\tilde{\boldsymbol{r}}_S|$, $|\tilde{\boldsymbol{r}}_P|$ are the amplitude ratio of the S component and P component of the incident light and the reflected light. φ_{rS} and φ_{rP} are the phase changes of the S component and P component light fields in the reflected light relative to the incident light when total reflection occurs. According to the above equation, when total reflection occurs, the intensity of reflected light is equal to the intensity of incident light, and the phase change of reflected light is complicated. The phase difference between them is determined by the following formula:

$$\Delta\varphi = \varphi_{rS} - \varphi_{rP} = 2\arctan\frac{\cos\theta_i\sqrt{\sin^2\theta_i - \boldsymbol{n}^2}}{\sin^2\theta_i} \tag{1-64}$$

Therefore, under a certain $\boldsymbol{n}$, the phase difference can be changed by properly controlling the incident angle, so as to change the polarization state of the reflected light, like the Fresnel prism.

When light is transmitted from an optically dense medium to an optically thinner medium and is total reflected at the interface, the intensity of the refracted light is zero. Here's the question, whether there's an optical field in a ***photophobic*** medium?

When the Fresnel's equation of t_S and t_P is extended to the complex field, it can be found that neither t_S nor t_P, P is equal to zero, that is, there are refracted light waves in the photophobic medium. When total reflection occurs, the light wave field will penetrate into a very thin layer (about the wavelength of light wave) of the second medium. This wave is called ***evanescent wave***.

Now, assuming that the dielectric interface is xOy plane and the incident plane is xOz plane, the transmitted wave field can be expressed as:

$$\boldsymbol{E}_t = \boldsymbol{E}_{0t}\exp[-\mathrm{i}(\omega_t t - \boldsymbol{k}_t \cdot \boldsymbol{r})] = \boldsymbol{E}_{0t}\exp[-\mathrm{i}(\omega_t t - \boldsymbol{k}_t x\sin\theta_t - \boldsymbol{k}_t z\cos\theta_t)] \tag{1-65}$$

Above equation can be rewritten as:

$$\boldsymbol{E}_t = \boldsymbol{E}_{0t}\exp\left[-\mathrm{i}\left(\omega_t t - \boldsymbol{k}_t x\sin\theta_t - \mathrm{i}\boldsymbol{k}_t z\sqrt{\left(\frac{n_1}{n_2}\sin\theta_i\right)^2 - 1}\right)\right] \tag{1-66}$$

$$\boldsymbol{E}_t = \boldsymbol{E}_{0t}\exp\left[-\boldsymbol{k}_t z\sqrt{\left(\frac{n_1}{n_2}\sin\theta_i\right)^2 - 1}\right]\exp\left[-\mathrm{i}\left(\omega_t t - \boldsymbol{k}_t x\frac{n_1}{n_2}\sin\theta_i\right)\right] \tag{1-67}$$

This is an inhomogeneous wave with amplitude attenuation along the z direction and propagation along the x direction of the interface, i.e., total reflection evanescent wave. Thus, the validity of the previous discussion can be explained: Only by taking the imaginary form of $\cos\theta_t$ and a positive sign can the objectively existing evanescent wave be obtained.

Incident wave spends a certain amount of energy to establish the evanescent wave electromagnetic field at the beginning of reaching the interface. When it reaches a stable state, it does not need to provide energy to it. Evanescent wave only propagates along the interface and does not enter the second medium. Therefore, $R_S=1$, $t_S\neq 0$ and $R_P=1$, $t_P\neq 0$ in total reflection do not violate the law of conservation of energy.

1.2.6 Interference

We now turn to the phenomena of interference and diffraction, where light is treated as a periodic wave, and ray optics provides a totally inadequate description. Optical interference involves the superposition of two or more electromagnetic waves in which the electric field vectors are added; the fields add vectorially. The waves are assumed to be nearly monochromic, and have to the same frequency. Two waves can only interfere if they exhibit mutual temporal coherence at a point in space where they interact.

Interference effects occur when two or more wavefronts are superposed, giving a resultant wave amplitude which depends on their relative phases. The principle of superposition of waves states that when two or more propagating waves of same type are incident on the same point, the total displacement at that point is equal to the vector sum of the displacements of the individual waves. If a crest of a wave meets a crest of another wave of the same frequency at the same point, then the magnitude of the displacement is the sum of the individual magnitudes—this is ***constructive interference***. If a crest of one wave meets a trough of another wave then the magnitude of the displacements is equal to the difference in the individual magnitudes—this is known as ***destructive interference***. The constructive interference and destructive interference is shown in Fig. 1-28.

Constructive interference occurs when the phase difference between the waves is a multiple of 2π, whereas destructive interference occurs when the difference is an odd multiple of π. If the difference between the phases is intermediate between these two extremes, then the magnitude of the displacement

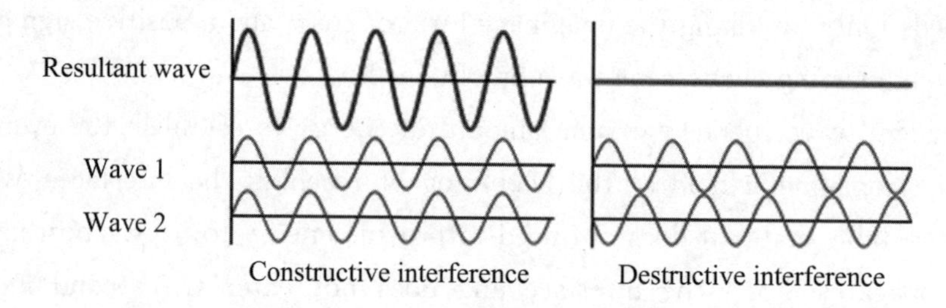

Fig. 1-28 Constructive interference and destructive interference

of the summed waves lies between the minimum and maximum values. ① When two waves with the same frequency with fields E_1 and E_2 interfere, that is, $E=E_1+E_2$. Consider two linearly polarized plane waves that originate from O_1 and O_2, as schematically shown in Fig. 1-29, so that the field oscillations at some arbitrary point of interest P is given by

$$E_1 = E_{O1}\sin(\omega t - kr_1 - \varphi_1)$$
$$E_2 = E_{O2}\sin(\omega t - kr_2 - \varphi_2) \quad (1\text{-}68)$$

where r_1 and r_2 are the distances from O_1 and O_2 to P. These waves have the same ω and k. Due to the process that generates the waves, there is a constant phase difference between given by $\varphi_2 - \varphi_1$. The resultant field at P will be the sum of these two waves, that is, $E=E_1+E_2$. Its irradiance depends on the time average of $E * E$, that is, $\overline{E \cdot E}$, so that:

$$\overline{E \cdot E} = \overline{(E_1 + E_2)\cdot(E_1+E_2)} = \overline{E_1^2} + \overline{E_2^2} + 2\,\overline{E_1 E_2} \quad (1\text{-}69)$$

It is clear that the interference effect is in the $2\,\overline{E_1 \cdot E_2}$ term. We can simplify the above equation a little further by assuming E_{O1} and E_{O2} are parallel with magnitudes E_{O1} and E_{O2}. Further, irradiance of the interfering waves are $I_1=\frac{1}{2}c\varepsilon_0 E_{O1}^2$ and $I_2=\frac{1}{2}c\varepsilon_0 E_{O2}^2$ so that the resultant irradiance is given by the sum of individual irradiances, I_1 and I_2, and has an additional third term I_{21}, that is

$$I = I_1 + I_2 + 2\,(I_1 I_2)^{1/2}\cos\delta \quad (1\text{-}70)$$

where the last term is usually written as $2(I_1 I_2)^{1/2}\cos\delta = I_{21}$, and δ is a phase difference given by

$$\delta = k(r_2 - r_1) + (\varphi_2 - \varphi_1) \quad (1\text{-}71)$$

① Kasap S O. Optoelectronics and Photonics: Principles & Practices[M]. 2nd ed. New Jersey: Prentice Hall, Inc., 2012:67-69.

Since we are using nearly ***monochromatic waves***, $\varphi_2 - \varphi_1$ is constant, and the interference therefor depends on the term $k(r_2 - r_1)$, which represents the phase difference between the two waves as a result of the optical path difference between the waves. As we move point P, $k(r_2 - r_1)$ will change the optical path difference between the two waves will change; and the interference will therefore also change.

Suppose $\varphi_2 - \varphi_1 = 0$, the two waves are emitted from a spatially coherent source. Then, if the path difference $k(r_2 - r_1)$ is zero, 2π or a multiple of 2π, that is, $2m\pi$, $m = 0, \pm 1, \pm 2, \cdots$, then the interference intensity I will be maximum; such interference is defined as constructive interference. If the path difference $k(r_2 - r_1)$ is π or 3π or an odd multiple of π, $(2m+1)\pi$, then the waves will be 180° out of phase, and the interference intensity will be minimum; such interference is defined as destructive interference; both constructive and destructive intensity are shown in Fig. 1-29. The maximum and minimum irradiance are given by

$$I_{\max} = I_1 + I_2 + 2(I_1 I_2)^{1/2}$$
$$I_{\min} = I_1 + I_2 - 2(I_1 I_2)^{1/2} \tag{1-72}$$

If the interfering beams have equal irradiance, then $I_{\max} = 4I_1$ and $I_{\min} = 0$.

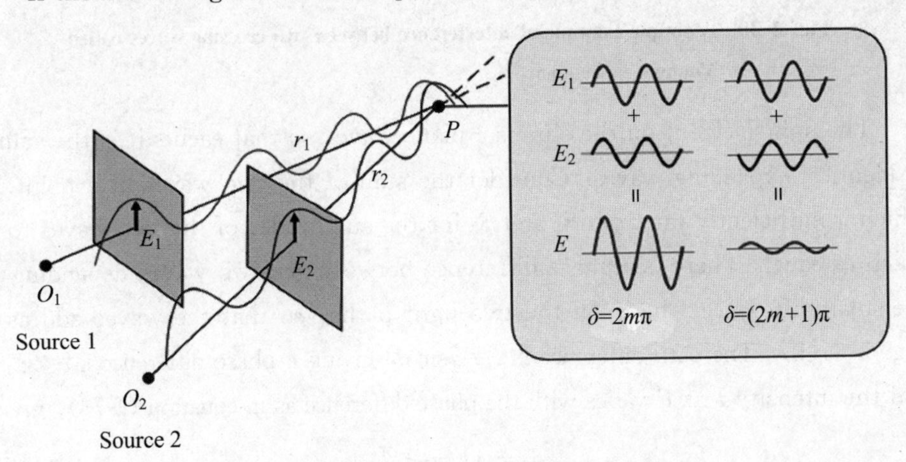

Fig. 1-29 Constructive and destructive intensity①

A simple example of interference between two crossing waves is Young's

① Smith F G, King T A, Wilkins D. Optics and Photonics: An Introduction[M]. 2nd ed. New York: John Wiley & Sons, Inc., 2007:188-190.

experiment, which provided the first demonstration of the wave nature of light. Two closely spaced narrow slits, S_1 and S_2 in Fig. 1-30, transmit two elements of a light wave from a single source. The two sets of waves spread by diffraction, then overlap and interfere. If they then illuminate a screen, there will be light and dark bands across the illuminated patch; these are the interference fringes. Fig. 1-30 shows that the geometry of the fringes becomes simpler at increasing distance from the slits, where the two sets of waves from S_1 and S_2 behave like two sets of nearly plane waves crossing at a small angle, like the plane waves of Fig. 1-30. We will consider the effect for monochromatic light.

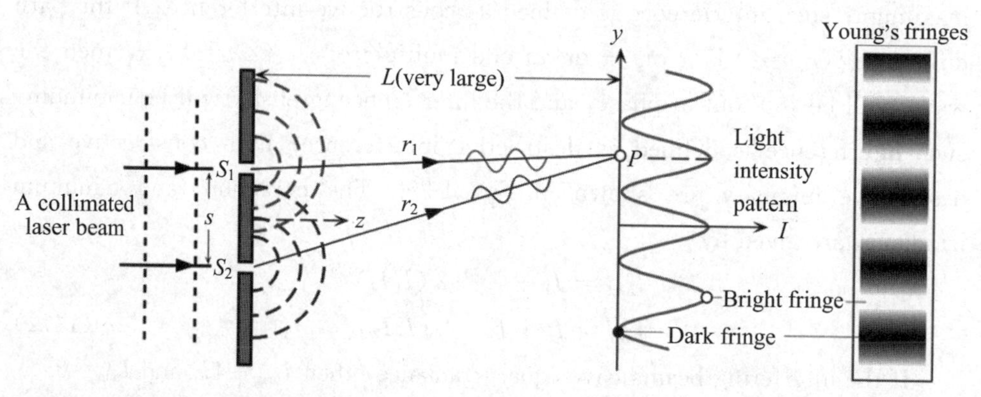

Fig. 1-30　A simple example of interference between two crossing waves called Young's experiment

The light incident on the slits is a plane wave, so that each slit is the source of identical expanding waves. Consider the sum of the two waves at a point P, which is sufficiently far from S_1 and S_2 for the amplitudes of the two waves to be taken as equal. There is a phase difference between the two waves depending on the small difference l between the two light paths, so that the waves add as in Fig. 1-31(b). The path difference is $l = d\sin\theta$, giving a phase difference $\varphi = 2\pi l/\lambda$, and the intensity I at P varies with the phase difference as in equation (1-73), giving

$$I = 4I_0\cos^2\frac{\pi l}{\lambda} \tag{1-73}$$

where I_0 is the intensity of each wave at P. The phase reference may be taken at O, half-way between the slits. The waves are then advanced and retarded on the reference by $\pi l/\lambda$, as in the phasor diagram of Fig. 1-31(b). Thus where constructive interference occurs the intensity is four times that due to one slit, or twice the intensity due to two slits if interference did not happen. The conditions for constructive and destructive interference are as follows:

$$\text{Constructive } l = d\sin\theta = N\lambda \tag{1-74}$$

$$\text{Destructive } l = d\sin\theta = \left(N+\frac{1}{2}\right)\lambda \tag{1-75}$$

where N is an integer. N is called the order of the interference; it is the number of whole wavelengths difference in the paths to points where constructive interference takes place. The bright and dark "fringes" which appear on a screen placed anywhere to the right of the double slit system can therefore each be labelled according to their order. The highest possible order is the integral part of d/λ. The fringes are spaced uniformly in $\sin\theta$ at intervals λ/d, and at a distance x the linear spacing, in the small-angle approximation, is $x\lambda/d$.

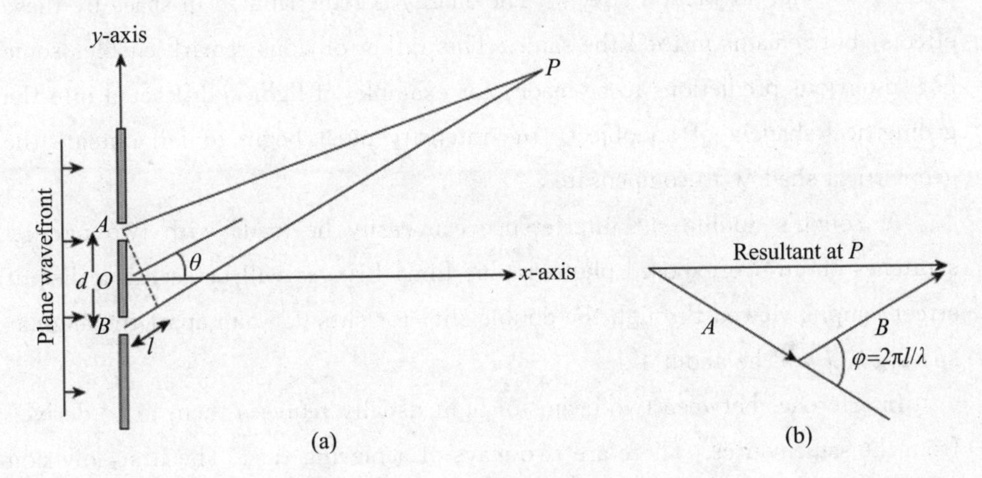

Fig. 1-31 Young's experiment

(a) Two wavelets spread out from the pair of slits, and interfere. Constructive or destructive interference occurs at P according to whether the path difference l is $N\lambda$ or $\left(N+\frac{1}{2}\right)\lambda$. (b) Phasor diagram for the sum at P, at a large distance from the slits. The phase reference (giving a horizontal phasor) is at O and the phase difference between the two wave is $\varphi=2\pi l/\lambda$.

The condition that the distance is sufficient to allow the use of the simple relation $l=d\sin\theta$ is important; it is equivalent to the condition that the phase of any elementary wave from the screen is a linear function of coordinates x and y in the plane of the screen. This is the condition for Fraunhofer diffraction, of which the present example is a special case. Under this condition, the whole of the screen, whatever the pattern of apertures, can be considered as a single diffracting object.

The amplitude of the interference pattern of a pair of slit sources is the function:

$$A(\theta) = A(0)\left|\cos\left(\frac{\pi d}{\lambda}\sin\theta\right)\right| \tag{1-76}$$

$A(0)$ is the amplitude of the diffracted wave when $\theta=0$. The intensity is the square of $A(\theta)$, giving cos-squared fringes:

$$I(\theta) = I(0)\cos^2\left(\frac{\pi d}{\lambda}\sin\theta\right) \tag{1-77}$$

Notice that the average intensity across several fringes is $I(0)/2=2I_0$, because the peaks of the upper half of the cos-squared fringes just fill the troughs of the lower half. This is an example of an important general principle of all interference and diffraction effects: The energy is redistributed in space by these effects, but remains in total the same. This rather obvious remark enables some not so obvious predictions to be made; for example, if light is diffracted into the geometrical shadow of an object, the intensity must begin to fall outside the geometrical shadow to compensate.

A Young's double slit interference can easily be made with two parallel scratches on an overexposed photographic film. Fringes will be seen if a distant street lamp is viewed through the double slit; for slits 0.5 mm apart the angular spacing λ/d will be about $4'$.

Interference between two beams of light usually requires them to be derived from the same source. There are two ways of achieving this. The first, division of wavefront, means utilizing spatially separate parts of the wavefront as distinct sources, as in Young's double slit or by using a ***diffraction grating***. The second, which is the main subject of this chapter, is to divide the amplitude of the wave by partial reflection, obtaining identical wavefronts which can be brought together by different paths. The most familiar example of interference by this division of amplitude is the pattern of coloured fringes seen in soap bubbles and thin oil films.

Example 1.2.3

Problem

Why soap bubble is chromatic? Soap bubble with color interference fringes see Fig. 1-32.

Fig. 1-32 Soap bubble with color interference fringes

Solution

Soap film is a transparent film and has an upper and lower surface. When sunlight hits the surface of a soap film, it is reflected off the top and bottom surfaces of the film, so that a beam of light becomes two interfering reflections.

Sunlight is made up of red, orange, yellow, green, blue, indigo and purple, seven kinds of monochromatic lights composed of visible light, because the soap film's thickness is not uniform everywhere, according to the theory of constructive interference and destructive interference, the position of different thickness on the membrane surface, both have different color of light, constructive interference conditions will have different colors of light when they meet destructive interference conditions.

As a result, the seven monochromatic rays of sunlight, at different thicknesses of the soap film, appear to intensify, weaken, and even cancel each other out. In terms of color expression, there will be some places on the soap film are more red, some places are blue, some places are green, and some places will show other colors.

In this way, the soap film breaks up the sunlight into colorful patterns. In addition, the flow of the liquid causes a slight change in the thickness of the film, and the colored markings on the film flow accordingly, so that the colored ripples seen are also flowing.

1.2.7 Diffraction

Diffraction is the spreading of waves from a wavefront limited in extent, occurring either when part of the wavefront is removed by an obstacle, or when all but a part of the wavefront is removed by an aperture or stop. The general theory which describes diffraction at large distances is due to Fraunhofer, and is referred to as ***Fraunhofer diffraction***.

The Fraunhofer theory of diffraction is concerned with the angular spread of light leaving an aperture of arbitrary shape and size; if the light then falls on a screen at a large distance, the pattern of illumination is described adequately by this angular distribution. But if the screen is close to the aperture, so that one might expect to see a sharp shadow at the edges, we see instead diffraction fringes, whose analysis involves a theory introduced by Fresnel. A famous prediction of Fresnel's theory was that the shadow of a circular object should have a central bright spot; the demonstration that this indeed exists was a powerful argument in establishing the wave theory of light. ①

1.2.7.1 *Single-slit Diffraction*

A long slit of infinitesimal width which is illuminated by light diffracts the light into a series of circular waves and the wavefront which emerges from the slit is a cylindrical wave of uniform intensity see Fig. 1-33.

A slit which is wider than a wavelength produces interference effects in the space downstream of the slit. These can be explained by assuming that the slit behaves as though it has a large number of point sources spaced evenly across the width of the slit. The analysis of this system is simplified if we consider light of a single wavelength. If the incident light is coherent, these sources all have the

① Smith F G, King T A, Wilkins D. Optics and Photonics: An Introduction[M]. 2nd ed. Now York: John Wiley & Sons, Inc., 2007:231.

same phase. Light incident at a given point in the space downstream of the slit is made up of contributions from each of these point sources and if the relative phases of these contributions vary by 2π or more, we may expect to find minima and maxima in the diffracted light. Such phase differences are caused by differences in the path lengths over which contributing rays reach the point from the slit.

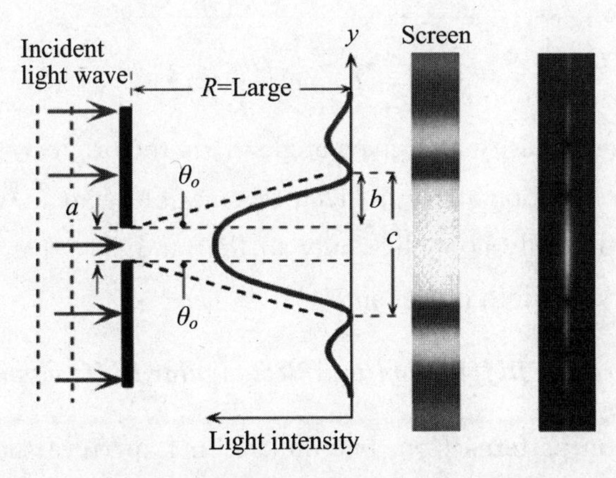

Fig. 1-33 Diffraction from a single slit[①]

We can find the angle at which a first minimum is obtained in the diffracted light by the following reasoning. The light from a source located at the top edge of the slit interferes destructively with a source located at the middle of the slit, when the path difference between them is equal to $\lambda/2$. Similarly, the source just below the top of the slit will interfere destructively with the source located just below the middle of the slit at the same angle. We can continue this reasoning along the entire height of the slit to conclude that the condition for destructive interference for the entire slit is the same as the condition for destructive interference between two narrow slits a distance apart that is half the width of the slit. The path difference is given by $d\sin(\theta)/2$ so that the minimum intensity occurs at an angle $\theta_{\min}$ given by

$$d\sin\theta_{\min} = \lambda \tag{1-78}$$

where d is the width of the slit, $\theta_{\min}$ is the angle of incidence at which the minimum intensity occurs, and λ is the wavelength of the light.

① Kasap S O. Optoelectronics and Photonics: Principles & Practices[M]. 2nd ed. New Jersey: Prentice Hall, Inc., 2012:76.

A similar argument can be used to show that if we imagine the slit to be divided into four, six, eight parts, etc. , minima are obtained at angles θ_n given by

$$d\sin\theta_n = n\lambda \tag{1-79}$$

where n is an integer other than zero.

There is no such simple argument to enable us to find the maxima of the diffraction pattern. The intensity profile can be calculated using the Fraunhofer diffraction equation as:

$$I(\theta) = I_0\,\mathrm{sinc}^2\left(\frac{d\pi}{\lambda}\sin\theta\right) \tag{1-80}$$

where, $I(\theta)$ is the intensity at a given angle, I_0 is the original intensity, and the unnormalized sinc function above is given by $\mathrm{sinc}(x)=\sin(x)/(x)$ if $x\neq0$, and $\mathrm{sinc}(0) = 1$. This analysis applies only to the far field, that is, at a distance much larger than the width of the slit.

1.2.7.2 *Circular Diffraction* and *Rectangular Diffraction*①

The diffraction patterns from two-dimensional aperture such as rectangular and circular apertures are more complicated to calculate but they use the same principle based on the multiple interference of waves emitted from all point sources in the apérture. The diffraction pattern of a rectangular aperture is shown in Fig. 1-34. It involves the multiplication of two individual single slit (sinc) functions, one slit of width a along the horizontal axis, and the other of width b along the vertical axis.

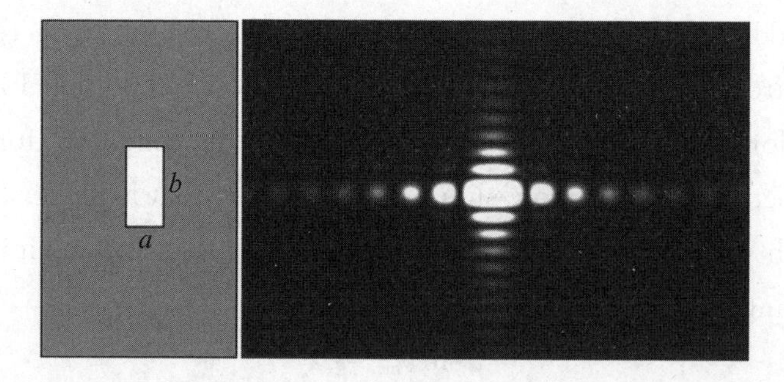

Fig. 1-34 The diffraction pattern of a rectangular aperture

---

① Kasap S O. Optoelectronics and Photonics: Principles & Practices[M]. 2nd ed. New Jersey: Prentice Hall, Inc. , 2012:74-80.

The diffraction pattern from a circular aperture, known as ***Airy rings***, was shown in Fig. 1-35. We can, as we did for the single slit, sum all waves emanating from every point in the circular aperture, taking into account their relative phases when they arrive at the screen to obtain the actual intensity pattern at the screen. The result is that the diffraction pattern is a Bessel function of the first kind, and not a simply rotated sinc function. The central white spot is called the Airy disk; its radius corresponds to the radius of the first dark ring. We can still imagine how diffraction occurs from a circular aperture, denoted as D. The angular position θ_o of the first dark ring is determined by the diameter D of the aperture and the wavelength λ, and is given by

$$\sin\theta_o = 1.22\frac{\lambda}{D} \tag{1-81}$$

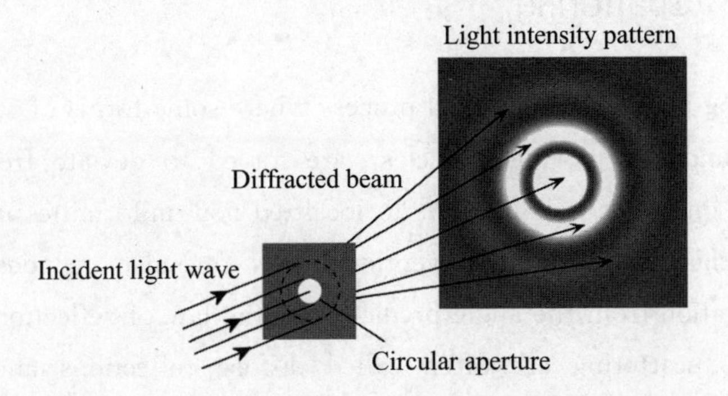

Fig. 1-35 The diffraction pattern from a circular aperture

Resolution of imaging systems is limited by diffraction effects. According to Fig. 1-36 and Fig. 1-37, As points S_1 and S_2 get closer, eventually the Airy patterns overlap so much that the resolution is lost. The Rayleigh criterion allows the minimum angular separation two of the point sources be determined.

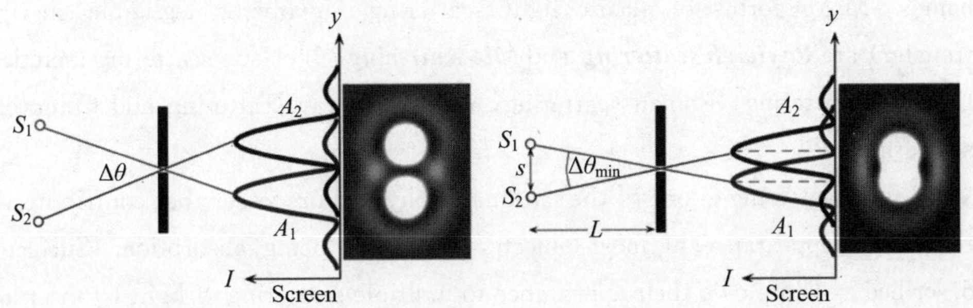

Fig. 1-36 Resolution of imaging system is limited by diffraction effects

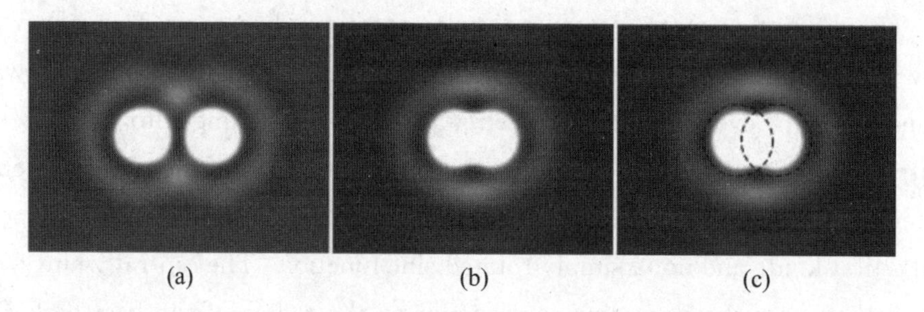

Fig. 1-37 Image of two point sources captured through a small circular aperture

(a) The two points are fully resolved since the diffraction patterns of the two sources are sufficiently separated. (b) The two images near the Rayleigh limit of resolution. (c) The first dark ring through the center of the bright Airy disk of the other pattern.

1.2.8 Scattering

Scattering is a general physical process where some forms of radiation, such as light, sound, or moving particles, are forced to deviate from a straight trajectory by one or more paths due to localized non-uniformities in the medium through which they pass. In conventional use, this also includes deviation of reflected radiation from the angle predicted by the law of reflection. Reflections that undergo scattering are often called diffuse reflections and unscattered reflections are called specular (mirror-like) reflections.

Electromagnetic waves are one of the best known and most commonly encountered forms of radiation that undergo scattering. Scattering of light and radio waves (especially in radar) is particularly important. Several different aspects of electromagnetic scattering are distinct enough to have conventional names. Major forms of elastic light scattering (involving negligible energy transfer) are ***Rayleigh scattering*** and ***Mie scattering***. Inelastic scattering includes Brillouin scattering, Raman scattering, inelastic X-ray scattering and Compton scattering.

Light scattering is one of the two major physical processes that contribute to the visible appearance of most objects, the other being absorption. Surfaces described as white owe their appearance to multiple scattering of light by internal or surface inhomogeneities in the object, for example by the boundaries of transparent microscopic crystals that make up a stone or by the microscopic fibers in a sheet of paper. More generally, the gloss (lustre or sheen) of the surface is determined by scattering. Highly scattering surfaces are described as being dull or having a matte finish, while the absence of surface scattering leads to a glossy appearance, as with polished metal or stone.

Spectral absorption, the selective absorption of certain colors, determines the color of most objects with some modification by elastic scattering. The apparent blue color of veins in skin is a common example where both spectral absorption and scattering play important and complex roles in the coloration. Light scattering can also create color without absorption, often shades of blue, as with the sky (Rayleigh scattering), the human blue iris, and the feathers of some birds. However, resonant light scattering in nanoparticles can produce many different highly saturated and vibrant hues, especially when ***surface plasmon resonance*** is involved.

Models of light scattering can be divided into three domains based on a dimensionless size parameter, α which is defined as $\alpha=\pi D_p/\lambda$, where πD_p is the circumference of a particle and λ is the wavelength of incident radiation. Based on the value of α, these domains are:

$\alpha \ll 1$: Rayleigh scattering (small particle compared to wavelength of light);

$\alpha \approx 1$: Mie scattering (particle about the same size as wavelength of light, valid only for spheres);

$\alpha \gg 1$: Geometric scattering (particle much larger than wavelength of light).

Rayleigh scattering is a process in which electromagnetic radiation (including light) is scattered by a small spherical volume of variant refractive index, such as a particle, bubble, droplet, or even a density fluctuation. Lord Rayleigh, from whom it gets its name, first modeled this effect successfully. In order for Rayleigh's model to apply, the sphere must be much smaller in diameter than the wavelength (λ) of the scattered wave; typically the upper limit is taken to be about 1/10 the wavelength. In this size regime, the exact shape of the scattering center is usually not very significant and can often be treated as a sphere of equivalent volume. The inherent scattering that radiation undergoes passing through a pure gas is due to microscopic density fluctuations as the gas molecules move around, which are normally small enough in scale for Rayleigh's model to apply. This scattering mechanism is the primary cause of the blue color of the Earth's sky on a clear day, as the shorter blue wavelengths of sunlight passing overhead are more strongly scattered than the longer red wavelengths according to Rayleigh's famous $1/\lambda^4$ relation. Along with absorption, such scattering is a major cause of the attenuation of radiation by the atmosphere. The degree of scattering varies as a function of the ratio of the particle diameter to the wavelength of the radiation, along with many other factors including polarization, angle, and coherence.

For larger diameters, the problem of electromagnetic scattering by spheres was first solved by Gustav Mie, and scattering by spheres larger than the Rayleigh range is therefore usually known as Mie scattering. In the Mie regime, the shape of the scattering center becomes much more significant and the theory only applies well to spheres and, with some modification, spheroids and ellipsoids. Closed-form solutions for scattering by certain other simple shapes exist, but no general closed-form solution is known for arbitrary shapes.

Both Mie and Rayleigh scattering are considered elastic scattering processes, in which the energy (and thus wavelength and frequency) of the light is not substantially changed. However, electromagnetic radiation scattered by moving scattering centers does undergo a Doppler shift, which can be detected and used to measure the velocity of the scattering center(s) in forms of techniques such as LIDAR and radar. This shift involves a slight change in energy.

At values of the ratio of particle diameter to wavelength more than about 10, the laws of geometric optics are mostly sufficient to describe the interaction of light with the particle, and at this point the interaction is not usually described as scattering.

For modeling of scattering in cases where the Rayleigh and Mie models do not apply such as irregularly shaped particles, there are many numerical methods that can be used. The most common one is finite-element method which solves Maxwell's equations to find the distribution of the scattered electromagnetic field. Sophisticated software packages exist which allow the user to specify the refractive index or indices of the scattering feature in space, creating a 2-dimensional or sometimes 3-dimensional model of the structure. For relatively large and complex structures, these models usually require substantial execution times on a computer.

When light passes through a uniform transparent medium (such as water or glass), it is difficult to see it from the side. If the medium is heterogeneous, such as a cloudy liquid with suspended particles, we can clearly see the path of the beam from the side as the inhomogeneity in the medium causes the light to scatter in all direction. This phenomenon is called light scattering.

Assume that the incident light intensity is I_0, the transmission light intensity is

$$I = I_0 e^{-(\alpha_a+\alpha_s)l} = I_0 e^{-\alpha l} \tag{1-82}$$

where α_a is attenuation coefficient, α_s is scattering coefficient, α is the total coefficient of the reduced light.

1. What is the scattering in nonuniform medium?

Nonuniformity of optical properties:

(1) A ***homogeneous substance*** is distributed with a large number of particles of other substances with a different refractive index.

(2) The irregular aggregation of the components of matter itself (particles). For examples, dust, smoke, fog, suspension, emulsion, ground glass, etc.

Features: The linearity of impurity particles is generally smaller than the wavelength of light, the distance between each other is greater than the wavelength, the arrangement is irregular, the vibration under the light has no fixed phase relationship, any point can be seen the light emits sub-wave superposition, which do not cancel out, forming scattered light.

2. What is the difference between scattering and reflection, diffusion and diffraction?

(1) Scattering: The "secondary wave" emission center is arranged differently, which is irregular, while in direct beam, the reflection and refraction are regular, and the object's linearity is much greater than the wave length.

(2) Reflection: The law of reflection only applies under the condition that the medium surface is an ideal smooth plane (mirror). (Note: Any material surface can never be a geometric plane, due to molecular thermal motion, the surface is constantly changing, but as long as the "convex" and "concave" part of the line is far less than the wavelength of light, can be considered as an ideal smooth plane.)

(3) Diffuse reflection occurs mainly because the mirror is not ideal. In this case, it can be considered that the reflected light beam is a superposition of the reflected light from many small "mirrors". The light reflected from these small "mirrors" still follows the law of reflection, but these small "mirrors" are

chaotic. So diffuse reflection and scattering are different.

(4) The difference between scattering and diffraction: The inhomogeneous region of diffraction (holes, slits, etc.) can be compared with wavelength. Scattering is a small region formed by a large number of irregular and heterogeneous small regions. The small region is generally smaller than the wavelength. Although there is diffraction in the small region, the diffraction phenomenon cannot be observed as a whole because of the incoherent overlap of the irregular arrangement.

3. What is the Rayleigh scattering?

Drip milk in the water and observe the muddy matter. Viewed from the front, i.e., in the direction perpendicular to the incident light, different phenomena can be found.

In Fig. 1-38, z is the scattered light band is bluish blue, short wave or more dispersed. x is the red light.

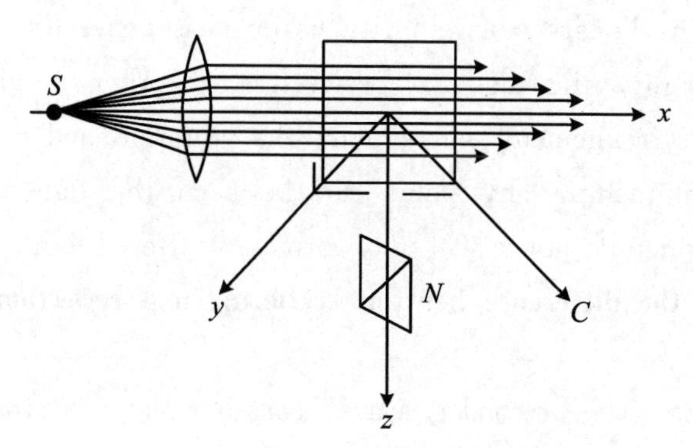

Fig. 1-38 Scattering in muddy water

Let the incident light, distribution is $f(\lambda)$, then the scattered light intensity distribution is $f(\lambda)\lambda^{-4}$.

The scattering of incident light by particles whose linearity is less than the wavelength of light is usually called Rayleigh scattering.

Explanation: The scattered light is superimposed by the subwaves generated by the forced oscillator:

$$\bar{S} = \bar{I} = \frac{1}{\mu C}\bar{E}^2 = \frac{\mu_0 e^2 A^2 \omega^4}{32\pi^2 CR^2}\sin^2\theta \tag{1-83}$$

Observe in the direction of x or θ, $\bar{I}$ is proportional to ω^4.

When the frequency of the scattered light is the same as that of the incident light, $\overline{I}$ is inversely proportional to λ^4.

This law shows that the scattered light is mainly composed of short wave, while the transmitted light is red because of the lack of short wave components.

Note: If the particle's size exceeds the wavelength, the phase difference between different incident light positions in a particle cannot be ignored, so the relationship between the scattering intensity and the incident light is not so simple (the power is less than four).

Therefore, when red light passes through the mist, it has a stronger penetration than blue light. Moreover, the infrared light has a stronger penetration than red light, it is suitable for transport photography or remote sensing technology.

Example 1.2.4

Problem

Why is the sky light blue? The scattering of sunlight see Fig. 1-39.

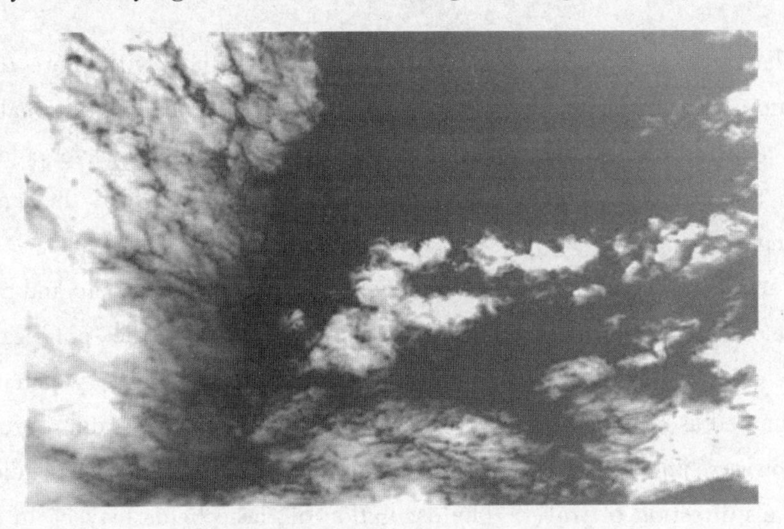

Fig. 1-39 The scattering of sunlight

Solution

Part of atmospheric scattering comes from suspended dust, and most of them is molecular scattering caused by density fluctuations. The effect of Rayleigh's inverse ratio law is more obvious.

Light blue and blue light scatters more than yellow and red light.

The brightness of the daytime sky is due entirely to the scattering of sunlight by the atmosphere. Otherwise the sun is a red ball of fire on a dark background.

The sun is red in the morning or evening at sunset, this is because the sun is almost parallel to the ground plane. Through the thick atmosphere, all the short wavelength scatter laterally, almost only red light wavelengths reach the observer, but at the moment, the sky is still light, the cloud illuminated by the sun is red too. However, noon sunlight through the thin atmosphere does not much scatter, so the sun is still white.

White clouds are made up of water droplets in the atmosphere, and the radius of the water droplets is not so small compared to the wavelength that Rayleigh scattering no longer applies. Thus, the scattering produced by water droplets does not have much to do with wavelength, which is why clouds appear white.

1.2.9 Polarization of Light

Polarization is a property of waves that can oscillate with more than one orientation. Electromagnetic waves, such as light, and gravitational waves exhibit polarization; sound waves in a gas or liquid do not have polarization because the medium vibrates only along the direction in which the waves are travelling.

In an electromagnetic wave such as light, both the electric field and magnetic field are oscillating but in different directions; by convention the "polarization" of light refers to the polarization of the electric field. Light which can be approximated as a plane wave in free space or in an isotropic medium propagates as a ***transverse wave*** — both the electric and magnetic fields are perpendicular to the wave's direction of travel. The oscillation of these fields may be in a single direction (***linear polarization***), or the field may rotate at the optical frequency (***circular polarization*** or ***elliptical polarization***). In that case the direction of the fields' rotation, and thus the specified polarization, may be either clockwise or counter-clockwise; this is referred to as the wave's chirality or handedness.

1.2.9.1 Plane Waves

The simplest manifestation of polarization to visualize is that of a ***plane wave***, which is a good approximation of most light waves (a plane wave is a wave with infinitely long and wide wavefronts). For plane waves Maxwell's equations, specifically Gauss's laws, impose the transversality requirement that the electric and magnetic field be perpendicular to the direction of propagation and to each other. Conventionally, when considering polarization, the electric field vector is described and the magnetic field is ignored since it is perpendicular to the electric field and proportional to it. The electric field vector of a plane wave may be arbitrarily divided into two perpendicular components labeled x and y (with z indicating the direction of travel). For a simple harmonic wave, where the amplitude of the electric vector varies in a sinusoidal manner in time, the two components have exactly the same frequency. However, these components have two other defining characteristics that can differ. First, the two components may not have the same amplitude. Second, the two components may not have the same phase, that is they may not reach their maxima and minima at the same time. Mathematically, the electric field of a plane wave can be written as:

$$\boldsymbol{E}(\boldsymbol{r},t) = \mathrm{Re}[(A_x, A_y \cdot \mathrm{e}^{\mathrm{i}\varphi}, 0)\mathrm{e}^{\mathrm{i}(kz-\omega t)}] \tag{1-84}$$

or alternatively,

$$\boldsymbol{E}(\boldsymbol{r},t) = [A_x \cdot \cos(kz-\omega t), A_y \cdot \cos(kz-\omega t+\varphi), 0] \tag{1-85}$$

where A_x and A_y are the amplitudes of the x and y directions and φ is the relative phase between the two components.

1.2.9.2 Polarization State

The shape traced out in a fixed plane by the electric vector as such a plane wave passes over it (Lissajous figure) is a description of the polarization state. The following figures show some examples of the evolution of the electric field vector (black), with time (the vertical axes), at a particular point in space, along with its x and y components, and the path traced by the tip of the vector in the plane: The same evolution would occur when looking at the electric field at a particular time while evolving the point in space, along the direction opposite to propagation.

In Fig. 1-40(a), the two orthogonal (perpendicular) components are in phase. In this case the ratio of the strengths of the two components is constant, so the direction of the electric vector (the vector sum of these two components) is constant. Since the tip of the vector traces out a single line in the plane, this special case is called linear polarization. The direction of this line depends on the relative amplitudes of the two components.

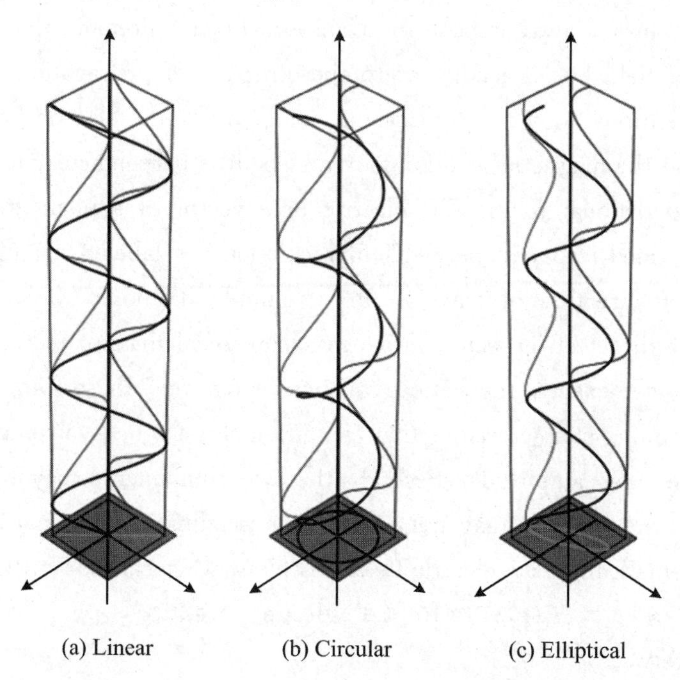

Fig. 1-40 Three different kinds of polarization state

In Fig. 1-40(b), the two orthogonal components have exactly the same amplitude and are exactly ninety degrees out of phase. In this case one component is zero when the other component is at maximum or minimum amplitude. There are two possible phase relationships that satisfy this requirement: The x component can be ninety degrees ahead of the y component or it can be ninety degrees behind the y component. In this special case, the electric vector traces out a circle in the plane, so this special case is called circular polarization. The direction the field rotates in depends on which of the two phase relationships exists. These cases are called right-hand circular polarization and left-hand circular polarization, depending on which way the electric vector rotates and the chosen convention.

Another case is when the two components are not in phase and either do not have the same amplitude or are not ninety degrees out of phase, though their phase offset and their amplitude ratio are constant. This kind of polarization is called elliptical polarization because the electric vector traces out an ellipse in the plane (the polarization ellipse). This is shown in Fig. 1-40(c).

1.2.9.3 Unpolarized Light

Most sources of electromagnetic radiation contain a large number of atoms or molecules that emit light. The orientation of the electric fields produced by these emitters may not be correlated, in which case the light is said to be ***unpolarized.*** If there is partial correlation between the emitters, the light is partially polarized. If the polarization is consistent across the spectrum of the source, partially polarized light can be described as a superposition of a completely unpolarized component, and a completely polarized one. One may then describe the light in terms of the degree of polarization, and the parameters of the polarization ellipse.

Example 1.2.5

Problem

How does 3D movies works? 3D film images see Fig. 1-41.

Fig. 1-41 3D film images

Solution

Images taken from two different perspectives are printed in two different colors on the same frame. When viewed with the naked eye, the blurred double image will appear, and the stereo effect can only be seen through the corresponding red and blue glasses. That is, the color is filtered red and blue to form parallax. At this time, the different images seen by the two eyes will overlap in the brain to present the 3D stereo effect.

New Words and Expressions

［1］ **electromagnetic wave**：电磁波。由同相振荡且互相垂直的电场与磁场在空间中以波的形式移动，其传播方向垂直于电场与磁场构成的平面，能有效地传递能量和动量。电磁辐射可以按照频率分类，从低频率到高频率，包括无线电波、微波、红外线、可见光、紫外光、X射线和伽马射线等。

［2］ **reflection**：反射。它是指光在两种物质分界面上改变传播方向又返回原来物质中的现象。

［3］ **refraction**：折射。它是指光从一种透明介质斜射入另一种透明介质时，传播方向一般会发生变化，这种现象叫作光的折射。

［4］ **interference**：干涉。两列或两列以上的波在空间中重叠时发生叠加而形成新波形的现象。

［5］ **polarization**：偏振。光波电矢量振动的空间分布对于光的传播方向失去对称性的现象叫作光的偏振，只有横波才能产生偏振现象。

［6］ **Thomas Young's double slit experiment**：托马斯杨氏双缝实验。这是一项著名的光学实验。在1807年，托马斯·杨总结出版了他的《自然哲学讲义》，里面综合整理了他在光学方面的工作，并在里面第一次描述了双缝实验：把一支蜡烛放在一张开了一个小孔的纸前面，这样就形成了一个点光源（从一个点发出的光源）。在纸的后面再放一张纸，不同的是第二张纸上开了两道平行的狭缝。从小孔中射出的光穿过两道狭缝投到屏幕上，就会形成一系列明暗交替的条纹，这就是现在众人皆知的双缝干涉条纹。

［7］ **diffraction**：衍射。在经典物理学中，波在穿过狭缝、小孔或圆盘之类的障碍物后会发生不同程度的弯散传播。假设将一个障碍物放置在光源和观察屏之间，则会有光亮区域与阴晦区域出现于观察屏，而且这些区域的边界并不清晰，是一种明暗相间的复杂图样，这种现象称为衍射。当波在其传播路径上遇到障碍物

时,都有可能发生这种现象。

[8] **photoelectric effect**: 光电效应。在光的照射下,某些物质内部的电子会被光子激发出来而形成电流,即光生电。

[9] **wavelength**: 波长。它是指沿着波的传播方向,在波的图形中相对平衡位置的位移时刻相同的相邻的两个质点之间的距离。

[10] **visible light**: 可见光。它是电磁波谱中人眼可以感知的部分,可见光谱没有精确的范围,一般人的眼睛可以感知的电磁波的波长为 400~700 nm。

[11] **electromagnetic spectrum**: 电磁波谱。它是在空间中传播着的交变电磁场,即电磁波。为了对各种电磁波有全面的了解,人们按照波长或频率、波数、能量的顺序把这些电磁波排列起来,这就是电磁波谱。

[12] **ultraviolet light**: 紫外线。紫外线是电磁波谱中波长为 10~400 nm 辐射的总称,肉眼无法看到紫外线。

[13] **X-rays**: X 射线,波长介于紫外线和伽马射线之间的电磁辐射。

[14] **gamma rays**: 伽马射线。它是原子核能级跃迁蜕变时释放出的射线,是波长短于 0.2 埃的电磁波,具有很强的穿透力。

[15] **Snell's law**: 斯涅尔定律。它是描述光的折射规律的定律,即光入射到不同介质的界面上会发生反射和折射。

[16] **refractive index**: 折射率。它是指光在真空中的速度与光在该材料中的速度的比率,材料的折射率越高,入射光发生折射的能力越强。

[17] **chromatic aberration**: 色差,又称色像差。它是透镜成像的一个严重缺陷,色差简单来说就是颜色的差别,发生在以多色光为光源的情况下,单色光不产生色差。不同波长的光,其通过透镜时的折射率各不相同,物方一个点,在像方则可能形成一个色斑。

[18] **dispersion**: 色散。复色光分解为单色光而形成光谱的现象叫作光的色散。色散可以利用棱镜或光栅等作为"色散系统"的仪器来实现。复色光进入棱镜后,由于它对各种频率的光具有不同的折射率,各种色光的传播方向有不同程度的偏折,因而在离开棱镜时就各自分散,形成光谱。

[19] **Abbe number**: 阿贝常数。

[20] **achromats**: 消色差透镜。它由两种不同材质的透镜组合而成,消色差透镜的用途是把两种不同颜色的光聚焦到同一个点,或称为修正色像差。

[21] **wavefront**: 波前。它是指波在介质中传播时,某时刻刚刚开始位移的质点构成的面。它代表某时刻波能量到达的空间位置,它是运动着的。

[22] **specular reflection**: 镜面反射。

[23] **total internal reflection (TIR)**: 全反射。它是指光由光密(即光在此介

质中的折射率大的)媒质射到光疏(即光在此介质中折射率小的)媒质的界面时,全部被反射回原媒质内的现象。

[24] **Brewster's angle**:布儒斯特角,又称偏振角。光以布儒斯特角入射时,反射光与折射光互相垂直。

[25] **circular polarization**:圆偏振。圆偏振光是指光矢量的振动方向不变,而大小随相位改变。

[26] **transverse wave**:横波。质点的振动方向与波的传播方向垂直。

[27] **linear polarization**:线偏振。线偏振光在光的传播方向上,光矢量只沿一个固定的方向振动。

[28] **elliptical polarization**:椭圆偏振。椭圆偏振光是指光的电场方向或光矢量末端在垂直于传播方向的平面上描绘出的轨迹。

[29] **plane wave**:平面波。它是指电磁波的电场与磁场都可以在一个平面内分析。

[30] **unpolarized**:非偏振。

[31] **constructive interference**:干涉相长。

[32] **destructive interference**:干涉相消。

[33] **monochromatic waves**:单色波。

[34] **diffraction grating**:衍射光栅。它是一种由密集、等间距平行刻线构成的非常重要的光学器件,分为反射和透射两大类。

[35] **Fraunhofer diffraction**:夫琅和费衍射,也称远场衍射。在光的衍射实验中,光源或观察屏离开衍射孔或缝为无限远。

[36] **single-slit diffraction**:单缝衍射。它是指光在传播过程中遇到障碍物,光波会绕过障碍物继续传播的一种现象。

[37] **circular diffraction**:圆孔衍射。

[38] **rectangular diffraction**:矩孔衍射。

[39] **Airy rings**:爱里斑。由于光的波动性,光通过小孔会发生衍射,出现明暗相间的条纹衍射图样,条纹间距随小孔尺寸的减小而变大。大约有84%的光能量集中在中央亮斑,其余16%的光能量分布在各级明环上。衍射图样的中心区域有最大的亮斑,称为爱里斑。

[40] **Maxwell's equations**:麦克斯韦方程组。它是英国物理学家詹姆斯·克拉克·麦克斯韦在19世纪建立的一组描述电场、磁场与电荷密度、电流密度之间关系的偏微分方程。

[41] **conduction current**:传导电流。

[42] **charge density**:电荷密度。

[43] **electromagnetic field**：电磁场。

[44] **matter equation**：物质方程。它描述了物质在电磁场影响下的特性的三个方程。它是通过麦克斯韦方程组求解各个场量时必不可少的方程。

[45] **boundary conditions**：边界条件。它是指在求解区域边界上所求解的变量或其导数随时间和地点的变化规律。边界条件是控制方程有确定的解的前提，对于任何问题，都需要给定边界条件。

[46] **Fresnel's equation**：菲涅耳公式(或称菲涅耳方程)，由奥古斯丁·让·菲涅尔导出。用来描述光在不同折射率的介质之间的行为。由公式推导出的光的反射称为"菲涅尔反射"。

[47] **photophobic**：避光的。

[48] **evanescent wave**：倏逝波。它是指光在发生全内反射时，光波不是绝对地在界面上被全部反射回第一介质，而是投入第二介质大约一个波长的深度，并沿着界面流过波长量级距离后重新返回第一介质，沿着反射光的方向射出。

[49] **scattering**：散射。

[50] **spectral absorption**：光吸收。

[51] **surface plasmon resonance**：表面等离子体共振。

[52] **Rayleigh scattering**：瑞利散射。

[53] **Mie scattering**：米氏散射。

[54] **homogeneous substance**：均匀介质。

1.3 Circuit Basis

An ***electronic circuit*** is composed of individual electronic components, such as ***resistors***, transistors, ***capacitors***, ***inductors*** and ***diodes***, connected by conductive wires or traces through which electric current can flow. The combination of components and wires allows various simple and complex operations to be performed: Signals can be amplified, computations can be performed, and data can be moved from one place to another. Circuits can be constructed of discrete components connected by individual pieces of wire, but today it is much more common to create interconnections by photolithographic techniques on a laminated substrate (a printed circuit board or PCB) and solder the components to these interconnections to create a finished circuit. In an integrated circuit or IC, the components and interconnections are formed on the same substrate, typically a semiconductor such as silicon or (less commonly) gallium arsenide.

Breadboards, perfboards or stripboards are common for testing new designs. They allow the designer to make quick changes to the circuit during the development. An electronic circuit can usually be categorized as an analog circuit, a digital circuit or a mixed-signal circuit (a combination of ***analog circuits*** and ***digital circuits***).

1.3.1 Basic Electronic Components

An electronic component is any basic discrete device or physical entity in an electronic system used to affect electrons or their associated fields. Electronic components, as shown in Fig. 1-42, are mostly industrial products, available in a singular form and are not to be confused with electrical elements, which are conceptual abstractions representing idealized electronic components.

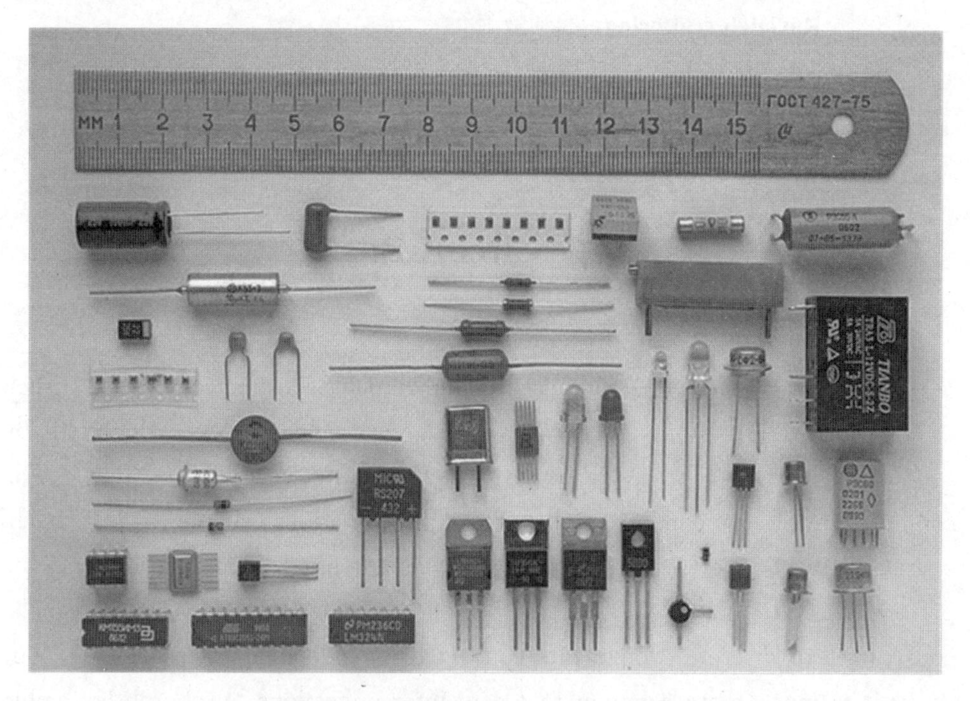

Fig. 1-42 Various electronic components

Electronic components have two or more electrical terminals (or leads) aside from antennas which may only have one terminal. These leads connect, usually soldered to a printed circuit board, to create an electronic circuit (a discrete circuit) with a particular function (e. g., an amplifier, radio receiver, or oscillator). Basic electronic components may be packaged discretely, as arrays or networks of like components, or integrated inside of packages such as semiconductor integrated circuits, hybrid integrated circuits, or thick film devices.

1.3.1.1 Resistor

A resistor, as shown in Fig. 1-43, is a passive two-terminal electrical component that implements electrical resistance as a circuit element. Resistors act to reduce current flow, and, at the same time, act to lower voltage levels within circuits. Resistors may have fixed resistances or variable resistances, such as those found in thermistors, varistors, trimmers, photoresistors and potentiometers.

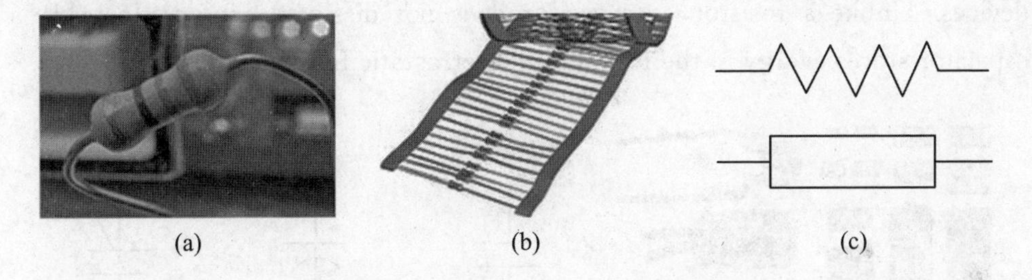

Fig. 1-43 A typical resistor and its symbol

The current through a resistor is in direct proportion to the voltage across the resistor's terminals. This relationship is represented by Ohm's law:

$$I=\frac{V}{R} \tag{1-86}$$

where I is the current through the conductor in units of amperes, V is the potential difference measured across the conductor in units of volts, and R is the resistance of the conductor in units of ohms (symbol: Ω).

The ratio of the voltage applied across a resistor's terminals to the intensity of current in the circuit is called its resistance, and this can be assumed to be a constant (independent of the voltage) for ordinary resistors working within their ratings.

Resistors are common elements of electrical networks and electronic circuits and are ubiquitous in electronic equipment. Practical resistors can be composed of various compounds and films, as well as resistance wires (wire is made of a high-resistivity alloy, such as nickel-chrome). Resistors are also implemented within integrated circuits, particularly analog devices, and can also be integrated into hybrid and printed circuits.

1.3.1.2 Capacitor

A capacitor (originally known as a condenser), as shown in Fig. 1-44, is a passive two-terminal electrical component used to store energy electrostatically in an electric field. The forms of practical capacitors vary widely, but all contain at least two electrical conductors (plates) separated by a dielectric (i.e., insulator). The conductors can be thin films of metal, aluminum foil or disks, etc. The "nonconducting" dielectric acts to increase the capacitor's charge capacity. A dielectric can be glass, ceramic, plastic film, air, paper, mica, etc. Capacitors are widely used as parts of electrical circuits in many common electrical devices. Unlike a resistor, a capacitor does not dissipate energy. Instead, a capacitor stores energy in the form of an electrostatic field between its plates.

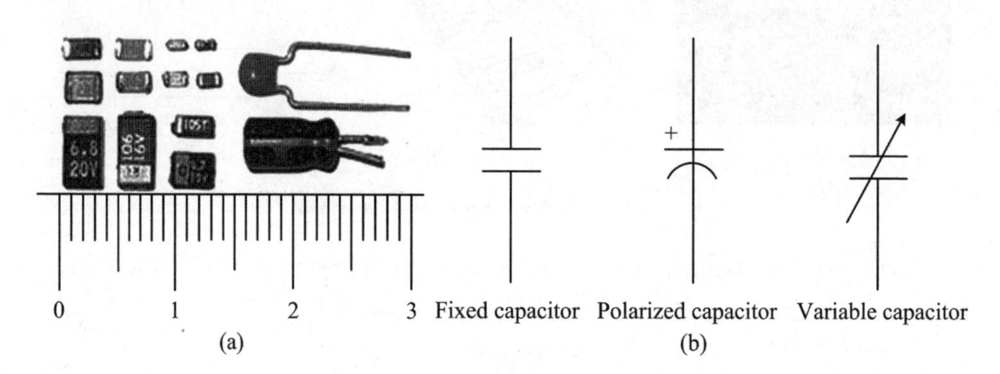

Fig. 1-44 Miniature low-voltage capacitors (next to a cm ruler) and electronic symbol of capacitor

When there is a potential difference across the conductors (e.g., when a capacitor is attached across a battery), an electric field develops across the dielectric, causing positive charge ($+Q$) to collect on one plate and negative charge ($-Q$) to collect on the other plate. If a battery has been attached to a capacitor for a sufficient amount of time, no current can flow through the capacitor. However, if an accelerating or alternating voltage is applied across the leads of the capacitor, a displacement current can flow.

An ideal capacitor is characterized by a single constant value for its capacitance (see Fig. 1-45). Capacitance is expressed as the ratio of the electric charge (Q) on each conductor to the potential difference (V) between them. The SI unit of capacitance is the farad (F), which is equal to one coulomb per volt (1 C/V). The most common subunits of capacitance in use today are the microfarad (μF), nanofarad (nF), picofarad (pF), and, in microcircuits, femtofarad (fF).

Fig. 1-45 Different types of capacitors

(a) A typical electrolytic capacitor. (b) Four electrolytic capacitors of different voltages and capacitance. (c) Solid-body, resin-dipped 10 μF 35 V tantalum capacitors. The "+" sign indicates the positive lead.

The capacitance is greater when there is a narrower separation between conductors and when the conductors have a larger surface area. In practice, the dielectric between the plates passes a small amount of leakage current and also has an electric field strength limit, known as the breakdown voltage. The conductors and leads introduce an undesired inductance and resistance.

Capacitors are widely used in electronic circuits for blocking direct current while allowing alternating current to pass. In analog filter networks, they smooth the output of power supplies. In resonant circuits, they tune radios to particular frequencies. In electric power transmission systems they stabilize voltage and power flow.

1.3.1.3 Inductor

In electromagnetism and electronics, inductance is the property of a conductor by which a change in current flowing through it "induces" (creates) a voltage (electromotive force) in both the conductor itself (self-inductance) and in any nearby conductors (mutual inductance).

These effects are derived from two fundamental observations of physics: First, that a steady current creates a steady magnetic field (Oersted's law); and second, that a time-varying magnetic field induces voltage in nearby conductors (Faraday's law of induction). According to Lenz's law, a changing electric current through a circuit that contains inductance, induces a proportional voltage, which opposes the change in current (self-inductance). The varying field in this circuit may also induce an e. m. f. in neighbouring circuits (mutual inductance).

To add inductance to a circuit, electrical or electronic components called inductors are used. An inductor, as shown in Fig. 1-46, also called a coil or a reactor, is a passive two-terminal electrical component which resists changes in electric current passing through it. It consists of a conductor such as a wire, usually wound into a coil. When a current flows through it, energy is stored temporarily in a magnetic field in the coil. When the current flowing through an inductor changes, the time-varying magnetic field induces a voltage in the conductor, according to Faraday's law of electromagnetic induction, which opposes the change in current that created it.

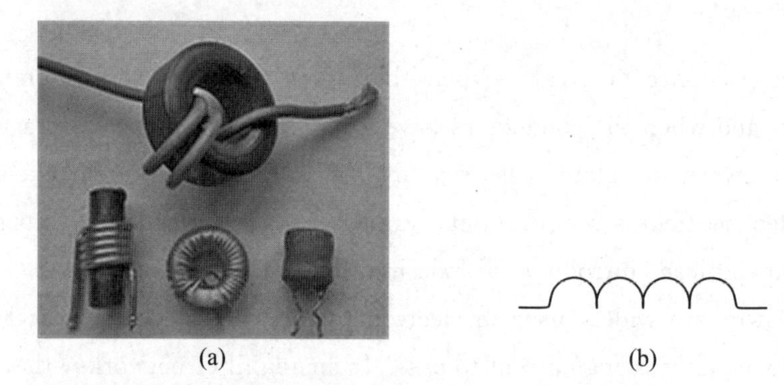

(a)　(b)

Fig. 1-46　A selection of low-value inductors and electronic symbol of inductors

An inductor is characterized by its inductance, the ratio of the voltage to the rate of change of current, which has units of henries (H). Inductors have values that typically range from 1 μH (10^{-6} H) to 1 H. Many inductors have a magnetic core made of iron or ferrite inside the coil, which serves to increase the magnetic field and thus the inductance. Along with capacitors and resistors, inductors are one of the three passive linear circuit elements that make up electric circuits.

Inductors are widely used in alternating current (AC) electronic equipment, particularly in radio equipment. They are used to block the flow of AC current while allowing DC to pass; inductors designed for this purpose are called chokes. They are also used in electronic filters to separate signals of different frequencies, and in combination with capacitors to make tuned circuits, used to tune radio and TV receivers.

1.3.2 Analog Circuit

Analog electronic circuits are those in which current or voltage may vary continuously with time to correspond to the information being represented. Analog circuitry is constructed from two fundamental building blocks: series and parallel circuits. In a ***series circuit***, the same current passes through a series of components. A string of Christmas lights is a good example of a series circuit: if one goes out, they all do. In a ***parallel circuit***, all the components are connected to the same voltage, and the current divides between the various components according to their resistance.

The basic components of analog circuits are wires, resistors, capacitors, inductors, diodes, and transistors. Recently, memristors have been added to the list of available components. Analog circuits are very commonly represented in schematic diagrams, in which wires are shown as lines, and each component has a unique symbol. Analog circuit analysis employs Kirchhoff's circuit laws: All the currents at a node (a place where wires meet) must add to 0, and the voltage around a closed loop of wires is 0. Wires are usually treated as ideal zero-voltage interconnections; any resistance or reactance is captured by explicitly adding a parasitic element, such as a discrete resistor or inductor. Active components such as transistors are often treated as controlled current or voltage sources: For example, a field-effect transistor can be modeled as a current source from the source to the drain, with the current controlled by the gate-source voltage.

When the circuit size is comparable to a wavelength of the relevant signal frequency, a more sophisticated approach must be used. Wires are treated as transmission lines, with (hopefully) constant characteristic impedance, and the impedances at the start and end determine transmitted and reflected waves on the line. Such considerations typically become important for circuit boards at frequencies above 1 GHz; integrated circuits are smaller and can be treated as lumped elements for frequencies less than 10 GHz or so.

An alternative model is to take independent power sources and induction as basic electronic units; this allows modeling frequency dependent negative resistors, gyrators, negative impedance converters, and ***dependent sources*** as secondary electronic components.

1.3.3 Digital Circuits

Digital electronics, or digital (electronic) circuits, represent signals by discrete bands of analog levels, rather than by a continuous range. All levels within a band represent the same signal state. Relatively small changes to the analog signal levels due to manufacturing tolerance, signal attenuation or parasitic noise do not leave the discrete envelope, and as a result are ignored by signal state sensing circuitry.

In most cases the number of these states is two, and they are represented by two voltage bands: one near a reference value (typically termed as "ground" or zero volts) and a value near the supply voltage, corresponding to the "false" ("0") and "true" ("1") values of the Boolean domain respectively.

Digital techniques are useful because it is easier to get an electronic device to switch into one of a number of known states than to accurately reproduce a continuous range of values.

Digital electronic circuits are usually made from large assemblies of ***logic gates***, simple electronic representations of Boolean logic functions.

New Words and Expressions

[1] **electronic circuit**：电子电路。

[2] **resistor**：电阻。它是一个限流元件，将电阻接在电路中后，电阻器的阻值是固定的，一般是两个引脚，它可限制通过它所连支路的电流大小。

[3] **capacitor**：电容。电容是电容器的简称，是电子设备中大量使用的电子元件之一，广泛应用于隔直、耦合、旁路、滤波、调谐回路、能量转换、控制电路等方面。

[4] **inductor**：电感器。它是能够把电能转化为磁能而存储起来的元件。

[5] **diode**：二极管。它是能够单向传导电流的电子器件。在半导体二极管内部有一个 pn 结和两个引线端子，这种电子器件按照外加电压的方向，具备单向电流的传导性。

[6] **analog circuit**：模拟电路。它是处理模拟信号的电子电路[模拟信号：时间和幅度都连续的信号(连续的含义是在某取值范围内可以取无穷多个数值)]。

[7] **digital circuit**：数字电路。它是用数字信号完成对数字量进行算术运算和逻辑运算的电路。

[8] **series circuit**：串联电路。它是电路中的元件或部件排列得使电流全部通过每一部件或元件而不分流的一种电路连接方式。

[9] **parallel circuit**：并联电路。电路、线路或元件为达到某种设计要求的功能的连接方式，特点是对两个同类或不同类的元件、电路、线路等首首相接，同时尾尾亦相连的一种连接方式。

[10] **logic gate**：逻辑门。它是集成电路上的基本组件，简单的逻辑门可由晶体管组成，这些晶体管的组合可以使代表两种信号的高低电平在通过它们之后产生高电平或者低电平的信号。

[11] **dependent sources**：受控源，又称非独立源。一般来说，一条支路的电压或电流受本支路以外的其他因素控制时统称为受控源。

References

[1] Smith F G, King T A, Wilkins D. Optics and Photonics: An Introduction[M]. 2nd ed. New York: John Wiley & Sons, Inc., 2007.

[2] Kasap S O. Optoelectronics and Photonics: Principles & Practices [M]. 2nd ed. New Jersey: Prentice Hall, Inc., 2012.

Questions

1. List one unique property of semiconductor and illustrate how it works.

2. Why is semiconductor the material of the optoelectronics but not metal or glass?

3. How does the electron-hole generate?

4. If we are told the wavelength $\lambda = 100$ mm, and the planck's constant $h=6.63\times10^{-34}$ J · s, solving the momentum p.

5. There is a double-sided mirror system, and the light incident paralleling to one of the mirror, after two times refractions, the light emitted paralleling to the other mirror, what is the angle of the two mirrors?

6. What is the light velocity within glass? Taking the refractive index of glass, $n=1.5$.

7. Let $n_1=1$, $\theta_1=45°$, and $n_2=1.33$. (see Fig. 1-47) What are θ_3 and θ_2?

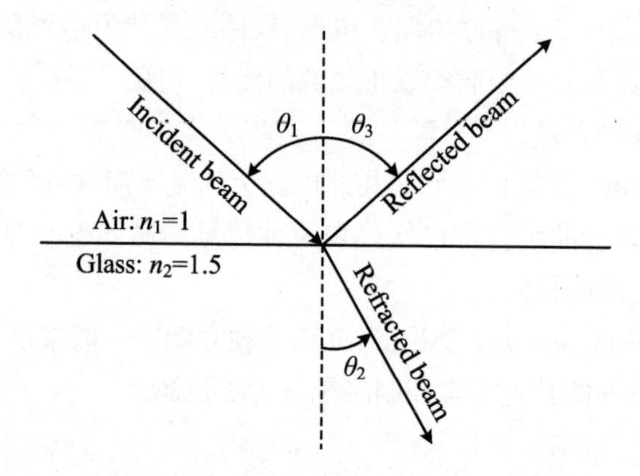

Fig. 1-47 Light travels from air to glass

8. Assume you have a glass rod surrounded by air, as shown in Fig. 1-48. Find the critical incident angle.

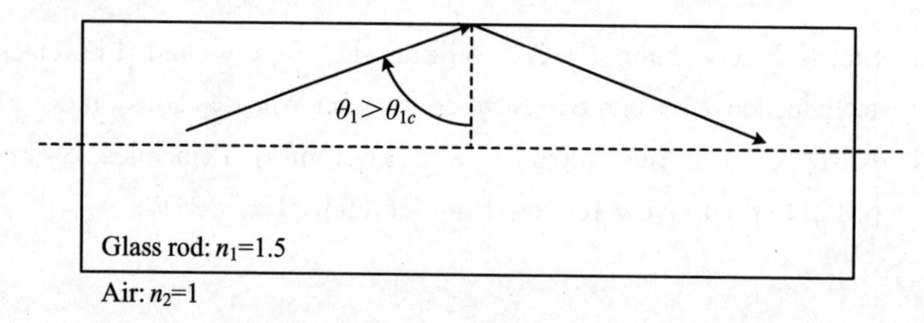

Fig. 1-48 Saving light inside a glass rod

9. Can you explain the following phenomenon (see Fig. 1-49) using what you have learned?

Fig. 1-49 A straw appears to be broken in the cup

Chapter 2
Electro-optic Information Transformation

Human get information through watching, hearing, touching and tasting, most of information is obtained by the visual. Visual information uses the light as carrier, and therefore we need to convert the signals into optical signals including that an optical signal is converted to an electrical signal. On the other hand, in the signal processing and transmission, sometimes the raw signal needs to convert into an optical signal. For example, in optical fiber communication systems, electrical signals need to be converted into optical signals and then coupled to an optical fiber for transmission. The device that helps the electrical signal convert to the optical signal is called a electro-optic conversion device. Currently the electro-optic conversion device is widely various and distinctive. This chapter starts with emission and absorption theory during light-matter interactions with typical instruments based on spontaneous emission (LED) and stimulated emission (LD). The dynamic equilibrium for the particles transition (population inversion) are highlighted. In addition, the basic working principle and application of other electro-optic conversion devices like LCD, OLED and PLED are also introduced.

2.1 Emission and Absorption During Light-matter Interactions

Light emission and absorption including ***spontaneous emission***, ***stimulated emission***, and ***stimulated absorption*** are associated with ***transitions*** of electrons.

During ***light-matter interactions***, emission/absorption happens at the same time. Generally, the spontaneous emission plays a major role in ordinary light sources, for example, the radiation from candles and the sun. While the stimulated emission/absorption can be observed in laser systems.

A large number of atomic or molecular energy levels exists, even for one diatomic molecular gas. If we consider transitions among all energy levels, actually, it brings a really huge task. To simplify, we assume that there are two states related with light-matter interactions, an ***excited state*** E_2 and a ***ground state*** E_1, which follow the law of radiation transition selection.

2.1.1 Spontaneous Emission

Based on Heisenberg uncertainty principle, one electron is not able to eternally locate at an excited state. It can jump down to the ground state through spontaneously transition, once its life time at the excited state finishes. The above process can release one photon with the energy of $h\nu$ (see Fig. 2-1).

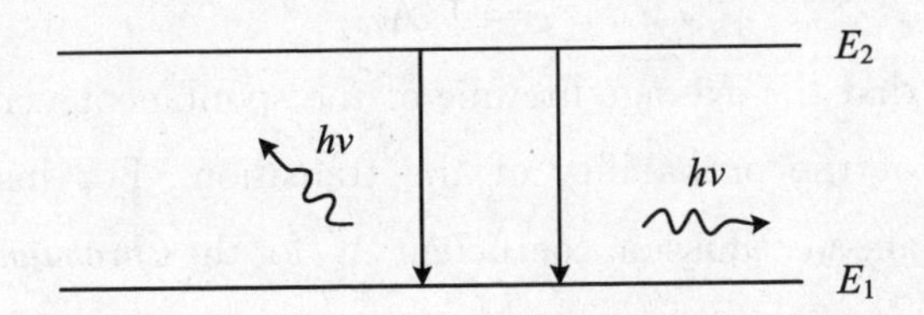

Fig. 2-1 Spontaneous emission

$$h\nu = E_2 - E_1 \tag{2-1}$$

Each transition for spontaneous emission is an independent process. It means that there is no common "stating time" for the spontaneous emission. In this sense, the phase, propagation direction, and polarization of the emitted photons by spontaneous emission are distributed randomly. Thus, radiations from the spontaneous emission are ***incoherent***.

In case of spontaneous emission, n_2 is used to indicate the number of electrons in a unit volume at the excited state E_2, $-\mathrm{d}n_2$ represents the reduction of the number of electrons at E_2 within a time interval of $\mathrm{d}t$. It writes:

$$-\mathrm{d}n_2 = A_{21} n_2 \mathrm{d}t \tag{2-2}$$

where the coefficient A_{21} is called Einstein spontaneous emission coefficient (or spontaneous emission coefficient). The above formula can be derived as:

$$A_{21} = -\frac{dn_2}{n_2 dt} \tag{2-3}$$

Physically, A_{21} means that the percentage of the particle number density for the particle that experiences spontaneous emission within a unit time. Here, we use the term "particle" since the transition could be done by either an electron or a molecule. When we integrate two sides of the equation (2-3), it derives:

$$n_2(t) = n_{20} e^{-A_{21} t} \tag{2-4}$$

where n_{20} is the particle number density at E_2, when $t = 0$. Equation (2-4) indicates that for spontaneous emission, if we assume no external energy supplements, the particle number density at an excited state will decay exponentially.

Thus, the lifetime of spontaneous emission can be obtained by averaging the time required for particles to complete the transitions. The average lift time (τ) is equal to the time taken for the particle number density at the excited state which drops to 1/e of the initial number density, which is

$$\tau = 1/A_{21} \tag{2-5}$$

We can find that the average lifetime of the spontaneous emission is equal to the reciprocality of the probability of the transition. For instance, in a ***ruby crystal***, the spontaneous emission coefficient A_{21} for the ***Chromium ion*** is of the order of $10^2\ s^{-1}$, which represents its average lifetime is 10^{-2} s. Equation (2-5) only considers the transition from E_2 to E_1. In fact, as shown by Fig. 2-2, there are many excited states that involve in the radiation. Assuming the transition probability from a higher-excited state to the ground state or a lower-excited state is A_{nm}, the average lifetime of the spontaneous emission writes:

$$\tau = 1/\sum_m A_{nm} \tag{2-6}$$

Here, the power of radiation by spontaneous emission can be obtained as well.

$$q_{21} = n_2 A_{21} h\nu \tag{2-7}$$

where the photon energy is $h\nu$. Here, the unit is W/m^3. One typical instrument based on spontaneous emission is light emitting diode (LED). The basic working principle and applications of this device will be discussed in Section 2. 2.

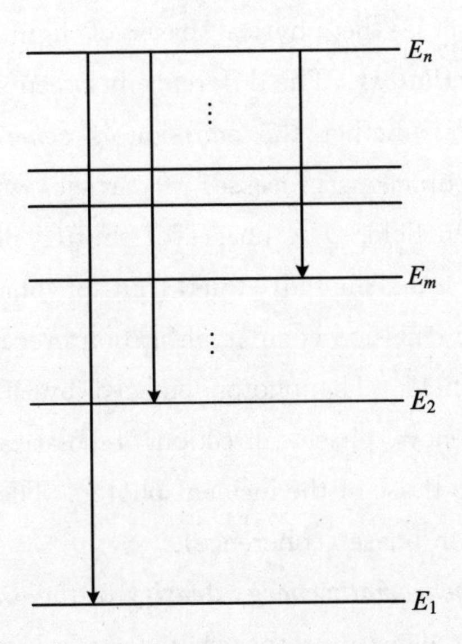

Fig. 2-2 Transitions among multiple energy levels

2.1.2 Stimulated Emission

If a laser-active atom or ion is at an excited state, it is possible that the photon emission is provoked by an incident photon. If the incident photon has a suitable photon energy ($h\nu = E_2 - E_1$), the stimulated emission will happen (see Fig. 2-3). In this case, another photon is emitted into the mode of the incident photon. In effect, the power of the incident radiation is amplified.

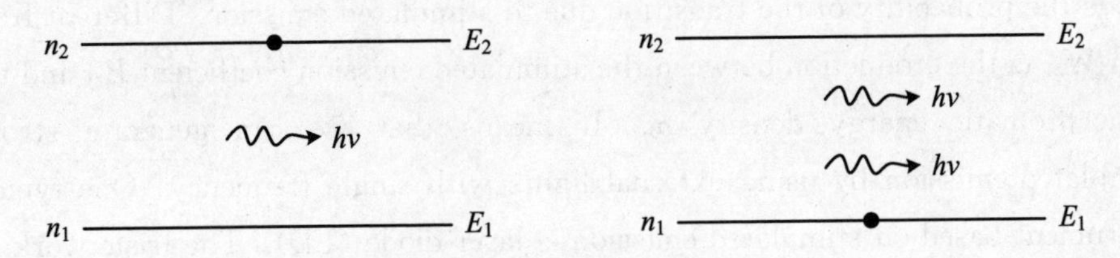

Fig. 2-3 Stimulated emission

Stimulated emission is characterized by,

(1) The stimulated emission happens only if the energy of the incident photon is equal to the energy between the bandgap $h\nu = E_2 - E_1$.

(2) The photon emitted by stimulated emission has the same frequency, phase, direction of polarization, and direction of propagation as those of the incident photon exactly.

Stimulated emission is the physical basis of light amplification in ***laser amplifiers*** and ***laser oscillators***. The difference between stimulated emission and spontaneous emission is whether the emission is ***coherent***. The spontaneous emission is through spontaneous processes of particles without being controlled by the external radiation field. The phases of emitted photons by spontaneous emission are irregularly distributed, thus, out of phase (incoherence). In contrast, the stimulated emission is an amplification process under the control of the external radiation field. The photon emitted by stimulated emission has exactly the same frequency, phase, direction of polarization, and direction of propagation according to those of the incident photon. Therefore, the photons by stimulated emission are in phase (coherence).

We define the ***monochromatic energy density*** of the external light field as ρ_ν, and the particle number density at the excited state E_2 as n_2. For stimulated emission, we have:

$$-\mathrm{d}n_2 = B_{21} n_2 \rho_\nu \mathrm{d}t \tag{2-8}$$

where the negative sign indicates that the particle number density at E_2 decreases. B_{21} is a proportional constant for the stimulated emission, which is called Einstein stimulated emission coefficient (or stimulated emission coefficient). The value of B_{21} is determined by the rate of transitions from E_2 to E_1. If we define $W_{21}=B_{21}\rho_\nu$, equation (2-8) writes:

$$W_{21} = B_{21}\rho_\nu = -\frac{\mathrm{d}n_2}{n_2 \mathrm{d}t} \tag{2-9}$$

W_{21} is the probability of the transition due to stimulated emission. Different from A_{21}, W_{21} is the production between the stimulated emission coefficient B_{21} and the monochromatic energy density ρ_ν. It means that we can generate strong stimulated emission by using external lights with single frequency. One typical instrument based on stimulated emission is laser diode (LD). The basic working principle and applications of LD device will be introduced in Section 2.3.

2.1.3 Stimulated Absorption

The stimulated absorption is the contrary process of stimulated emission. The schematic diagram for stimulated absorption is shown in Fig. 2-4. An electron at ground state E_1 jumps to an excited state E_2 by absorbing a photon with the energy equal to the energy gap $E=h\nu=E_2-E_1$. For the stimulated absorption, we have:

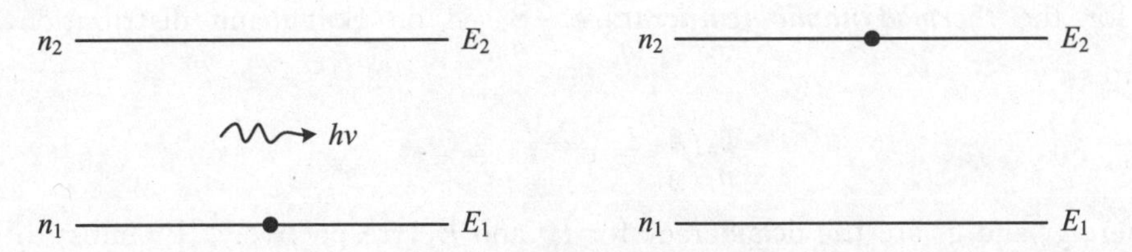

Fig. 2-4 Stimulated absorption

$$dn_2 = B_{12} n_1 \rho_\nu dt \tag{2-10}$$

where B_{12} is Einstein stimulated absorption coefficient. If we define $W_{12}=B_{12}\rho_\nu$, equation (2-10) writes:

$$W_{12} = B_{12}\rho_\nu = \frac{dn_2}{n_1 dt} \tag{2-11}$$

W_{12} is the probability of the transition by stimulated absorption. More specifically, W_{12} is the probability for a particle (electron) at the ground state E_1 involved in the transition $E_1 \rightarrow E_2$ due to stimulated absorption. We can generate stronger stimulated absorption by using external lights with single frequency.

2.1.4 Dynamic Equilibrium Among Spontaneous Emission, Stimulated Emission, and Stimulated Absorption

In fact, during light-matter interactions, spontaneous emission, stimulated emission, and stimulated absorption occur at the same time and closely related. Assuming the incident light with monochromatic energy density ρ_ν, if we count the number of electrons within a very short time period of dt, a ***dynamic equilibrium*** happens among three processes, which is

$$A_{21} n_2 dt + B_{21} n_2 \rho_\nu dt = B_{12} n_1 \rho_\nu dt \tag{2-12}$$

In equation (2-12), $A_{21} n_2 dt$, $B_{21} n_2 \rho_\nu dt$ and $B_{12} n_1 \rho_\nu dt$ indicate the number of electrons involved in spontaneous emission, stimulated emission, and stimulated absorption within a unit time period of dt, respectively. Furthermore, owing to ***conservation of energy***, the number of particles (electrons) jumps down from excited state to ground state through spontaneous emission and stimulated emission must be equal to the number of particles (electrons) promoted from ground state to excited state by stimulated absorption.

In order to provide a detailed physical picture of the dynamic equilibrium during light-matter interactions, we assume that spontaneous emission, stimulated emission, and stimulated absorption happen in an atomic system with T for the ***thermodynamic temperature***. Based on Boltzmann distribution, it writes:

$$\frac{n_2/g_2}{n_1/g_1} = e^{-\frac{E_2-E_1}{kT}} = e^{-\frac{h\nu}{kT}} \tag{2-13}$$

where g_2 and g_1 are the degeneracy for E_2 and E_1, respectively. By substituting equation (2-13) to equation (2-11), we derive:

$$(B_{21}\rho_\nu + A_{21})\frac{g_2}{g_1}e^{-\frac{h\nu}{kT}} = B_{12}\rho_\nu \tag{2-14}$$

In this sense, the monochromatic energy density in thermal equilibrium writes:

$$\rho_\nu = \frac{A_{21}}{B_{21}} \frac{1}{\frac{B_{12}g_1}{B_{21}g_2}e^{\frac{h\nu}{kT}} - 1} \tag{2-15}$$

According to Planck's ***black body radiation law***, the formula for monochromatic energy density of black body is

$$\rho_\nu = \frac{8\pi h\nu^3}{c^3}\frac{1}{e^{\frac{h\nu}{kT}} - 1} \tag{2-16}$$

Comparing equation (2-15) with equation (2-16), we derive:

$$A_{21}/B_{21} = 8\pi h\nu^3/c^3 \tag{2-17}$$

$$g_1 B_{12} = g_2 B_{21} \tag{2-18}$$

Equation (2-17) and equation (2-18) are the basic relationship between Einstein coefficients (A_{21}, B_{12} and B_{21}). Equation (2-17) and equation (2-18) are independent with any specific processes, meaning generally applicable. If the degeneracy of the ground state and excited states are equal ($g_1 = g_2$), equation (2-18) writes:

$$B_{12} = B_{21} \tag{2-19}$$

Besides, if we define the refractive index as μ. The speed of light in the transparent medium is c/μ. In this case, equation (2-17) writes:

$$A_{21}/B_{21} = 8\pi\mu^3 h\nu^3/c^3 \tag{2-20}$$

New Words and Expressions

[1] **spontaneous emission**：自发辐射。它是指在没有任何外界作用下，激发态原子自发地从高能级向低能级跃迁，同时辐射出一个光子的过程。

[2] **stimulated emission**：受激辐射。它是指处于激发态的发光原子在外来辐射场的作用下，向低能态或基态跃迁时，辐射光子的过程。

[3] **stimulated absorption**：受激吸收。它是指处于低能级的原子，受到外来光子的激励下，在满足能量恰好等于低、高两能级之差 ΔE 时，该原子就吸收这部分光子能量的过程。

[4] **transitions**：跃迁。

[5] **light-matter interactions**：光与物质相互作用。

[6] **monochromatic energy density**：单色能量密度。

[7] **dynamic equilibrium**：动态平衡。多种效应共同作用达到平衡。

[8] **incoherent**：非相干的。

[9] **coherent**：相干的。

[10] **conservation of energy**：能量守恒。

[11] **ruby crystal**：红宝石。红宝石激光器是1960年发明的第一台红宝石激光器。

[12] **Chromium ion**：铬离子。

[13] **ground state**：基态。材料的最低能级。

[14] **excited state**：激发态。相对于基态，指材料的高能级。

[15] **laser amplifiers**：激光放大器。

[16] **laser oscillators**：激光振荡器。

[17] **thermodynamic temperature**：热力学温度。以绝对零度为起点的温度。

[18] **black body radiation law**：普朗克黑体辐射定律。此定律描述，在任意温度 T 下，从一个黑体中发射出的电磁辐射的辐射率与频率彼此之间的关系。

2.2 Typical Instrument Based on Spontaneous Emission：Light Emitting Diode（LED）

LED is a solid state semiconductor devices，which can help electricity be directly converted into light. LED is a typical instrument based on spontaneous emission which we have initially introduced in the previous section. LED is the heart of a semiconductor chip，the chip is attached to one end of a ***stent***，is the negative side，the other end of the power of the ***cathode***，the entire chip package to be ***epoxy resin***. Semiconductor chip is composed of two parts，part of the p-type semiconductor, it is hole-dominated, the other side is the n-type semiconductor, which is mainly electronic. But linking the two semiconductors, among them is the formation of a "p-n junction". When a positive bias voltage is applied to a p-n junction, the forward bias produces an electric field opposite to the built-in electric field in the barrier region. This reduces the electric field intensity in the barrier region and results in the injection of minority carriers from both sides of the junction. Finally, the non-equilibrium carriers higher than the concentration at the equilibrium state will recombine near the junction and produce photons, and this is the principle of LED luminescence. The wavelength of light that is the color of light, which depends on the p-n junction with different materials.

2.2.1 History and Development of LED

History of artificial lighting sources is related to the development of human civilization. In 1879, Thomas Edison invented the incandescent lamp. Since then, with the rapid development of lighting sources, fluorescent lamps and high-pressure sodium lamps have emerged one after another, acting as an important lighting tool. These lights sources cannot meet the high demand of energy saving and environmental protection for the growing human civilization.

People are paying more and more attention to the research and application of semiconductor materials and technology. In the 1940s, the research of semiconductor physics and p-n junction flourished. In 1947, the transistor was born in the bell telephone Laboratory of the United States. Shockley, Bardeenan and Brattaln won the Nobel Prize in physics in 1956. People begin to realize that p-n junctions can be used in light emitting devices. In 1951, K. Lehovec et al. explained the electroluminescence phenomenon of SiC. That is, after carriers (i. e. , current carriers) are injected into the junction region, electrons and holes recombine to produce luminescence. However, the measured photon energy is lower than the bandgap energy of SiC. They believe that the recombination process may be dominated by impurities or lattice defects. In 1955 and 1956, J. R. Haynes of Bell laboratory confirmed that the electroluminescence observed in germanium and silicon was due to the radiation recombination of electrons and holes in p-n junctions.

In 1957, H. Kroemer predicted that heterojunction has higher injection efficiency than homojunction, and proposed many ideas for the application of heterojunction in solar cells. In 1960, R. L. Anderson made high-quality heterojunction for the first time, and proposed the theoretical model and energy band diagram of the system. H. Kroemer and Z. I. Alferov won the Nobel Prize in physics in 2000 for their groundbreaking contributions to the invention of heterocrystalline tubes and laser diodes (LD). Since then, GaAs has attracted much attention, and the preparation technology of p-n junction based on GaAs has developed rapidly. GaAs is a direct bandgap semiconductor material. The recombination of electrons and holes does not require the participation of phonons, so it is very suitable for making light emitting devices. The bandgap of GaAs is 1.4 eV and the corresponding emission wavelength is in the ***infrared*** region.

In 1962, Nick Holonyak Jr. invented the first p-n red light-emitting diode (LED) using GaAsP as luminescent material. It was commercialized in 1968 as an indicator lamp because of its long life, anti-shock and anti-seismic characteristics. In the next forty years, other semiconductor materials have been used, for example, GaP (green, 550 nm), GaAsP (orange and yellow, 650 nm), GaAlAs (red, 680 nm), AlInGaP (yellow-orange, 590～620 nm). In the 1990s, the use of quaternary AlInGaP/GaAs lattice matching materials increased the luminous efficiency of LEDs to several tens of lm/W (lm: lumens, the unit representing luminous flux). The efficiency of the orange-red LED obtained by HP reaches 100 lm/W. The luminous wavelength is limited to red and yellow-green for a long time. It is difficult to achieve a blue LED. Until around the 1990s, Isamu Akasaki, Hiroshi Amano, and Shuji Nakamura (2014 Nobel Prize in physics) solved the challenge related to the growth of high quality GaN film and p-type doping, and realized the high-efficiency nitride blue LED. The epitaxy methods of group Ⅲ nitrides mainly include metal organic compounds vapor phase epitaxy (MOCVD), hydride vapor phase epitaxy (HVPE), and molecular beam epitaxy (MBE). The invention of GaN blue light and shorter wavelength LED makes solid white light source possible. Therefore, the invention of GaN based blue LED is known as "the second lighting revolution".

The international competition for the research and industrialization of high-efficiency white LEDs has rapidly emerged, and has continued to this day. The luminous efficiency has been continuously improved. At present, it has exceeded 300 lm/W, and the electro-optical conversion rate has reached more than 50%.

The service life of LED lamps is up to 100000 hours, while that of incandescent lamps is only 1000 hours and that of fluorescent lamps is also around 1000 hours. Therefore, the use of LED lamps can greatly save resources. LED is a cold light source, without invisible infrared and ultraviolet light, and its energy consumption is only 1/8 of that of incandescent lamps. We may as well estimate that the national power generation in 2017 was 6275. 8 billion kilowatt-hour, of which 1/5 was consumed by lighting, that is, about 1. 2 trillion kilowatt-hour. Assuming that half of them are consumed by incandescent lamps, it is 600 billion kilowatt-hour. If incandescent lamps are replaced by LEDs, 480 billion kilowatt-hour of electric energy will be saved. At present, more than 1. 5 billion people in the world do not have access to power grid. LED, as shown in Fig. 2-5, is especially suitable for users powered by solar energy, and is expected to bring light to people in the dark and improve their lives.

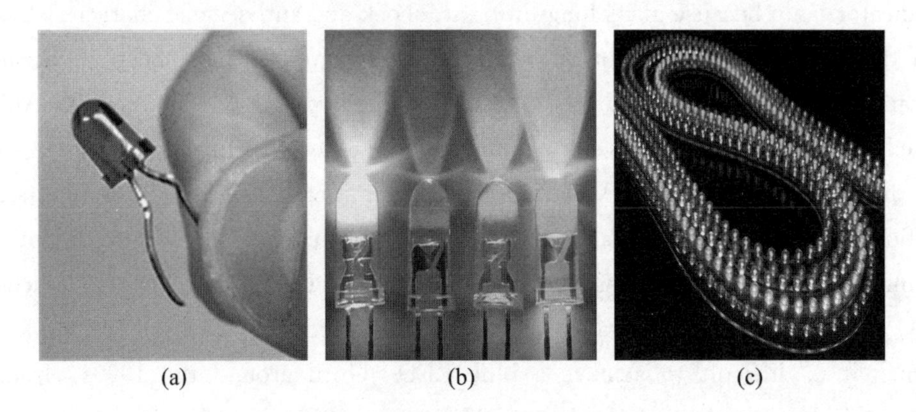

(a)　(b)　(c)

Fig. 2-5　Colorful LED in our daily life

2.2.2 Working Principle of LED

LED is a semiconductor optoelectronic device. In contrast with the traditional light sources, LED has a list of advantages, such as high efficiency, long lifetime, not easy to break, fast reaction speed and high reliability. The basic structure is a p-n junction. The basic principle is that electrons and holes in the semiconductors recombine and emit photons under forward bias.

In order to fabricate efficient LED, the radiation recombination probability of electrons and holes in semiconductor must be increased as much as possible, and the non-radiation recombination probability must be reduced. LED has different optical and electronical characteristics from the transitional incandescent lamp and other light sources. Output from a LED is based on the light radiation (ultraviolet, visible or infrared) produced by the electron excitation transition, which excludes any radiation caused by pure temperature of material. This section takes p-n junction as the theoretical model, mainly introduces the working principle of LED.

As is described in last chapter, a semiconductor with extra electrons is called n-type material, since it has extra negatively-charged particles. In n-type material, free electrons move from a negatively-charged area to a positively-charged area.

A semiconductor with extra holes is called p-type material, since it effectively has extra positively-charged particles. Electrons can move from a negatively-charged area to a positively-charged area. As a result, the holes themselves appear to move from a positively-charged area to a negatively-charged area.

A diode comprises a section of n-type material bonded to a section of p-type material, with ***electrodes*** on each end. This arrangement conducts electricity in only one direction. When no voltage is applied to the diode, electrons from the n-type material fill holes from the p-type material along the junction between the layers, forming a ***depletion zone***, as shown in Fig. 2-6. In the depletion zone, the semiconductor material is returned to its original insulating state — all of the holes are filled, so there are no free electrons or spaces for electrons, and charge carriers can't flow.

To get rid of the depletion zone, you have to make electrons moving from the n-type area to the p-type area and holes moving in the reverse direction. To do this, you connect the n-type side of the diode to the negative end of a circuit and the p-type side to the positive end, as shown in Fig. 2-7. The free electrons in the n-type material are repelled by the negative electrode and drawn to the positive electrode. The holes in the p-type material move the other way. When the voltage difference between the electrodes is high enough, the electrons in the depletion zone are boosted out of their holes and begin moving freely again. The depletion zone disappears, and charge moves across the diode.

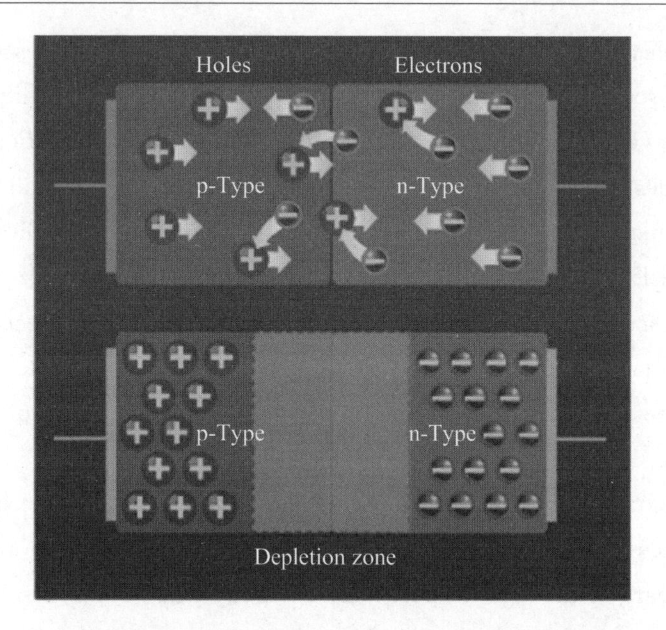

Fig. 2-6　Formation of depletion zone

At the junction, free electrons from the n-type material fill holes from the p-type material. This creates an insulating layer in the middle of the diode called the depletion zone.

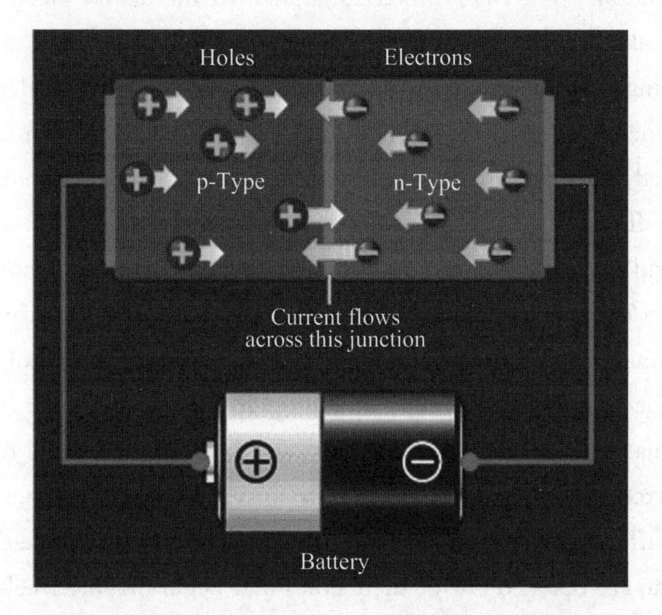

Fig. 2-7　Disappearance of depletion zone

When the negative end of the circuit is hooked up to the n-type layer and the positive end is hooked up to p-type layer, electrons and holes start moving and the depletion zone disappears.

If you try to run current by the other way, with the p-type side connected to the negative end of the circuit and the n-type side connected to the positive end of the circuit, as shown in Fig. 2-8, current will not flow. The negative electrons in the n-type material are attracted to the positive electrode. While, the positive holes in the p-type material are attracted to the negative electrode. Thus, no current flows across the junction because both holes and electrons are moving in the wrong direction. The depletion zone is increased.

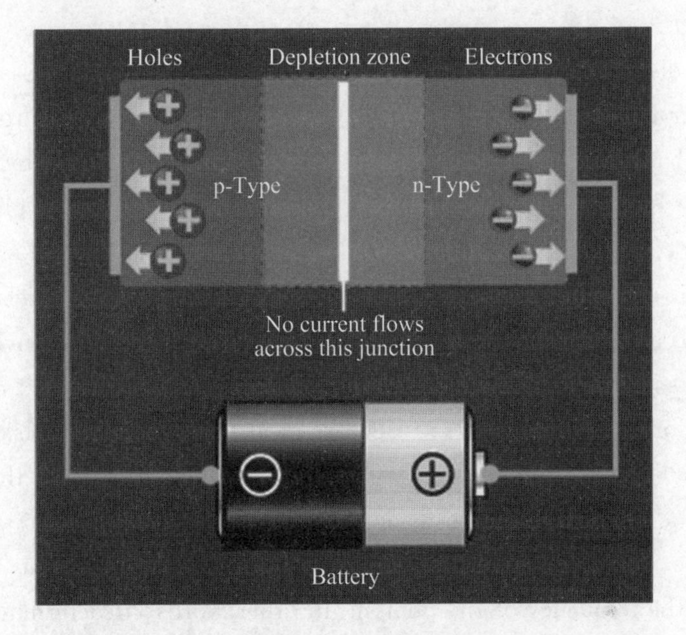

Fig. 2-8 Expansion of depletion zone

When the positive end of the circuit is hooked up to the n-type layer and the negative end is hooked up to the p-type layer, free electrons collect on one end of the diode and holes collect on the other. The depletion zone becomes larger.

The interaction between electrons and holes in this process has an interesting effect — it generates light. The LED does not emit light under reverse bias. It can be found that LED has the characteristics of "ON" under ***forward voltage*** and "OFF" under reverse bias. In the next section, we'll find out exactly why this is.

Light is a form of energy that can be released by an atom. It is made up of many small particle-like packets that have energy and momentum but no mass. These particles, called photons, are the most basic units of light.

Photons are released as a result of moving electrons. In an atom, electrons move in orbitals around the nucleus. Electrons in different orbitals have different amounts of energy. Generally speaking, electrons with greater energy move in orbitals farther away from the nucleus.

For an electron to jump from a lower orbital to a higher orbital, something has to boost its energy level. Conversely, an electron releases energy when it drops from a higher orbital to a lower one. This energy is released in the form of a photon. A greater energy drop releases a higher-energy photon, which is characterized by a higher frequency (Check out how light works for a full explanation).

As we discussed in the last section, free electrons moving across a diode can fall into empty holes from the p-type layer. This involves a drop from the conduction band to a lower orbital, thus the electrons release energy in the form of photons. This happens in any diode, but you can only see the photons when the diode is composed of certain material. The atoms in a standard silicon diode, for example, are arranged in such a way that the electron drops a relatively short distance. As a result, the photon's frequency is so low that it is invisible to the human—it is in the infrared portion of the light spectrum. This isn't necessarily a bad thing, for example: Infrared LEDs are ideal for remote controls.

Visible light-emitting diodes (VLEDs), such as the ones that light up numbers in a digital clock, are made of materials characterized by a wider gap between the conduction band and the lower orbitals. The size of the gap determines the frequency of the photon. In other words, it determines the color of the light. Internal structure of LED is shown in Fig. 2-9. Depending on the materials used in LEDs, they can be built to shine in infrared, ultraviolet, and all the colors of the visible spectrum in between. Note that the LEDs are specially constructed to release a large number of photons outward. Additionally, they are housed in a plastic bulb that concentrates the light in a particular direction.

2.2.3 Advantages and Disadvantages of LED

LEDs have several advantages over conventional incandescent lamps. For one thing, they don't have a filament which will burn out, so they last much longer. Additionally, their small plastic bulb makes them a lot more durable. They also fit more easily into modern electronic circuits.

But the main advantage of LED is efficiency. In conventional incandescent

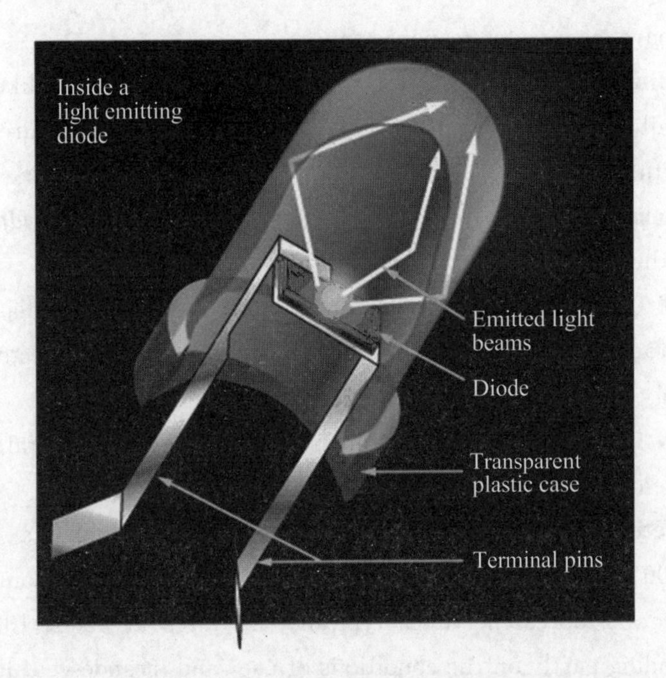

Fig. 2-9 Internal structure of LED

bulbs, the light-production process involves generating a lot of heat (the filament must be warmed). This is completely a waste of energy, unless you're using the lamp as a heater, because a huge portion of the available electricity isn't going towards producing visible light. LEDs generate very little heat, relatively speaking. A much higher percentage of the electrical power is going directly to generate light, which cuts down on the electricity demands considerably.

2.2.3.1 Advantages

(1) **Efficiency**: LEDs emit more light per watt than incandescent light bulbs. Their efficiency is not affected by shape and size, unlike fluorescent light bulbs or tubes.

(2) **Color**: LEDs can emit light of an intended color without using any color filters as traditional lighting methods needed. This is more efficient and can lower initial costs.

(3) **Size**: LEDs can be very small (smaller than 2 mm) and are easily populated onto printed circuit boards.

(4) **On/Off time**: LEDs light up very quickly. A typical red indicator LED will achieve full brightness within a microsecond. LEDs used in communication

devices can have even faster response times.

(5) **Cycling**: LEDs are ideal for frequent on-off cycling, unlike fluorescent lamps that fail faster when cycled often, or HID lamps that require a long time before restarting.

(6) **Dimming**: LEDs can very easily be dimmed either by ***pulse-width modulation*** or lowering the forward current.

(7) **Cool light**: In contrast to most light sources, LEDs radiate very little heat in the form of IR that can cause damage to sensitive objects or fabrics. Wasted energy is dispersed as heat through the base of the LED.

(8) **Slow failure**: LEDs mostly fail by dimming over time, rather than the abrupt failure of incandescent bulbs.

(9) **Lifetime**: LEDs can have a relatively long useful life. One report estimates 35000 ~ 50000 hours of useful life, though time to complete failure may be longer. Fluorescent tubes typically are rated at about 10000 ~ 15000 hours, depending partly on the conditions of use, and incandescent light bulbs at 1000 ~ 2000 hours.

(10) **Shock resistance**: LEDs, being solid state components, are difficult to damage with external shock, unlike fluorescent and incandescent bulbs, which are fragile.

(11) **Focus**: The solid package of the LED can be designed to focus its light. Incandescent and fluorescent sources often require an external reflector to collect light and direct it in a usable manner.

2.2.3.2 Disadvantages

(1) **High initial price**: LEDs are currently more expensive, price per lumen, on an initial capital cost basis, than most conventional lighting technologies. As of 2010, the cost per thousand lumens (kilolumen) was about $18. The price is expected to reach $2/kilolumen by 2015. The additional expense partially stems from the relatively low lumen output and the drive circuitry and power supplies needed.

(2) **Temperature dependence**: LED performance largely depends on the ambient temperature of the operating environment. Over-driving an LED in high ambient temperatures may result in overheating the LED package, eventually leading to device failure. An adequate heat sink is needed to maintain long

lifetime. This is especially important in automotive, medical, and military uses where devices must operate over a wide range of temperatures, and need low failure rates.

(3) **Voltage sensitivity**: LEDs must be supplied with the voltage above the ***threshold*** and a current below the rating. This can involve series resistors or current-regulated power supplies.

(4) **Light quality**: Most cool-white LEDs have spectra that differ significantly from a black body radiator like the sun or an incandescent light. The spike at 460 nm and dip at 500 nm can cause the color of objects to be perceived differently under cool-white LEDs illumination than sunlight or incandescent sources, due to metamerism, red surfaces being rendered particularly badly by typical phosphor-based cool-white LEDs. However, the color rendering properties of common fluorescent lamps are often inferior to what is now available in state-of-art white LEDs.

(5) **Area light source**: Single LEDs do not approximate a point source of light giving a spherical light distribution, but rather a ***Lambertian distribution***. Therefore, LEDs are difficult to apply to the uses needing a spherical light field. LEDs cannot provide divergence below a few degrees. In contrast, lasers can emit beams with divergences of 0.2 degrees or less.

(6) **Electrical polarity**: Unlike incandescent light bulbs, which illuminate regardless of the electrical polarity, LEDs will only light with correct electrical polarity.

(7) **Blue hazard**: There is a concern that blue LEDs and cool-white LEDs are now capable of exceeding safe limits of the so-called blue-light hazard as defined in eye safety specifications such as ANSI/IESNA RP-27.1-05: Recommended Practice for Photobiological Safety for Lamp and Lamp Systems.

(8) **Blue pollution**: Because cool-white LEDs emit proportionally more blue light than conventional outdoor light sources such as high-pressure sodium vapor lamps, the strong wavelength dependence of Rayleigh scattering means that cool-white LEDs can cause more light pollution than other light sources. The International Dark-Sky Association discourages using white light sources with correlated color temperature above 3000 K.

(9) **Droop**: The efficiency of LEDs tends to decrease as the current increases.

2.2.4 Applications of LED

In general, all the LED products can be divided into two major parts, the public lighting and the indoor lighting. LED uses fall into four major categories:

(1) Visual signals where light goes more or less directly from the source to the human eye, to convey a message or meaning.

(2) Illumination where light is reflected from objects to give visual response of these objects.

(3) Measuring and interacting with processes involving no human vision.

(4) Narrow band light sensors where LEDs operate in a ***reverse-bias*** mode and respond to incident light, instead of emitting light.

For more than 70 years, until the LED, practically all lighting was incandescent and fluorescent with the first fluorescent light only being commercially available after the 1939 World's Fair.

2.2.4.1 Indicators and Signs

The low energy consumption, low maintenance and small size of modern LEDs has led to use as status indicators and displays on a variety of equipment and installations. Large-area LED displays are used as stadium displays and as dynamic decorative displays. Thin, lightweight message displays are used at airports and railway stations, and are used as destination displays for trains, buses, trams, and ferries.

One-color light is well suited for traffic lights and signals, exit signs, emergency vehicle lighting, ships' navigation lights or lanterns and LED-based Christmas lights. In cold climates, LED traffic lights may remain snow covered. Red or yellow LEDs are used in indicator and alphanumeric displays in environments where night vision must be retained: aircraft cockpits, submarine and ship bridges, astronomy observatories, and in the field, e. g. night time animal watching and military field use.

Because of their long life and fast switching times, LEDs have been used in brake lights for cars high-mounted brake lights, trucks, and buses, and in turn signals for some time, but many vehicles now use LEDs for their rear light clusters. The use in brakes improves safety, due to a great reduction in the time needed to light fully, or faster rise time, up to 0.5 second faster than an incandescent bulb. This gives drivers more time to react. In a dual intensity circuit (i.e., rear markers and brakes) if the LEDs are not pulsed at a fast enough frequency, they can create a phantom array, where ghost images of the LED will appear if the eyes quickly scan across the array. White LED headlamps are starting to be used. Using LEDs has styling advantages because LEDs can form much thinner lights than incandescent lamps with parabolic reflectors. Due to the relative cheapness of low output LEDs, they are also used in many temporary uses such as glowsticks and the photonic textile.

2.2.4.2 Lighting

With the development of high-efficiency and high-power LEDs, it has become possible to use LEDs in lighting and illumination. Replacement light bulbs have been made, as well as dedicated fixtures and LED lamps. LEDs are used as street lights and in other architectural lighting where color changing is used. The mechanical robustness and long lifetime is used in automotive lighting on cars, motorcycles, and bicycle lights.

LED street lights are employed on poles and in parking garages. In 2007, the Italian village Torraca was the first place to convert its entire illumination system to LEDs.

LEDs are used in aviation lighting. Airbus has used LED lighting in their Airbus A320 Enhanced since 2007, and Boeing plans its use in the 787. LEDs are also being used now in airport and heliport lighting. LED airport fixtures currently include medium-intensity runway lights, runway centerline lights, taxiway centerline and edge lights, guidance signs, and obstruction lighting.

LEDs are also suitable for backlighting for LCD televisions and lightweight laptop displays and light source for DLP projectors (see LED TV). RGB LEDs raise the color gamut by as much as 45%. Screens for TV and computer displays can be made thinner using LEDs for backlighting.

LEDs are used increasingly in aquarium lights. In particular for reef aquariums, LED lights provide an efficient light source with less heat output to help maintain optimal aquarium temperatures. LED-based aquarium fixtures also have the advantage of being manually adjustable to emit a specific color-spectrum for ideal coloration of corals, fish, and invertebrates while optimizing photosynthetically active radiation (PAR), which raises growth and sustainability of photosynthetic life such as corals, anemones, clams, and macroalgae. These fixtures can be electronically programmed to simulate various lighting conditions throughout the day, reflecting phases of the sun and moon for a dynamic reef experience. LED fixtures typically cost up to five times as much as similarly rated fluorescent or high-intensity discharge lighting designed for reef aquariums and are not as high output to date.

The lack of IR or heat radiation makes LEDs ideal for stage lights using banks of RGB LEDs that can easily change color and decrease heating from traditional stage lighting, as well as medical lighting where IR-radiation can be harmful. In energy conservation, LEDs lower heat output also means air conditioning (cooling) systems have less heat to dispose of, reducing carbon dioxide emissions.

LEDs are small, durable and need little power, so they are used in hand held devices such as flashlights. LED strobe lights or camera flashes operate at a safe, low voltage, instead of the 250+ volts commonly found in xenon flashlamp-based lighting. This is especially useful in cameras on mobile phones, where space is at a premium and bulky voltage-raising circuitry is undesirable.

LEDs are used for infrared illumination in night vision uses including security cameras. A ring of LEDs around a video camera, aimed forward into a retroreflective background, allows chroma keying in video productions.

LEDs are now used commonly in all market areas from commercial to home use: standard lighting, AV, stage, theatrical, architectural, public installations, and wherever artificial light is used.

LEDs are increasingly finding uses in medical and educational applications, for example as mood enhancement, and new technologies such as AmBX, exploiting LED versatility. NASA has even sponsored research for the use of LEDs to promote health for astronauts.

2.2.4.3 Smart Lighting

Light can be used to transmit broadband data, which is already implemented in IrDA standards using infrared LEDs. Because LEDs can cycle on and off millions of times per second, they can be wireless transmitters and access points for data transport. Lasers can also be modulated in this manner.

2.2.4.4 Sustainable Lighting

Efficient lighting is needed for sustainable architecture. In 2009, a typical 13-watt LED lamp emitted 450～650 lumens, which is equivalent to a standard 40-watt incandescent bulb. In 2011, LEDs have become more efficient, so that a 6-watt LED can easily achieve the same results. A standard 40-watt incandescent bulb has an expected lifespan of 1000 hours, whereas an LED can continue to operate with reduced efficiency for more than 50000 hours, 50 times longer than the incandescent bulb.

1. Energy Consumption

In the US, one kilowatt-hour of electricity will cause 1.34 pounds (610 g) of CO_2 emission. Assuming the average light bulb is on for 10 hours a day, one 40-watt incandescent bulb will cause 196 pounds (89 kg) of CO_2 emission per year. The 6-watt LED equivalent will only cause 30 pounds (14 kg) of CO_2 over the same time span. A building's carbon footprint from lighting can be reduced by 85% by exchanging all incandescent bulbs for new LEDs.

2. Economically Sustainable

LED light bulbs could be a cost-effective option for lighting a home or office space because of their very long lifetimes. Consumer use of LEDs as a replacement for conventional lighting system is currently hampered by the high cost and low efficiency of available products. 2009 DOE testing results showed an average efficacy of 35 lm/W, below that of typical CFLs, and as low as 9 lm/W, worse than standard incandescent. However, as of 2011, there are LED bulbs available as efficient as 150 lm/W and even inexpensive low-end models typically exceed 50 lm/W. The high initial cost of the commercial LED bulb is due to the expensive sapphire substrate, which is key to the production process. The sapphire apparatus must be coupled with a mirror-like collector to reflect light that would otherwise be wasted.

2.2.4.5 Light Sources for Machine Vision Systems

Machine vision systems often require bright and homogeneous illumination, so features of interest are easier to process. LEDs are often used for this purpose, and this is likely to remain one of their major uses until price drops low enough to make signaling and illumination uses more widespread. Barcode scanners are the most common example of machine vision, and many low-cost ones use red LEDs instead of lasers. Optical computer mice are also another example of LEDs in machine vision, as it is used to provide an even light source on the surface for the miniature camera within the mouse. LEDs constitute a nearly ideal light source for machine vision systems for several reasons:

The size of the illuminated field is usually comparatively small and machine vision systems are often quite expensive, so the cost of the light source is usually a minor concern. However, it might not be easy to replace a broken light source placed within complex machinery, and here the long service life of LEDs is a benefit.

LED elements tend to be small and can be placed with high density over flat or even shaped substrates (PCBs, etc.) so that bright and homogeneous sources that direct light from tightly controlled directions on inspected parts can be designed. This can often be obtained with small, low-cost lenses and diffusers, helping to achieve high light densities with control over lighting levels and homogeneity. LED sources can be shaped in several configurations (spot lights for reflective illumination; ring lights for coaxial illumination; back lights for contour illumination; linear assemblies; flat, large format panels; dome sources for diffused, omnidirectional illumination).

LEDs can be easily strobed (in the microsecond range and below) and synchronized with imaging. High-power LEDs are available allowing well-lit images even with very short light pulses. This is often used to obtain crisp and sharp "still" images of quickly moving parts.

LEDs come in several different colors and wavelengths, allowing easy use of the best color for each need, where different color may provide better visibility of features of interest. Having a precisely known spectrum allows tightly matched filters to be used to separate informative ***bandwidth*** or to reduce disturbing effects of ambient light. LEDs usually operate at comparatively low working temperatures, simplifying heat management and dissipation. This allows using plastic lenses, filters, and diffusers. Waterproof units can also easily be designed, allowing use in harsh or wet environments (food, beverage, oil industries).

2.2.4.6 Other Applications

The light from LEDs can be modulated very quickly so they are used extensively in optical fiber and Free Space Optics communications. This include remote controls, such as for TVs, VCRs, and LED Computers, where infrared LEDs are often used. Opto-isolators use an LED combined with a photodiode or ***phototransistor*** to provide a signal path with electrical isolation between two circuits. This is especially useful in medical equipment where the signals from a low-voltage sensor circuit (usually battery-powered) in contact with a living organism must be electrically isolated from any possible electrical failure in a recording or monitoring device operating at potentially dangerous voltages. An ***optoisolator*** also allows information to be transferred between circuits not sharing a common ground potential.

Many sensor systems rely on light as the signal source. LEDs are often ideal as a light source due to the requirements of the sensors. LEDs are used as movement sensors, for example, in optical computer mice. The Nintendo Wii's sensor bar uses infrared LEDs. Pulse oximeters use them for measuring oxygen saturation. Some flatbed scanners use arrays of RGB LEDs rather than the typical cold-cathode fluorescent lamp as the light source. Having independent control of three illuminated colors allows the scanner to calibrate itself for more accurate color balance, and there is no need for warm-up. Further, its sensors only need be monochromatic, since at any one time the page being scanned is only lit by one color of light. Touch sensing: Since LEDs can also be used as photodiodes, they can be used for both photo emission and detection. This could be used in for example a touch-sensing screen that register reflected light from a finger or stylus.

Many materials and biological systems are sensitive to or dependent on light. Grow lights use LEDs to increase photosynthesis in plants and bacteria, and viruses can be removed from water and other substances using UV LEDs for sterilization. Other uses are as UV curing devices for some ink and coating methods, and in LED printers.

Plant growers are interested in LEDs because they are more energy-efficient, emit less heat (can damage plants close to hot lamps), and can provide the optimum light frequency for plant growth and bloom periods compared to currently used grow lights: HPS (high-pressure sodium), MH (metal halide) or

CFL/low-energy. However, LEDs have not replaced these grow lights due to higher price. As mass production and LED kits develop, the LED products will become cheaper.

LEDs have also been used as a medium-quality voltage reference in electronic circuits. The forward voltage drop (e. g., about 1.7 V for a normal red LED) can be used instead of a Zener diode in low-voltage regulators. Red LEDs have the flattest I/V curve above the knee. Nitride-based LEDs have a fairly steep I/V curve and are useless for this purpose. Although LED forward voltage is far more current-dependent than a good Zener, Zener diodes are not widely available below voltages of about 3 V.

New Words and Expressions

[1] **stent**：支架。

[2] **cathode**：阴极。它是指得到电子的极。

[3] **epoxy resin**：环氧树脂。它是泛指分子中含有两个或两个以上环氧基团的有机高分子化合物，是一类重要的热固性塑料，广泛用于黏合剂、涂料等。

[4] **bandwidth**：带宽。信号的带宽是指该信号所包含的各种不同频率成分所占据的频率范围。

[5] **forward voltage**：正向电压。

[6] **infrared**：红外线。它是波长介乎微波与可见光之间的电磁波，其波长为760 nm～1 mm，是波长比红光长的非可见光。

[7] **electrodes**：电极。

[8] **depletion zone**：耗尽区。它是指pn结中在漂移运动和扩散作用的双重影响下载流子数量非常少的一个高电阻区域。

[9] **pulse-width modulation**：脉宽调制。

[10] **threshold**：阈值。

[11] **Lambertian distribution**：朗伯分布。

[12] **reverse-bias**：反向偏置。

[13] **phototransistor**：光电三极管。它是在光电二极管的基础上发展起来的光电器件，本身具有放大功能。

[14] **optoisolator**：光隔离器。它是一种只允许单向光通过的无源光器件，它的作用是防止光路中由于各种原因产生的后向传输光对光源以及光路系统产生的不良影响。

2.3 Typical Instrument Based on Stimulated Emission: Laser Diode (LD)

A laser diode, or LD, is a typical instrument based on stimulated emission which we have initially introduced in Section 2. 1. 2. LD can be taken as an electrically pumped semiconductor laser (see Fig. 2-10) in which the active medium is formed by a p-n junction of a semiconductor diode, which is similar to that found in a light-emitting diode.

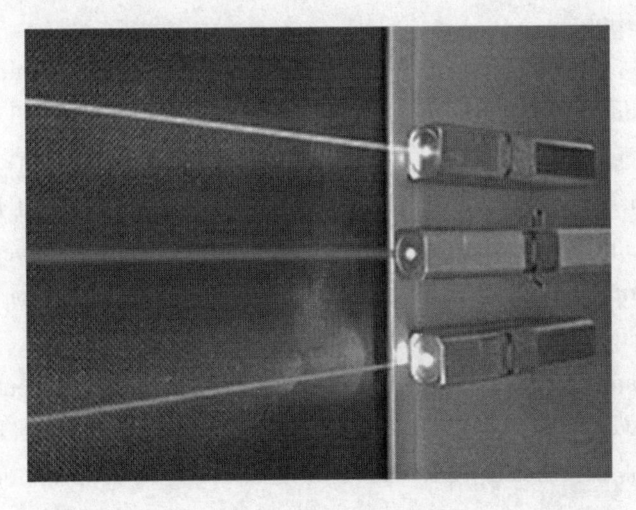

Fig. 2-10 Semiconductor lasers (445 nm, 520 nm, 635 nm)

The laser diode is the most common type of laser produced. Laser diodes have a very wide range of uses that include, but are not limited to, fiber optic communications, barcode readers, laser pointers, CD/DVD/Blu-ray reading and recording, laser printing, scanning and increasingly directional lighting sources.

2.3.1 History of LD

The acronym laser means ***light amplification*** by the ***stimulated emission of radiation***. The first working ruby laser was developed in 1960 by the American scientist Theodore Meiman. The theoretical and practical foundations for this development were made by the American Charles Townes and the Russians Alexander Prokhorov and Nikolay Basov, who shared the Nobel Prize for physics in 1964 for their work.

Interestingly, the laser is not a light amplifier, as the term suggests, but, rather, a light generator. This was true for the first laser, it is true for today's devices. However, since the term exists and is well accepted, so be it. It is also interesting to note that in the technical literature, derivative words like "lasing" and "to lase" have become common.

The laser is a device that amplifies (or, as we now know, "generates") light by means of the stimulated emission of radiation. How a laser produces light amplification and what the words stimulated emission of radiation mean are our next consideration.①

Coherent light emission from a semiconductor (gallium arsenide) diode (the first laser diode) was demonstrated in 1962 by two US groups led by Robert N. Hall at the General Electric Research Center and by Marshall Nathan at the IBM T. J. Watson Research Center. The priority is given to General Electric group who have obtained and submitted their results earlier; they also went further and made a ***resonant cavity*** for their diode. The first visible wavelength laser diode was demonstrated by Nick Holonyak, Jr. later in 1962.

Other teams at MIT Lincoln Laboratory, Texas Instruments, and RCA Laboratories were also involved in and received credit for their historic initial demonstrations of efficient light emission and lasing in semiconductor diodes in 1962 and thereafter. GaAs lasers were also produced in early 1963 in the Soviet Union by the team led by Nikolay Basov.

In the early 1960s liquid phase epitaxy (LPE) was invented by Herbert Nelson of RCA Laboratories. By layering the highest quality crystals of varying compositions, it enabled the demonstration of the highest quality ***heterojunction*** semiconductor laser materials for many years. LPE was adopted by all the leading laboratories, worldwide and used for many years. It was finally supplanted in the 1970s by molecular beam epitaxy and organometallic chemical vapor deposition.

Diode lasers of that era operated with threshold current densities of

---

① Mynbaev D K, Scheiner L L. Fiber-Optic Communications Technology[M]. New Jersey: Prentice Hall, Inc., 2000:35.

1000 A/cm^2 at 77 K temperatures. Such performance enabled continuous-lasing to be demonstrated in the earliest days. However, when operated at room temperature, about 300 K, threshold current densities were two orders of magnitude greater, or 100000 A/cm^2 in the best devices. The dominant challenge for the remainder of the 1960s was to obtain low threshold current density at 300 K and thereby to demonstrate continuous-wave lasing at room temperature from a diode laser.

The first diode lasers were ***homojunction*** diodes. That is, the material (and thus the bandgap) of the ***waveguide*** core layer and that of the surrounding clad layers, were identical. It was recognized that there was an opportunity, particularly afforded by the use of liquid phase epitaxy using aluminum gallium arsenide, to introduce heterojunctions. Heterostructures consist of layers of semiconductor crystal having varying ***bandgap*** and refractive index. Heterojunctions (formed from heterostructures) had been recognized by Herbert Kroemer, while working at RCA Laboratories in the mid-1950s, as having unique advantages for several types of electronic and optoelectronic devices including diode lasers. LPE afforded the technology of making heterojunction diode lasers.

The first heterojunction diode lasers were single-heterojunction lasers. These lasers utilized aluminum gallium arsenide p-type injectors situated over n-type gallium arsenide layers grown on the substrate by LPE. An admixture of aluminum replaced gallium in the semiconductor crystal and raised the bandgap of the p-type injector over that of the n-type layers beneath. It worked: the 300 K threshold currents went down by 10× to 10000 amperes per square centimeter. Unfortunately, this was still not in the needed range and these single-heterostructure diode lasers did not function in continuous wave operation at room temperature.

The innovation that met the room temperature challenge was the double heterostructure laser. The trick was to quickly move the wafer in the LPE apparatus between different "melts" of aluminum gallium arsenide (p-type and n-type) and a third melt of gallium arsenide. It had to be done rapidly since the gallium arsenide core region needed to be significantly under 1 μm in thickness. This may have been the earliest true example of "***nanotechnology***". The first laser diode to achieve continuous wave operation was a double heterostructure demonstrated in 1970 essentially simultaneously by Zhores Alferov and collaborators (including Dmitri Z. Garbuzov) of the Soviet Union, and Morton Panish and Izuo Hayashi working in the United States. However, it is widely accepted that Zhores Alferov and team reached the milestone first. For their accomplishment and that of their co-workers, Alferov and Kroemer shared the Nobel Prize in physics in 2000.

2.3.2 Principle of Action

2.3.2.1 Stimulated Emission

We distinguish between two types of radiation: spontaneous and stimulated. "Spontaneous" means that radiation occurs without external cause. That's exactly what happens in an LED: Excited electrons from the conduction band fall, without any external inducement, to the valence band, which results in spontaneous radiation.

A different process occurs if you let an external photon hit an excited electron. Their interaction includes an electron transition and the radiation of a new photon. Now the induced emission is stimulated by an external photon. Thus, this radiation is called stimulated radiation. ①

In the classical view, the energy of an electron orbiting an atomic nucleus is larger for orbits further from the nucleus of an atom. However, ***quantum mechanical effects*** force electrons to take on discrete positions in orbitals. Thus, electrons are found in specific energy levels of an atom, two of which are shown in Fig. 2-11.

When an electron absorbs energy either from light (photons) or heat (phonons), it receives that incident quantum of energy. But transitions are only allowed in between discrete energy levels such as the two shown above. This leads to emission lines and absorption lines.

When an electron is excited from a lower to a higher energy level, it will not stay that way forever. An electron in an excited state may decay to a lower energy state which is not occupied, according to a particular time constant characterizing

① Mynbaer D K, Scheiner L L. Fiber-Optic Communications Technology[M]. New Jersey: Prentice Hall, Inc., 2000:36.

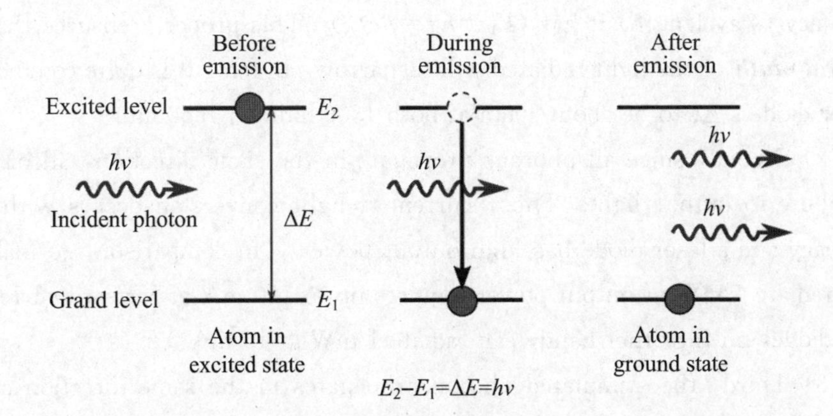

Fig. 2-11 Stimulated emission

that transition. When such an electron decays without external influence, emitting a photon, that is called "spontaneous emission". The phase associated with the photon that is emitted is random. A material with many atoms in such an excited state may thus result in radiation which is very spectrally limited (centered around one wavelength of light), but the individual photons would have no common phase relationship and would emanate in random directions. This is the mechanism of fluorescence and thermal emission.

An external electromagnetic field at a frequency associated with a transition can affect the quantum mechanical state of the atom. As the electron in the atom makes a transition between two stationary states (neither of which shows a dipole field), it enters a transition state which does have a dipole field, and which acts like a small electric dipole, and this dipole oscillates at a characteristic frequency. In response to the external electric field at this frequency, the probability of the atom entering this transition state is greatly increased. Thus, the rate of transitions between two stationary states is enhanced beyond that due to spontaneous emission. Such a transition to the higher state is called absorption, and it destroys an incident photon (the photon's energy goes into powering the increased energy of the higher state). A transition from the higher to a lower energy state, however, produces an additional photon; this is the process of stimulated emission.

Stimulated emission has four main properties:

(1) First, an external photon forces a photon with similar energy (E_P) to be emitted. In other words, the external photon stimulates radiation with the same

frequency (wavelength) it has ($E_P = hf = hc/\lambda$). This property ensures that the ***spectral width*** of the light radiated will be narrow. In fact, it is quite common for a laser diode's $\Delta\lambda$ to be about 1 nm at both 1300 nm and 1550 nm.

(2) Second, since all photons propagate in the same direction, all of them contribute to output light. Thus, current-to-light conversion occurs with high efficiency and a laser diode has high output power. (In comparison, to make an LED radiate 1 mW of output power requires up to 150 mA of forward current; a laser diode, on the other hand, can radiate 1 mW at 10 mA.)

(3) Third, the stimulated photon propagates in the same direction as the photon that stimulated it; hence, the stimulated light will be well directed. If you compare the beam of a laser pointer — available in any stationery store — with any type of lamp, you'll appreciate the difference between spontaneous and stimulated emission in terms of the way each direct light.

(4) Fourth, since a stimulated photon is radiated only when an external photon triggers this action, both photons are said to be synchronized, that is, time-aligned. This means that both photons are in phase and so the stimulated emission is coherent. ①

2.3.2.2 *Positive Feedback*

To radiate stimulated light with essential power, we need not one photon but millions. Here is how we can stimulate such radiation: We place a mirror at one end of an active layer. Two photons — one external and one stimulated — are then reflected back and directed to the active layer again. These two photons now work as external radiation and stimulate the emission of two other photons. The four photons are reflected by a second mirror, which is positioned at the other end of the active layer. When these photons pass the active layer, they stimulate emission of another four photons. These eight photons are reflected back into the active layer by the first mirror and this process continues ad infinitum, see Fig. 2-12.

---

① Mynbaev D K, Scheiner L L. Fiber-Optic Communications Technology[M]. New Jersey: Prentice Hall, Inc., 2003:36-40.

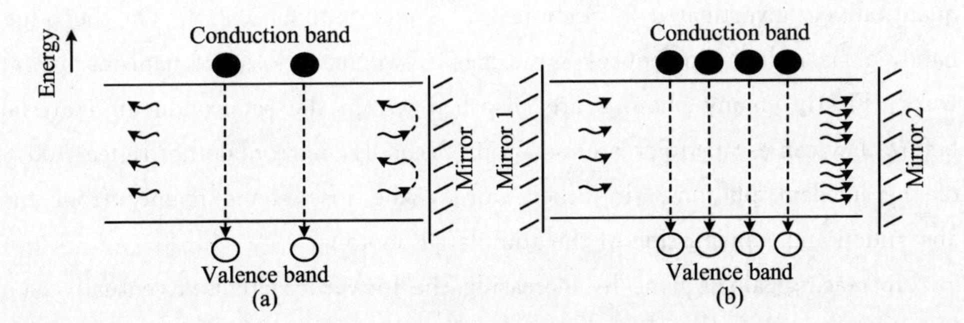

Fig. 2-12 Optical amplification

Thus, the two mirrors provide positive optical feedback — positive because the feedback adds the output (stimulated photons) to the input (external photons). If the output is subtracted from the input, the feedback is called negative. These two mirrors, then, constitute a resonator. ①

2.3.2.3 Population Inversion

Note how fast the number of stimulated photons rises. To sustain this dynamic process, we need an incalculable number of excited electrons available at the conduction band. We know that using external energy — forward current for an LED — makes it possible to excite a number of electrons. But in lasers depletion of the conduction band occurs much faster than it does in LEDs; hence, we need to excite electrons at a much higher rate than we did in the LED process. In fact, for laser action (lasing) we need to have more electrons at the higher-energy conduction band than at the lower-energy valence band. This situation is called population inversion because, normally, the valence band is much more heavily populated than the conduction band. To create this population inversion, high-density forward current is passed through the small active area.

Population inversion is a necessary condition to create a lasing effect because the greater the number of excited electrons, the greater the number of stimulated photons that can be radiated. What's more, the emission intensity will be higher as well. In other words, the number of excited electrons determines the gain of a semiconductor diode. The detailed physical picture of this part will be

---

① Mynbaev D K, Scheiner L L. Fiber-Optic Communications Technology[M]. New Jersey: Prentice Hall, Inc., 2000:39.

quantitatively investigated in Section 2. 3. 3 and Section 2. 3. 4. On the other hand, a laser diode introduces some loss. Two main loss mechanisms are at work: Firstly, many photons are absorbed within the semiconductor material before they can escape to create radiation. Secondly, mirrors do not reflect 100% of the incident photons. In other words, the loss stems mainly from the absorption and transmission of the stimulated photons.

Increasing gain is done by increasing the forward current. Eventually gain becomes equal to loss, a situation called the threshold condition. (The corresponding forward current is called the threshold current.) At this threshold condition, a semiconductor diode starts to act like a laser. As we continue to increase the forward current (that is, the gain), the number of emitted stimulated photons continues to increase, which means the intensity of the output light also continues to increase. What we have, then, is a semiconductor diode that radiates monochromatic, well-directed, highly intense, coherent light. ①

2.3.3 Population Inversion in Three-levels and Four-levels Systems

As we introduced in the previous section, we have qualitatively understood that population inversion is essential to achieve "gain" in a laser resonator. With the excitation from the power supply, the particles (electrons/molecules) initially located at the ground state are promoted to the excited states. ***Lasing action*** happens when one of the excited states has more population as compared with that of the ground state or lower-excited state. In this section, the energy diagrams including pumping, spontaneous emission, stimulated emission, and stimulated absorption in a gain medium are discussed. The rate equations are employed to quantitatively depict the transitions. Then, according to rate equations, the difference between the particle number density at the upper and lower level is calculated in order to demonstrate ***population inversion*** in a gain medium.

As shown by the energy diagram in Fig. 2-13, population inversion can be

① Mynbaev D K, Scheiner L L. Fiber-Optic Communications Technology[M]. New Jersey: Prentice Hall, Inc., 2000:40.

resulted in both ***three-levels system*** and ***four-levels system***. A three-levels system means that three energy levels including the ground state E_1, lower excited state E_2, and higher excited state E_3 are involved in the transitions giving rise to lasing action [see Fig. 2-13 (a)]. In fact, E_3 could be a coupled level by densely distributed excited states close to the continuum other than a single energy level. Materials with three-levels systems were used as gain media of lasers in the olden days. For instance, the first ruby laser invented in 1960 was a laser with a three-levels system.

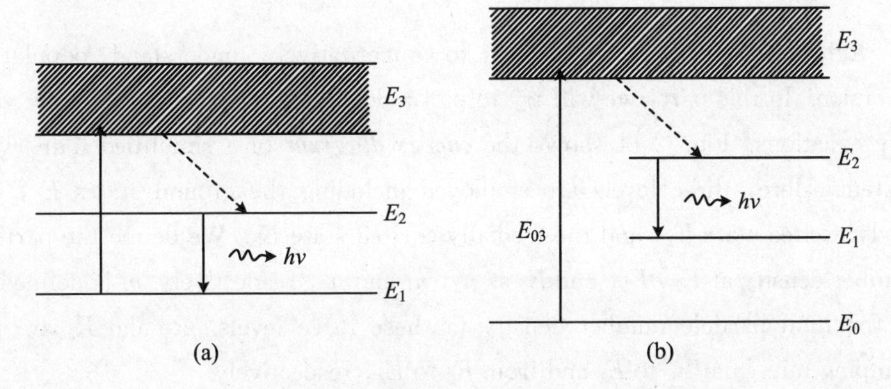

Fig. 2-13 Energy diagrams for a three-levels system (a) and a four-levels system (b)

The transitions of a three-levels system is shown in Fig. 2-13(a). Particles (electrons) from the ground states E_1 are firstly promoted to the higher excited state E_3. Due to limited lifetime at E_3, particles will be transferred to the lower excited state E_2 through non-radiative transitions. Population inversion is resulted between lower excited state E_2 and the ground state E_1, once more than half of the populations at E_1 are redistributed at E_2.

On the other hand, in a four-levels system [see Fig. 2-13 (b)], the ground sate E_0, and three excited states E_1, E_2 and E_3 are involved in the transitions for lasing action. To simplify, we name E_1, E_2 and E_3 in Fig. 2-13(b) as lowly, medially, highly excited state, respectively. The particles at ground state E_0 are firstly promoted to the highly excited state E_3 and then jump down to E_2 through non-radiative transitions. The population inversion happens between medially-excited state E_2 and the lowly-excited state E_1. Particles at the medially-excited state usually have a relatively longer lifetime as compared with those at the lowly-excited state in order to benefit the lasing action.

The lower level in a four-levels system for stimulated emission is an excited state (lowly-excited state), which is initially empty. It is compared with the lower level in a three-levels system, which is the ground state and initially highly occupied. Therefore, it is much easier to create population inversion in a four-levels system than that in a three-levels system. In fact, nowadays, most of lasers have gain media with four-levels systems.

2.3.4 From Rate Equations to Population Inversion

Rate equations are the keys to quantitatively understand population inversion. In this part, we will try to plot a detailed physical picture by deriving rate equations. Fig. 2-14 shows the ***energy diagram*** of a simplified four-levels system. Here, three levels are employed including the ground states E_0, the lowly-excited state E_1, and the medially-excited state E_2. We define the particle number density at E_2, E_1, and E_0 as n_2, n_1 and n_0, respectively. n is defined to be the total particle number density in these three levels. R_1 and R_2 are the pumping rate from E_0 to E_1 and from E_0 to E_2, respectively.

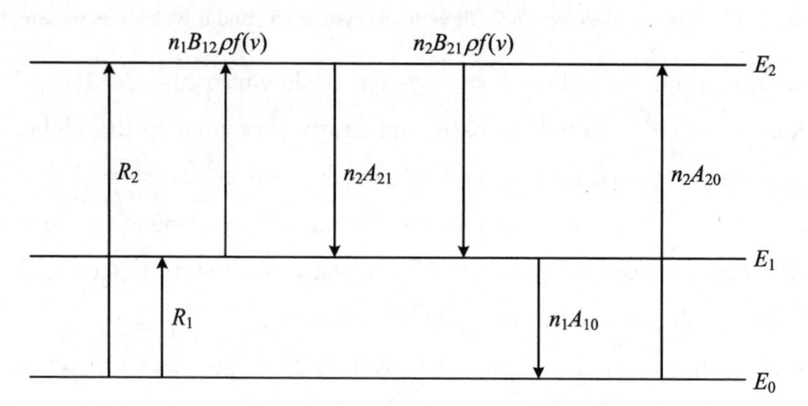

Fig. 2-14 Simplified energy diagram for a four-levels system

When the laser starts working, the particles (electrons) at the ground state E_0 are firstly promoted to the excited states (E_2 and E_1). We assume that the oscillations are initiated by ***monochromatic lights***, which have much narrower bandwidth as compared with the absorption band of the gain medium. Due to the limited lifetimes for particles at excited states, the particles at E_2 will jump down to E_1 and E_0 through stimulated/spontaneous emission, so as to reduce the particle number density at E_2. On the other hand, the particle number density at E_2 can be increased through stimulated absorption from E_0. In this sense, for the medially-excited state E_2, the instantaneous particle number density change at E_2 can be expressed as:

$$\frac{dn_2}{dt} = R_2 - n_2 A_2 - (n_2 B_{21} - n_1 B_{12}) \rho f(\nu) \tag{2-21a}$$

here, $A_2 = A_{21} + A_{20}$. Similarly, for the lowly-excited state E_1, we have:

$$\frac{dn_1}{dt} = R_1 + n_2 A_{21} - n_1 A_1 + (n_2 B_{21} - n_1 B_{12}) \rho f(\nu) \tag{2-21b}$$

The total particle number density n is

$$n = n_0 + n_1 + n_2 \tag{2-21c}$$

Equations (2-21a) ~ Equations (2-21c) are rate equations which describe the dynamic equilibrium by emissions/absorptions. Starting from rate equations, in principle, the number of particles in each energy level can be derived.

Population inversion happens once we switch on a laser. The initial radiation circulates inside a resonator with its power gradually amplified. Normally, it takes some time for a laser, from small signal oscillation to stable operation. When a laser works stably, its output parameters (e. g. power, pulse duration, spectral width, and repetition rate) keep constant. In this case, the equilibrium is achieved by pumping, stimulated emission, stimulated absorption, spontaneous emission, thermal effects, nonlinear effects, and etc.

In this part, we are supposed to derive population inversion when the laser system is stably operating. It is the most common and useful condition for the users of lasers. When a laser operates stably, its output power keeps constant. It means that the number of particles at each level will not change. In this case, the dynamic equilibrium writes:

$$\frac{dn_0}{dt} = \frac{dn_1}{dt} = \frac{dn_2}{dt} = 0 \tag{2-22}$$

If we assume that the degeneracy for energy levels E_2 and E_1 are equal, meaning $g_1 = g_2$. Thus,

$$B_{21} = B_{12} \tag{2-23}$$

In practical cases, for spontaneous emission, the probability of transitions from E_2 to E_1 is much larger than that from E_2 to E_0 ($A_{21} > A_{20}$). Thus, we derive:

$$A_2 \approx A_{21} \tag{2-24}$$

In this case, rate equations can be simplified:

$$\begin{aligned} R_2 - n_2 A_2 - (n_2 - n_1) B_{21} \rho f(\nu) &= 0 \\ R_1 - n_1 A_1 + n_2 A_2 + (n_2 - n_1) B_{21} \rho f(\nu) &= 0 \end{aligned} \tag{2-25}$$

We can add the above two equations to derive:

$$R_1 + R_2 = n_1 A_1 = \frac{n_1}{\tau_1} \tag{2-26}$$

By substituting the equation (2-26) into equation (2-25), we have:

$$\begin{aligned} n_2 &= \frac{R_2 + (R_1 + R_2)\tau_1 B_{21} \rho f(\nu)}{A_2 + B_{21} \rho f(\nu)} \\ &= \frac{R_2 + (R_1 + R_2)\tau_1 B_{21} \rho f(\nu)}{\frac{1}{\tau_2} + B_{21} \rho f(\nu)} \\ &= \frac{R_2 \tau_2 + (R_1 + R_2)\tau_1 \tau_2 B_{21} \rho f(\nu)}{1 + \tau_2 B_{21} \rho f(\nu)} \end{aligned} \tag{2-27}$$

where τ_2 and τ_1 are the lifetimes for particles at E_2 and E_1, respectively. When a laser is operating stably, the population inversion (population difference) between the medially-excite E_2 and lowly-excited states E_1 writes:

$$\begin{aligned} \Delta n = n_2 - n_1 &= \frac{R_2 \tau_2 + (R_1 + R_2)\tau_1 \tau_2 B_{21} \rho f(\nu)}{1 + \tau_2 B_{21} \rho f(\nu)} - (R_1 + R_2)\tau_1 \\ &= \frac{R_2 \tau_2 - (R_1 + R_2)\tau_1}{1 + \tau_2 B_{21} \rho f(\nu)} = \frac{\Delta n_0}{1 + \tau_2 B_{21} \rho f(\nu)} \end{aligned} \tag{2-28}$$

Equation (2-28) provides the value for the population inversion (population difference between upper and lower levels). From equation (2-28), we can find that the value of $R_2 \tau_2 - (R_1 + R_2)\tau_1$ must be positive to achieve population inversion ($\Delta n > 0$). In this sense, in the gain medium of a diode laser, particles at the upper level should have a relative longer lifetime, meaning large value of τ_2. Particles at lower level should have a shorter lifetime, meaning small value of τ_1.

2.3.5 Laser-diode Light: An Analysis[①]

A laser diode radiates light that can be characterized as follows:

(1) ***Monochromatic.*** The spectral width of the radiated light is very narrow. Indeed, the line width for a laser diode can be in tenths or even hundredths of a

① Mynbaev D K, Scheiner L L. Fiber-Optic Communications Technology[M]. New Jersey: Prentice Hall, Inc., 2000:41.

nanometer.

(2) Well directed. A laser diode radiates a narrow, well-directed beam that can be easily launched into an optical fiber.

(3) Highly intense and power-efficient. A laser diode can radiate hundreds of milliwatts of output power. A new type of laser diode, the VCSEL, radiates 1 mW at 10 mA of forward current, making current-to-light conversion 10 times more efficient than it is in the best LEDs.

(4) Coherent. Light radiated by a laser diode is coherent; that is, all oscillations are in phase. This property is important for the transmission and detection of an information signal.

As you can see, these characteristics are very similar to those of stimulated emission. Note that only the combination of an active medium and a resonator, which together form a laser, produces light with these remarkable properties.

2.3.6 Design of LD

A laser consists of a ***gain medium***, a mechanism to supply energy to it, and something to provide optical feedback, see Fig. 2-15. The gain medium is a material with properties that allow it to amplify light by stimulated emission. Light of a specific wavelength that passes through the gain medium is amplified (increases in power).

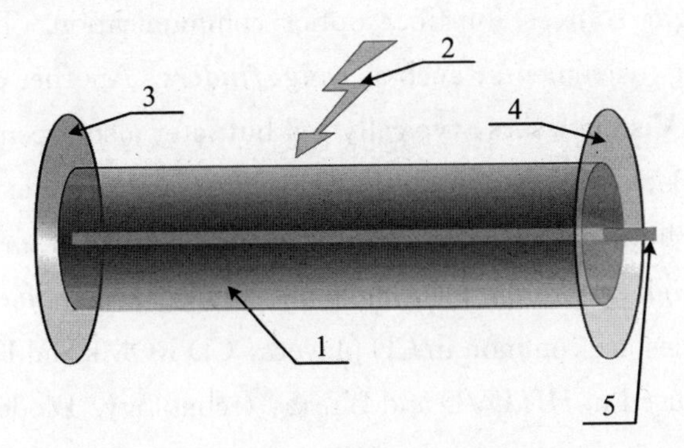

Fig. 2-15 Components of a typical laser

1. Gain medium; 2. Laser pumping energy; 3. High reflector; 4. Output coupler; 5. Laser beam.

For the gain medium to amplify light, it needs to be supplied with energy.

This process is called pumping. The energy is typically supplied as an electrical current, or as light at a different wavelength. Pump light may be provided by a flash lamp or by another laser.

The most common type of laser uses feedback from an optical cavity — a pair of mirrors on either end of the gain medium. Light bounces back and forth between the mirrors, passing through the gain medium and being amplified each time. Typically, one of the two mirrors, the output coupler, is partially transparent. Some of the light escapes through this mirror. Depending on the design of the cavity (whether the mirrors are flat or curved), the light coming out of the laser may spread out or form a narrow beam. This type of device is sometimes called a laser oscillator in analogy to electronic oscillators, in which an electronic amplifier receives electrical feedback that causes it to produce a signal.

Most practical lasers contain additional elements that affect properties of the emitted light, such as the polarization, the wavelength, and the shape of the beam.

2.3.7 Applications of LD

Laser diodes are numerically the most common laser type, with 2004 sales of approximately 733 million units, as compared to 131000 of other types of lasers.

Laser diodes find wide use in telecommunication as easily modulated and easily coupled light sources for fiber optics communication. They are used in various measuring instruments, such as ***rangefinders***. Another common use is in barcode readers. Visible lasers, typically red but later also green, are common as laser pointers. Both low and high-power diodes are used extensively in the printing industry both as light sources for scanning (input) of images and for very high-speed and ***high-resolution*** printing plate (output) manufacturing. Infrared and red laser diodes are common in CD players, CD-ROMs and DVD technology. Violet lasers are used in HD DVD and Blu-ray technology. Diode lasers have also found many applications in laser absorption spectrometry (LAS) for high-speed, low-cost assessment or monitoring of the concentration of various species in gas phase. High-power laser diodes are used in industrial applications such as heat treating, cladding, seam welding and for pumping other lasers, such as diode-pumped solid-state lasers.

Uses of laser diodes can be categorized in various ways. Most applications could be served by larger solid-state lasers or optical parametric oscillators, but the low cost of mass-produced diode lasers makes them essential for mass-market applications. Diode lasers can be used in a great many fields; since light has many different properties (power, wavelength, spectral and beam quality, polarization, etc.), it is useful to classify applications by these basic properties.

Many applications of diode lasers primarily make use of the "directed energy" property of an optical beam. In this category, one might include the laser printers, barcode readers, image scanning, illuminators, designators, optical data recording, combustion ignition, laser surgery, industrial sorting, and industrial machining. Some of these applications are well-established while others are emerging.

Laser medicine: Medicine and especially dentistry have found many new uses for diode lasers. The shrinking size of the units and their increasing user friendliness makes them very attractive to clinicians for minor soft tissue procedures. The 800～980 nm units have a high absorption rate for hemoglobin and thus make them ideal for soft tissue applications, where good hemostasis is necessary.

Uses which may make use of the coherence of diode-laser-generated light include interferometric distance measurement, holography, coherent communications, and coherent control of chemical reactions.

Uses which may make use of "narrow spectral" properties of diode lasers include range-finding, telecommunications, infra-red countermeasures, spectroscopic sensing, generation of radio-frequency or terahertz waves, atomic clock state preparation, quantum key cryptography, frequency doubling and conversion, water purification (in the UV), and photodynamic therapy (where a particular wavelength of light would cause a substance such as porphyrin to become chemically active as an anti-cancer agent only where the tissue is illuminated by light).

Uses where the desired quality of laser diodes is their ability to generate ultra-short pulses of light by the technique known as "mode-locking" include clock distribution for high-performance integrated circuits, high-peak-power sources for laser-induced breakdown ***spectroscopy*** sensing, arbitrary waveform generation for radio-frequency waves, photonic sampling for ***analog-to-digital***

conversion, and optical ***code-division-multiple-access*** systems for secure communication.

New Words and Expressions

[1] **light amplification**：光放大。

[2] **stimulated emission of radiation**：受激辐射。它是指处于激发态的发光原子在外来辐射场的作用下，向低能态或基态跃迁时，辐射光子的现象。

[3] **resonant cavity**：谐振腔。它是指光波在其中来回反射从而提供光能反馈的空腔，是激光器的必要组成部分。

[4] **heterojunction**：异质结。由两种不同的半导体相接触所形成的界面区域。

[5] **homojunction**：同质结。由两种相同的半导体相接触所形成的界面区域。

[6] **waveguide**：波导。它是指一种在微波或可见光波段中传输电磁波的装置。

[7] **bandgap**：带隙。导带的最低点和价带的最高点的能量之差。

[8] **nanotechnology**：纳米技术。它是指用单个原子、分子制造物质的科学技术，研究结构尺寸为 0.1～100 nm 的材料的性质和应用。

[9] **quantum mechanical effects**：量子力学效应。

[10] **spectral width**：谱宽。它是指辐射频谱分布曲线上的两上半最大强度点之间的频率宽度。

[11] **positive feedback**：正反馈。它是指受控部分发出反馈信息，其方向与控制信息一致，可以促进或加强控制部分的活动。

[12] **monochromatic**：单色性。

[13] **gain medium**：增益介质。它是指用来实现粒子数反转并产生光的受激辐射放大作用的物质体系。

[14] **rangefinders**：测距仪。

[15] **high-resolution**：高分辨率。

[16] **spectroscopy**：光谱学。

[17] **analog-to-digital conversion**：模拟数字转换。

[18] **code-division-multiple-access**：码分多址（一种扩频多址数字式通信技术）。

[19] **lasing action**：形成激光。

［20］ **three-levels system**：三能级系统。它是指在增益介质中有三个能级参与受激辐射跃迁。红宝石激光是典型的三能级系统。

［21］ **four-levels system**：四能级系统。它是指在增益介质中有四个能级参与受激辐射跃迁。四能级系统更容易实现粒子数反转，在形成激光方面比三能级系统更有优势。

［22］ **energy diagram**：能级图。

［23］ **population inversion**：高能级粒子数布局大于低能级粒子数布局。在外界激励源的作用下，将下能级的粒子抽运到上能级，使激光上下能级之间产生粒子数反转。

［24］ **rate equations**：速率方程组。它用来研究在泵浦、受激辐射、受激吸收、自发辐射动态平衡下各能级上的粒子数密度变化。

［25］ **monochromatic light**：单色光。

2.4 Liquid Crystal Display (LCD)

A liquid crystal display is a flat panel display, electronic visual display, or video display that uses the light modulating properties of liquid crystals. Liquid crystals do not emit light directly.

LCDs are available to display arbitrary images (as in a general-purpose computer display) or fixed images which can be displayed or hidden, such as preset words, digits, and 7-segment displays as in a digital clock. They use the same basic technology, except that arbitrary images are made up of a large number of small ***pixels***, while other displays have larger elements.

LCDs are used in a wide range of applications including computer monitors, televisions, instrument panels, aircraft cockpit displays and signage. They are common in consumer devices such as video players, gaming devices, clocks, watches, calculators and telephones, and have replaced ***cathode ray tube*** (CRT) displays in most applications. They are available in a wider range of screen sizes than CRT and ***plasma displays***, and since they do not use phosphors, they do not suffer image burn-in. LCDs are, however, susceptible to image persistence.

The LCD screen is more energy efficient and can be disposed of more safely than a CRT. Its low electrical power consumption enables it to be used in battery-powered electronic equipment. It is an electronically modulated optical device made up of any number of segments filled with liquid crystals and arrayed in front of a light source (backlight) or reflector to produce images in color or monochrome. Liquid crystals were first discovered in 1888. By 2008, worldwide sales of televisions with LCD screens exceeded annual sales of CRT units; the CRT became obsolete for most purposes.

In this chapter the basics of liquid crystal science, how liquid crystal displays are made and a selection of typical display modes will be described. At first, we will introduce basic knowledge of liquid crystal.

2.4.1 Introduction to Liquid Crystal

Liquid crystals (LCs) are matter in a state that has properties between those of conventional liquid and those of solid crystal. For instance, a liquid crystal may flow like a liquid, but its molecules may be oriented in a crystal-like way. There are many different types of ***liquid crystal phases***, which can be distinguished by their different optical properties (such as ***birefringence***). When viewed under a microscope using a polarized light source, different liquid crystal phases will appear to have distinct textures. The contrasting areas in the textures correspond to domains where the liquid-crystal molecules are oriented in different directions. Within a domain, however, the molecules are well ordered. LC materials may not always be in a liquid-crystal phase (just as water may turn into ice or steam).

Liquid crystals can be divided into thermotropic, lyotropic and metallotropic phases. Thermotropic and ***lyotropic liquid crystals*** consist of organic molecules. Thermotropic LCs exhibit a phase transition into the liquid-crystal phase as temperature is changed. Lyotropic LCs exhibit phase transitions as a function of both temperature and concentration of the liquid-crystal molecules in a solvent (typically water). Metallotropic LCs are composed of both organic and inorganic molecules; their liquid-crystal transition depends not only on temperature and concentration, but also on the inorganic-organic composition ratio.

2.4.2 Liquid Crystal Phases

The various liquid-crystal phases (called ***mesophases***) can be characterized by the type of ordering. One can distinguish positional order (whether molecules are arranged in any sort of ordered lattice) and orientational order (whether molecules are mostly pointing in the same direction), moreover, order can be either short-range (only between molecules close to each other) or long-range (extending to larger, sometimes macroscopic, dimensions). Most thermotropic LCs will have an ***isotropic*** phase at high temperature. That is that heating will eventually drive them into a conventional liquid phase characterized by random and isotropic molecular ordering (little to no long-range order), and fluid-like flow behavior. Under other conditions (for instance, lower temperature), a LC might inhabit one or more phases with significant anisotropic orientational structure and short-range orientational order while still having an ability to flow.

The ordering of liquid crystalline phases is extensive on the molecular scale. This order extends up to the entire domain size, which may be on the order of micrometers, but usually does not extend to the macroscopic scale as often occurs in classical crystalline solids. However, some techniques, such as the use of boundaries or an applied electric field, can be used to enforce a single ordered domain in a macroscopic liquid crystal sample. The ordering in a liquid crystal might extend along only one dimension, with the material being essentially disordered in the other two directions.

2.4.2.1 *Thermotropic Liquid Crystals*

Thermotropic phases are those that occur in a certain temperature range. If the temperature rises too high, thermal motion will destroy the delicate cooperative ordering of the LC phase, pushing the material into a conventional isotropic liquid phase. At too low temperature, most LC materials will form a conventional crystal. Many thermotropic LCs exhibit a variety of phases as the temperature is changed. For instance, a particular type of LC molecule (called mesogen) may exhibit various smectic and nematic (and finally isotropic) phases as the temperature is increased. An example of a compound displaying thermotropic LC behavior is para-azoxyanisole.

2.4.2.2 Nematic Phase

One of the most common LC phase is the nematic phase, as shown in Fig.

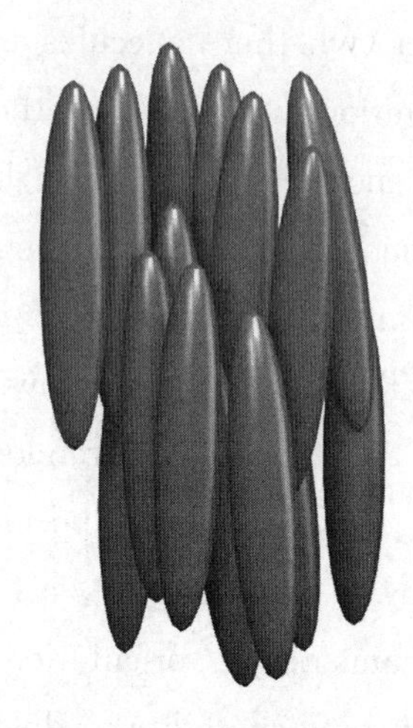
Fig. 2-16 Alignment in a nematic phase

2-16. The word nematic comes from the Greek "νήμα" (nema), which means "thread". This term originates from the thread-like topological defects observed in nematics, which are formally called ***"disclinations"***. Nematics also exhibit so-called " hedgehog " ***topological defects***. In a nematic phase, the calamitic or rod-shaped organic molecules have no positional order, but they self-align to have long-range directional order with their long axes roughly parallel. Thus, the molecules are free to flow and their center of mass positions are randomly distributed as in a liquid, but still maintain their long-range directional order. Most nematics are ***uniaxial***: They have one axis that is longer and preferred, with the other two being equivalent (can be approximated as cylinders or rods). However, some liquid crystals are ***biaxial*** nematics, meaning that in addition to orienting their long axis, they also orient along a secondary axis. Nematics have fluidity similar to that of ordinary (isotropic) liquids but they can be easily aligned by an external magnetic or electric field. Aligned nematics have the optical properties of uniaxial crystals and this makes them extremely useful in liquid crystal displays.

2.4.2.3 Smectic Phases

The smectic phases, as shown in Fig. 2-17, which are found at lower temperatures than the nematic, form well-defined layers that can slide over one another in a manner similar to that of soap. The word "smetic" originates from the Latin word "smecticus", meaning cleaning, or having soap like properties. The smectics are thus positionally ordered along one direction. In the smectic A phase, the molecules are oriented along the layer normal, while in the smectic C phase, they are tilted away from the layer normal. These phases are liquid-like within the layers. There are many different smectic phases, all characterized by different types and degrees of positional and orientational order.

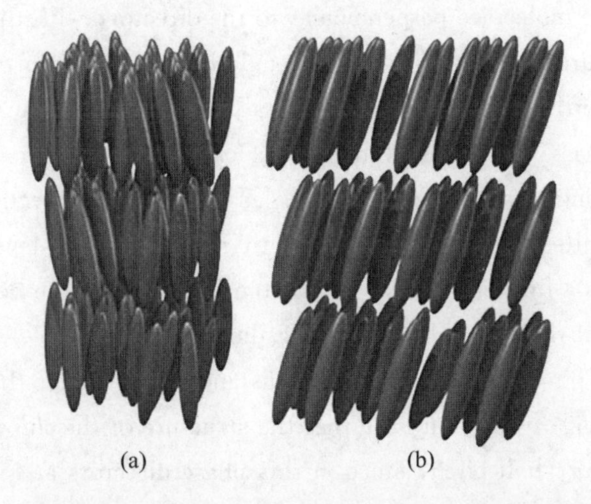

Fig. 2-17 Schematic of alignment in the smectic phases

The smectic A phase (a) has molecules organized into layers. In the smectic C phase (b), the molecules are tilted inside the layers.

2.4.2.4 Chiral Phases

The chiral phase exhibits chirality (handedness), as shown in Fig. 2-18. The phases are often called the cholesteric phases, because they were first observed for cholesterol derivatives. Only chiral molecules (i. e., those that have no internal planes of symmetry) can give rise to such a phase. This phases exhibit

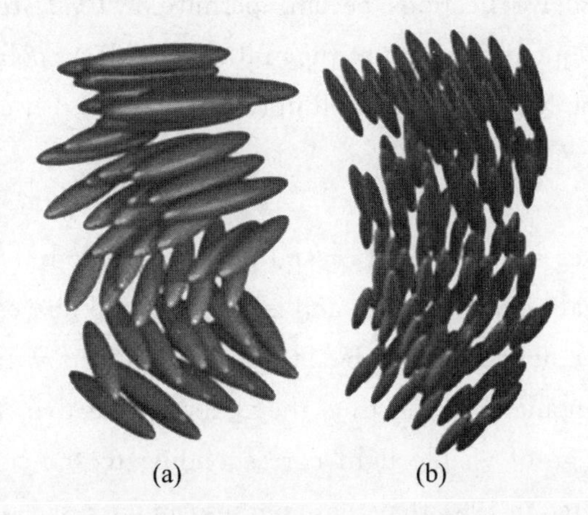

Fig. 2-18 Schematic of ordering in chiral liquid crystal phases

(a) The chiral nematic phase, also called the cholesteric phase. (b) The smectic C * phase.

a twisting of the molecules perpendicular to the director, with the molecular axis parallel to the director. The finite twist angle between adjacent molecules is due to their asymmetric packing, which results in longer-range chiral order. In the smectic C * phase (an asterisk denotes a chiral phase), the molecules have positional ordering in a layered structure (as in the other smectic phases), with the molecules tilted by a finite angle with respect to the layer normal. The chirality induces a finite azimuthal twist from one layer to the next, producing a spiral twisting of the molecular axis along the layer normal.

The chiral pitch, p, refers to the distance over which the LC molecules undergo a full 360° twist (but note that the structure of the ***chiral nematic phase*** repeats itself every half-pitch, since in this phase directors at 0° and ±180° are equivalent). The pitch, p, typically changes when the temperature is altered or when other molecules are added to the LC host (an achiral LC host material will form a chiral phase if doped with a chiral material), allowing the pitch of a given material to be tuned accordingly. In some liquid crystal systems, the pitch is of the same order as the wavelength of visible light. This causes these systems to exhibit unique optical properties, such as ***Bragg reflection*** and low-threshold laser emission, and these properties are exploited in a number of optical applications. For the case of Bragg reflection, only the lowest-order reflection is allowed if the light is incident along the ***helical axis***, whereas for oblique incidence higher-order reflections become permitted. ***Cholesteric liquid crystals*** also exhibit the unique property that they reflect circularly polarized light when it is incident along the helical axis and elliptically polarized if it comes in obliquely.

2.4.2.5 Blue Phases

The blue phases are the liquid crystal phases that appear in the temperature range between chiral nematic phases and isotropic liquid phases. The blue phases have a regular three-dimensional cubic structure of defects with lattice periods of several hundred nanometers, and thus they exhibit selective Bragg reflections in the wavelength range of visible light corresponding to the cubic lattice. It was theoretically predicted in 1981 that these phases can possess icosahedral symmetry similar to ***quasicrystals***.

Although the blue phases are of interest for fast light modulators or tunable photonic crystals, they exist in a very narrow temperature range, usually less than a few kelvin. Recently the stabilization of the blue phases over a temperature range of more than 60 K including room temperature (260～326 K) has been demonstrated. The blue phases stabilized at room temperature allow electro-optical switching with response times of the order of 10^{-4} s.

2.4.3 External Influences on Liquid Crystals

Scientists and engineers are able to use liquid crystals in a variety of applications because external perturbation can cause significant changes in the macroscopic properties of the liquid crystal system. Both electric and magnetic fields can be used to induce these changes. The magnitude of the fields, as well as the speed at which the molecules align are important characteristics industry deals with. Special surface treatments can be used in liquid crystal devices to force specific orientations of the director.

2.4.3.1 Electric and Magnetic Field Effects

The ability of the director to align along an external field is caused by the electric nature of the molecules. Permanent electric dipoles result when one end of a molecule has a net positive charge while the other end has a net negative charge. When an external electric field is applied to the liquid crystal, the dipole molecules tend to orient themselves along the direction of the field.

Even if a molecule does not form a permanent dipole, it can still be influenced by an electric field. In some cases, the field produces slight rearrangement of electrons and protons in molecules such that an induced electric dipole results. While not as strong as permanent dipoles, orientation with the external field still occurs. The effects of magnetic fields on liquid crystal molecules are analogous to electric fields. Because magnetic fields are generated by moving electric charges, permanent magnetic dipoles are produced by electrons moving about atoms. When a magnetic field is applied, the molecules will tend to align with or against the field.

2.4.3.2 Surface Preparations

In the absence of an external field, the director of a liquid crystal is free to point in any direction. It is possible, however, to force the director to point in a specific direction by introducing an outside agent to the system. For example, when a thin polymer coating (usually a polyimide) is spread on a glass substrate and rubbed in a single direction with a cloth, it is observed that liquid crystal molecules in contact with that surface align with the rubbing direction. The currently accepted mechanism for this is believed to be an epitaxial growth of the liquid crystal layers on the partially aligned polymer chains in the near surface layers of the polyimide.

2.4.3.3 *Fredericks Transition*

The competition between orientation produced by surface anchoring and by electric field effects is often exploited in liquid crystal devices. Consider the case in which liquid crystal molecules are aligned parallel to the surface and an electric field is applied perpendicular to the cell. At first, as the electric field increases in magnitude, no change in alignment occurs. However, at a threshold magnitude of electric field, deformation occurs. Deformation occurs where the director changes its orientation from one molecule to the next. The occurrence of such a change from an aligned to a deformed state is called a Fredericks transition and can also be produced by the application of a magnetic field of sufficient strength.

The Fredericks transition is fundamental to the operation of many liquid crystal displays, because the director orientation (and thus the properties) can be controlled easily by the application of a field.

2.4.4 Overview of LCD

Each pixel of an LCD typically consists of a layer of molecules aligned between two transparent electrodes and two polarizing filters, the axes of transmission of which are (in most of the cases) perpendicular to each other. With actual liquid crystal between the polarizing filters, light passing through the first filter would be blocked by the second (crossed) polarizer, see Fig. 2-19.

The surface of the electrodes that are in contact with the liquid crystal material are treated so as to align the liquid crystal molecules in a particular direction. This treatment typically consists of a thin polymer layer that is unidirectionally rubbed using, for example, a cloth. The direction of the liquid crystal alignment is then defined by the direction of rubbing. Electrodes are made

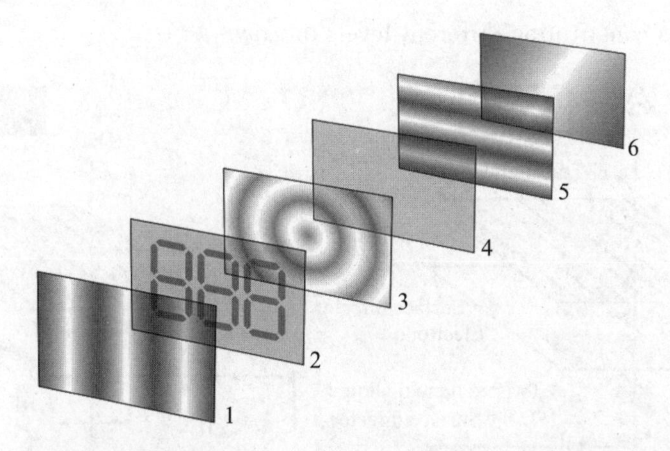

Fig. 2-19 Reflective twisted *nematic liquid crystal* display

1. Polarizing filter film with a vertical axis to polarize light as it enters. 2. Glass substrate with ITO electrodes. The shapes of these electrodes will determine the shapes that will appear when the LCD is turned on. Vertical ridges etched on the surface are smooth. 3. Twisted nematic liquid crystal. 4. Glass substrate with common electrode film (ITO) with horizontal ridges to line up with the horizontal filter. 5. Polarizing filter film with a horizontal axis to block/pass light. 6. Reflective surface to send light back to viewer. In a backlit LCD, this layer is replaced with a light source.

of the transparent conductor indium tin oxide (ITO). The liquid crystal display is intrinsically a "passive" device, it is a simple light valve. The managing and control of the data to be displayed is performed by one or more circuits commonly denoted as LCD drivers.

According to Fig. 2-20, before an electric field is applied, the orientation of the liquid crystal molecules is determined by the alignment at the surfaces of electrodes. In a twisted nematic device (still the most common liquid crystal device), the surface alignment directions at the two electrodes are perpendicular to each other, and so the ***molecules*** arrange themselves in a helical structure, or twist. This induces the rotation of the polarization of the incident light, and the device appears gray, as shown in Fig. 2-20 (a). If the applied voltage is large enough, the liquid crystal molecules in the center of the layer are almost completely untwisted and the polarization of the incident light is not rotated as it passes through the liquid crystal layer. This light will then be mainly polarized perpendicular to the second filter, and thus be blocked and the pixel will appear black, as shown in Fig. 2-20 (b). By controlling the voltage applied across the liquid crystal layer in each pixel, light can be allowed to pass through in varying amounts thus constituting different levels of gray.

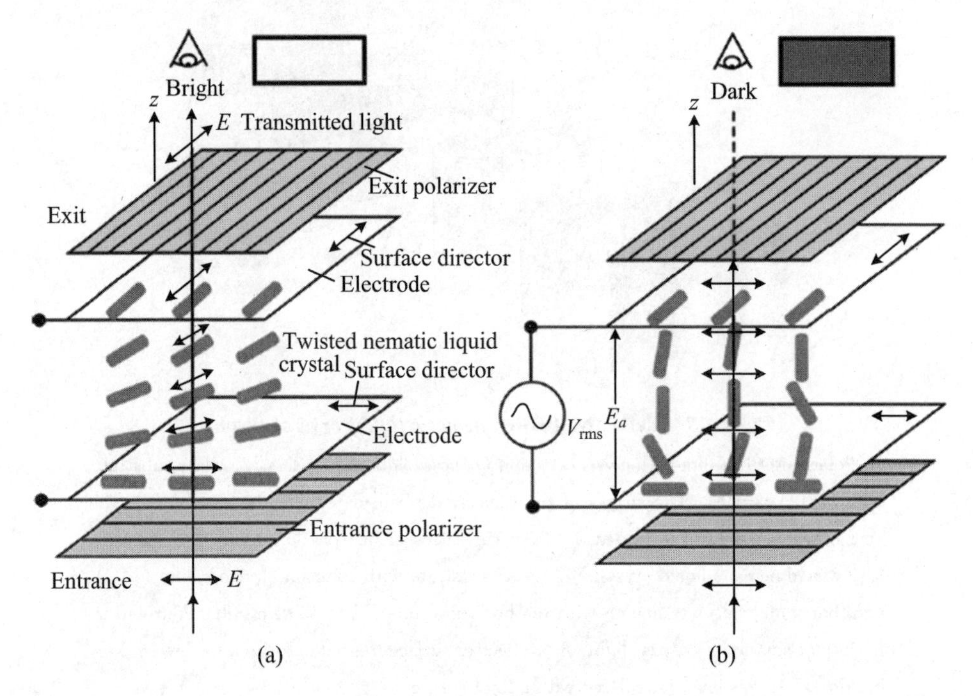

Fig. 2-20　Transmission based LCD

(a) In the absence of a field. The liquid crystal has the twisted nematic phase and the light passing through it has its polarization rotated by 90°. The light is transmitted through both polarizers. The viewer sees a bright image. (b) When a voltage and hence a field E_a is applied. The molecules in the liquid crystal align with the field E_a and are unable to rotate the polarization of the light passing through it; Light therefore cannot pass through the exit polarizer. The light is extinguished. And the viewer see dark image.

The optical effect of a twisted nematic device in the voltage-on state is far less dependent on variations in the device thickness than that in the voltage-off state. Because of this, these devices are usually operated among crossed polarizers such that they appear bright with no voltage (the eye is much more sensitive to variations in the dark state than the bright state). These devices can also be operated among parallel polarizers, in which case the bright and dark states are reversed. The voltage-off dark state in this configuration appears blotchy, however, because of small variations of thickness across the device.

Both the liquid crystal material and the alignment layer material contain ionic compounds. If an electric field of one particular ***polarity*** is applied for a long period of time, this ionic material is attracted to the surfaces and degrades the device performance. This is avoided either by applying an alternating current or by reversing the polarity of the electric field as the device is addressed (the response of the liquid crystal layer is identical, regardless of the polarity of the applied field).

Displays for a small number of individual digits and/or fixed symbols (as in digital watches and pocket calculators) can be implemented with independent electrodes for each segment. In contrast, full alphanumeric and/or variable graphics displays are usually implemented with pixels arranged as a matrix consisting of electrically connected rows on one side of the LC layer and columns on the other side, which makes it possible to address each pixel at the intersections. The general method of matrix addressing consists of sequentially addressing one side of the matrix, for example by selecting the rows one-by-one and by applying the picture information on the other side at the columns row-by-row.

2.4.5 Illumination of Liquid Crystal Display

Since LCD panels produce no light of their own, they require external light to produce a visible image. In a "transmissive" type of LCD, this light is provided at the back of the glass "stack" and is called the backlight. While passive-matrix displays are usually not backlit (e.g., calculators, wristwatches), active-matrix displays almost always are common implementations of LCD backlight technology.

2.4.5.1 *CCFL (Cold Cathode Fluorescent Lamp)*

The LCD panel is lit either by two cold cathode fluorescent lamps placed at opposite edges of the display or by an array of parallel CCFLs behind larger displays. A diffuser then spreads the light out evenly across the whole display. For many years, this technology had been used almost exclusively. Unlike white LEDs, most CCFLs have an even-white spectral output resulting in better color gamut for the display. However, CCFLs are less energy efficient than LEDs and require a somewhat costly inverter to convert whatever DC voltage the device uses (usually 5 V or 12 V) to about 1000 V needed to light a CCFL. The thickness of the inverter transformers also limit how thin the display can be.

2.4.5.2 WLED Array

The LCD panel is lit by a full array of white LEDs placed behind a diffuser which behind the panel. LCD displays that use this implementation will usually have the ability to dim the LEDs in the dark areas of the image being displayed, effectively increasing the contrast ratio of the display. As of 2012, this design gets most of its use from upscale, larger-screen LCD televisions.

2.4.5.3 RGB-LED

Similar to the WLED array, except the panel is lit by a full array of RGB-LEDs. While displays lit with white LEDs usually have a poorer color gamut than CCFL lit displays, panels lit with RGB-LEDs have very wide color gamuts. This implementation is most popular on professional graphics editing LCD displays. As of 2012, LCD displays in this category usually cost more than $1000. Today, most LCD screens are being designed with an LED backlight instead of the traditional CCFL backlight.

2.4.6 Future Prospects of LCD①

Liquid crystal displays are now the ultimate for flat screen displays; the scope of applications they can address is extremely broad. Their reign started by creating markets that the then best technology (CRTs) could not address and gradually started to replace the CRT with better products in the applications where it had the most problems. Now the liquid crystal display competes head on in the CRT's main market segment (monitors and TVs). Yet still after 30 years of liquid crystal display technology, the CRT remains overall the major display mode, because in most cases where the two compete it is much cheaper, but it cannot get cheaper while liquid crystal displays probably can.

There are now competing flat panel technologies (such as organic light-emitting diodes — OLEDs) that offer fast addressing times and are not as limited by temperature as liquid crystal displays. Will these replace liquid crystal displays

---

① Dakin J P, Brown R. Handbook of Optelectronics: Volume 1[M]. Boca Raton: CRC Press, 2010:987-988.

in the near future? Bearing in mind the CRT story, this seems unlikely even without technical arguments but simply due to the massive infrastructure and investment in liquid crystal displays. It is more probable that OLEDs will start by having success in areas where liquid crystal displays are not that good such as in wide temperature range applications. There is debate about which of the two uses less power for portable applications. However, the final choice will be that of the consumer and which display gives the best value. Most large liquid crystal display makers are also developing OLED displays as they did with liquid crystal displays when they only made CRTs. It is possible that new faster light-emitting displays can open markets that liquid crystal displays cannot address so easily, thus we may see new applications for flat panel displays.

Looking at the liquid crystal displays themselves, there are more exciting new modes being taken to full colour demonstration than ever before. These attempt to overcome the speed issues and have ever lower power consumption using bistable low voltage modes in reflection. Thus, we shall see improved monitors and television screens; the much quoted wall mounted television is here but most houses are not designed for it; they are designed for CRT televisions. However, flat screen monitors are being used and liked in the desk environment; it cannot be long before flat screen television liquid crystal displays replace portable or second TVs. In large size televisions, they will also compete with plasma screens. Projection liquid crystal displays operate in a tough market segment but many believe they can survive. Plastic displays sound attractive, but when considered seriously, no one has a killer application that must have plastic rather than glass; until this occurs, the market will be small and thus the plastic devices will be expensive.

New Words and Expressions

[1] **pixels**：像素。

[2] **cathode ray tube**：阴极射线管。它是将电信号转变为光学图像的一类电子束管。

[3] **plasma displays**：等离子显示器。

[4] **liquid crystal phase**：液晶相。

[5] **polarity**：极性。

［6］ **molecules**：分子。

［7］ **nematic liquid crystal**：向列液晶。

［8］ **isotropic**：各向同性。它是指物体的物理、化学等方面的性质不会因方向的不同而有所变化的特性。

［9］ **thermotropic liquid crystal**：热致液晶。

［10］ **lyotropic liquid crystal**：溶致液晶。

［11］ **mesophase**：中间相。

［12］ **Fredericks transition**：弗雷德里克斯过渡。

［13］ **topological defects**：拓扑缺陷。它是拓扑学中的一种规律。

［14］ **disclination**：扭曲向列型。

［15］ **uniaxial**：单轴的。

［16］ **biaxial**：双轴的。

［17］ **chiral nematic phase**：手性向列相。

［18］ **Bragg reflection**：布拉格反射。

［19］ **helical axis**：螺旋轴。

［20］ **cholesteric liquid crystals**：胆甾相液晶。

［21］ **quasicrystals**：准晶。它是一种介于晶体和非晶体之间的固体。

［22］ **birefringence**：双折射。光束入射到各向异性的晶体，分解为两束光而沿不同方向折射的现象。

［23］ **CCFL（Cold Cathode Fluorescent Lamp）**：冷阴极荧光灯管。它具有高功率、高亮度、低能耗等优点，广泛应用于显示器、照明等领域。

2.5 Organic Light Emitting Diode（OLED）

Imagine having a high-definition TV that is 80 inches wide and less than a quarter-inch thick, consumes less power than most TVs on the market today and can be rolled up when you're not using it. What if you could have a "heads up" display in your car? How about a display monitor built into your clothing? These devices may be possible in the near future with the help of a technology called ***organic light-emitting diodes（OLEDs）***.

An OLED is a light-emitting diode (LED) in which the emissive ***electroluminescent*** layer is a film of organic compound which emits light in response to an electric current. This layer of organic semiconductor is situated between two electrodes. Generally, at least one of these electrodes is transparent. OLEDs are used to create digital displays in devices such as television screens, computer monitors, portable systems such as mobile phones, handheld games consoles and PDAs. A major area of research is the development of white OLED devices for use in solid-state lighting applications, as shown in Fig. 2-21.

(a) (b)

Fig. 2-21 Applications of OLED

There are two main families of OLED: those based on small molecules and those employing polymers. Adding mobile ions to an OLED creates a light-emitting electrochemical cell or LEC, which has a slightly different mode of operation. OLED displays can use either ***passive-matrix*** (PMOLED) or ***active-matrix*** (AMOLED) addressing schemes. Active-matrix OLEDs (AMOLEDs) require a thin-film transistor backplane to switch each individual pixel on or off, but allow for higher resolution and larger display sizes.

An OLED display works without a backlight. Thus, it can display deep black levels and can be thinner and lighter than a liquid crystal display (LCD). In low ambient light conditions such as a dark room, an OLED screen can achieve a higher contrast ratio than an LCD, whether the LCD uses cold cathode fluorescent lamps or LED backlight.

2.5.1 Components of OLEDs

Like an LED, an OLED is a solid-state semiconductor device that is 100 ~ 500 nm thick or about 200 times smaller than a human hair. OLEDs can have either two layers or three layers of organic material; in the latter design, the third layer helps transport electrons from the cathode to the emissive layer. We'll be focusing on the two-layer design.

An OLED consists of the following several parts (see Fig. 2-22):

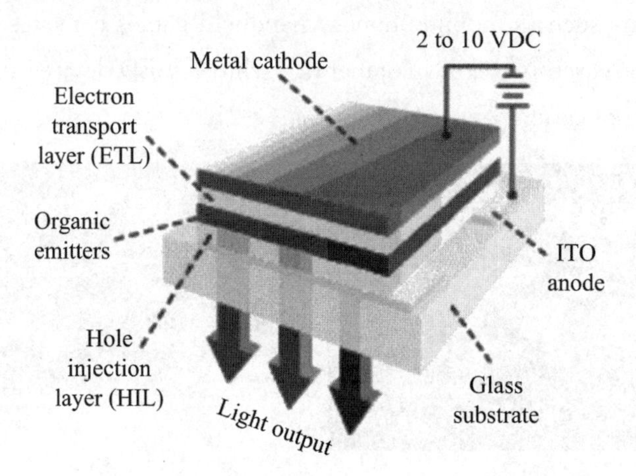

Fig. 2-22 OLED structure

(1) ***Substrate*** (clear plastic, glass, foil) — The substrate supports the OLED.

(2) ***Anode*** (transparent) — The anode removes electrons (adds electron "holes") when a current flows through the device.

(3) Organic layers — These layers are made of organic molecules or polymers.

(4) Conducting layer — This layer is made of organic plastic molecules that transport "holes" from the anode. One conducting polymer used in OLEDs is ***polyaniline***.

(5) Emissive layer — This layer is made of organic plastic molecules (different ones from the conducting layer) that transport electrons from the cathode; this is where light is made. One polymer used in the emissive layer is polyfluorene.

(6) Cathode (may or may not be transparent depending on the type of OLED) — The cathode injects electrons when a current flows through the device.

The biggest part of manufacturing OLEDs is applying the organic layers to the substrate. This can be done in the following three ways:

(1) Vacuum deposition or vacuum ***thermal evaporation*** (VTE) — In a vacuum chamber, the organic molecules are gently heated (evaporated) and allowed to condense as thin films onto cooled substrates. This process is

expensive and inefficient.

(2) ***Organic vapor phase deposition*** (OVPD) — In a low-pressure, hot-walled reactor chamber, a carrier gas transports evaporated organic molecules onto cooled substrates, where they condense into thin films. Using a carrier gas increases the efficiency and reduces the cost of making OLEDs.

(3) ***Inkjet printing*** — With inkjet technology, OLEDs is sprayed onto substrates just like inks are sprayed onto paper during printing. Inkjet technology greatly reduces the cost of OLED manufacturing and allows OLEDs to be printed onto very large films for large displays like 80-inch TV screens or electronic billboards.

2.5.2 Working Principle of OLED

OLEDs emit light in a similar manner to LEDs, through a process called ***electrophosphorescence.***

The process is as follows (see Fig. 2-23):

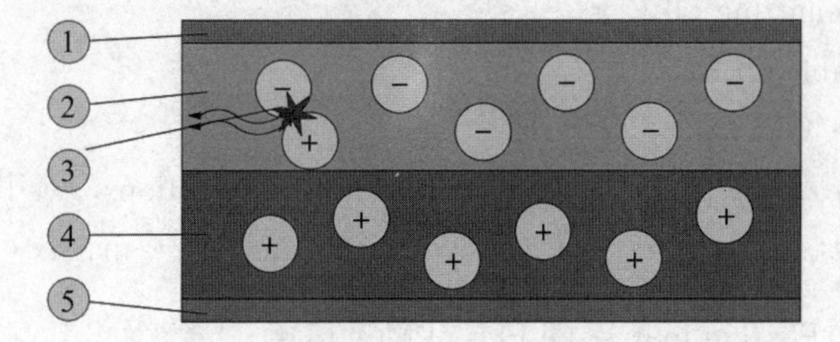

Fig. 2-23 Schematic of a bilayer OLED

1. Cathode (−); 2. Emissive layer; 3. Emission of radiation; 4. Conductive layer; 5. Anode (+).

(1) The battery or power supply of the device containing the OLED applies a voltage across the OLED.

(2) An electrical current flows from the cathode to the anode through the organic layers (an electrical current is a flow of electrons). The cathode gives electrons to the emissive layer of organic molecules. The anode removes electrons from the conductive layer of organic molecules. (This is the equivalent to giving electron holes to the conductive layer.)

(3) At the boundary between the emissive and the conductive layers, electrons find electron holes. When an electron finds an electron hole, the

electron fills the hole (it falls into an energy level of the atom that's missing an electron). When this happens, the electron gives up energy in the form of a photon of light.

(4) The OLED emits light.

(5) The color of the light depends on the type of organic molecule in the emissive layer. Manufacturers place several types of organic films on the same OLED to make color displays.

(6) The intensity or brightness of the light depends on the amount of electrical current applied: The more current, the brighter the light.

2.5.3 Types of OLEDs

There are several types of OLEDs:

(1) Passive-matrix OLEDs;

(2) Active-matrix OLEDs;

(3) Transparent OLEDs;

(4) Top-emitting OLEDs;

(5) Foldable OLEDs;

(6) White OLEDs.

Each type has different uses. In the following sections, we'll discuss each type of OLEDs. Let's start with passive-matrix and active-matrix OLEDs.

2.5.3.1 Passive-matrix OLEDs (PMOLEDs)

PMOLEDs have strips of cathode, organic layers and strips of anode. The anode strips are arranged perpendicular to the cathode strips. The intersections of the cathode and anode make up the pixels where light is emitted. External circuitry applies current to selected strips of anode and cathode, determining which pixels get turned on and which pixels remain off. Again, the brightness of each pixel is proportional to the amount of applied current.

PMOLEDs are easy to make, but they consume more power than other types of OLEDs, mainly due to the power needed for the external circuitry. PMOLEDs are most efficient for text and icons and are best suited for small screens (2-inch to 3-inch diagonal) such as those you find in cell phones, PDAs and MP3 players. Even with the external circuitry, passive-matrix OLEDs consume less battery power than the LCDs that currently power these devices.

2.5.3.2 Active-matrix OLEDs (AMOLEDs)

AMOLEDs have full layers of cathode, organic molecules and anode, but the anode layer overlays a ***thin film transistor*** (***TFT***) array that forms a matrix. The TFT array itself is the circuitry that determines which pixels get turned on to form an image.

AMOLEDs consume less power than PMOLEDs, because the TFT array requires less power than external circuitry, so they are efficient for large displays. AMOLEDs also have faster refresh rates suitable for video. The best uses for AMOLEDs are computer monitors, large-screen TVs and electronic signs or billboards.

2.5.3.3 Transparent OLEDs

Transparent OLEDs have only transparent components (substrate, cathode and anode) and, when turned off, are up to 85 percent as transparent as their substrate. When a transparent OLED display is turned on, it allows light to pass in both directions. A transparent OLED display can be either active-matrix or passive-matrix. This technology can be used for heads-up displays.

2.5.3.4 Top-emitting OLEDs

Top-emitting OLEDs have a substrate that is either opaque or reflective. They are best suited to active-matrix design. Manufacturers may use top-emitting OLED displays in smart cards.

2.5.3.5 Foldable OLEDs

Foldable OLEDs have substrates made of very flexible metallic foils or plastics. Foldable OLEDs are very lightweight and durable. Their use in devices such as cell phones and PDAs can reduce breakage, a major cause for return or repair. Potentially, foldable OLED displays can be attached to fabrics to create "smart" clothing, such as outdoor survival clothing with an integrated computer chip, cell phone, GPS receiver and OLED display sewn into it.

2.5.3.6 White OLEDs

White OLEDs emit white light that is brighter, more uniform and more energy efficient than that emitted by fluorescent lights. White OLEDs also have the true-color qualities of ***incandescent lighting***. Because OLEDs can be made in large sheets, they can replace fluorescent lights that are currently used in homes and buildings. Their use could potentially reduce energy costs for lighting.

2.5.4 Advantages of OLEDs

The different manufacturing process of OLEDs lends itself to several advantages over flat panel displays made with LCD technology.

2.5.4.1 Lower Cost in the Future

OLEDs can be printed onto any suitable substrate by an inkjet printer or even by screen printing, theoretically making them cheaper to produce than LCD or plasma displays. However, fabrication of the OLED substrate is more costly than that of a TFT LCD, until mass production methods lower cost through scalability. Roll-to-roll vapour-deposition methods for organic devices do allow mass production of thousands of devices per minute for minimal cost, although this technique also induces problems in that devices with multiple layers can be challenging to make because of registration, lining up the different printed layers to the required degree of accuracy.

2.5.4.2 Lightweight and Flexible Plastic Substrates

OLED displays can be fabricated on flexible plastic substrates leading to the possible fabrication of flexible organic light-emitting diodes for other new applications, such as roll-up displays embedded in fabrics or clothing. As the substrate used can be flexible such as polyethylene terephthalate (PET), the displays may be produced inexpensively.

2.5.4.3 Wider Viewing Angles and Improved Brightness

OLEDs can enable a greater artificial contrast ratio (both ***dynamic range*** and static, measured in purely dark conditions) and a wider viewing angle compared to LCDs because OLED pixels emit light directly. OLED pixel colors appear correct and unshifted, even as the viewing angle approaches 90° from normal.

2.5.4.4 Better Power Efficiency and Thickness

LCDs filter the light emitted from a backlight, allowing a small fraction of light through. So, they cannot show true black. However, an inactive OLED element does not produce light or consume power, thus allowing true blacks. Dismissing the backlight also makes OLEDs lighter because some substrates are not needed. This allows electronics potentially to be manufactured more cheaply, but, first, a larger production scale is needed, because OLEDs still somewhat are niche products. When looking at top-emitting OLEDs, thickness also plays a role when talking about index match layers (IMLs). Emission intensity is enhanced when the IML thickness is 1.3～2.5 nm. The refractive value and the matching of the optical IMLs property, including the device structure parameters, also enhance the emission intensity at these thicknesses.

2.5.4.5 Response Time

OLEDs also can have a faster response time than standard LCD screens. Whereas LCD displays are capable of 1～16 ms response time offering a refresh rate of 60～480 Hz, an OLED theoretically can have a response time less than 0.01 ms, enabling a refresh rate up to 100000 Hz. OLEDs also can be run as a ***flicker*** display, similar to a CRT, in order to eliminate the sample-and-hold effect that creates motion blur on OLEDs.

2.5.5 Disadvantages of OLEDs

2.5.5.1 Current Costs

OLED manufacture currently requires process steps that make it extremely expensive. Specifically, it requires the use of low-temperature ***polysilicon*** backplanes; LTPS backplanes, in turn, require laser annealing from an (***amorphous silicon***) start, so this part of the manufacturing process for AMOLEDs starts with the process costs of standard LCD, and then adds an expensive, time-consuming process that cannot currently be used on large-area glass substrates.

2.5.5.2 Lifespan

The biggest technical problem for OLEDs was the limited lifetime of the organic materials. One 2008 technical report on an OLED TV panel found that "after 1000 hours the blue luminance degraded by 12%, the red by 7% and the green by 8%". In particular, blue OLEDs historically have had a lifetime of around 14000 hours to half original brightness (five years at 8 hours a day) when used for flat-panel displays. This is lower than the typical lifetime of LCD, LED or PDP technology. Each currently is rated for about 25000~40000 hours to half brightness, depending on manufacturer and model. Degradation occurs because of the accumulation of nonradiative recombination centers and luminescence quenchers in the emissive zone. It is said that the chemical breakdown in the semiconductors occurs in four steps: ① recombination of charge carriers through the absorption of UV light, ② hemolytic dissociation, ③ subsequent radical addition reactions that form π radicals, and ④ disproportionation between two radicals resulting in hydrogen-atom transfer reactions. However, some manufacturers' displays aim to increase the lifespan of OLED displays, pushing their expected life past that of LCD displays by improving light out coupling, thus achieving the same brightness at a lower drive current. In 2007, experimental OLEDs were created which can sustain 400 cd/m^2 of luminance for over 198000 hours for green OLEDs and 62000 hours for blue OLEDs.

2.5.5.3 Color Balance Issues

Additionally, as the OLED material used to produce blue light degrades significantly more rapidly than the materials that produce other colors, blue light output will decrease relative to the other colors of light. This variation in the differential color output will change the color balance of the display and is much more noticeable than a decrease in overall luminance. This can be avoided partially by adjusting color balance, but this may require advanced control circuits and interaction with the users, which is unacceptable for some users. More commonly, though, manufacturers optimize the size of the R, G and B ***subpixels*** to reduce the current density through the subpixel in order to equalize lifetime at full luminance. For example, a blue subpixel may be 100% larger than the green subpixel. The red subpixel may be 10% smaller than the green subpixel.

2.5.5.4 Efficiency of Blue OLEDs

Improvements to the efficiency and lifetime of blue OLEDs are vital to the success of OLEDs as replacements for LCD technology. Considerable research has been invested in developing blue OLEDs with high external quantum efficiency as well as a deeper blue color. External quantum efficiency values of 20% and 19% have been reported for red (625 nm) and green (530 nm) diodes, respectively. However, blue diodes (430 nm) have only been able to achieve maximum external quantum efficiencies in the range of 4%～6%.

2.5.5.5 Water Damage

Water can damage the organic materials of the displays. Therefore, improved sealing processes are important for practical manufacturing. Water damage especially may limit the longevity of more flexible displays.

2.5.6 Polymer Light-emitting Diode (PLED)

Polymers are substances formed by a chemical reaction in which two or more molecules combine to form larger molecules. PLEDs are thin film displays that are created by sandwiching an undoped conjugated polymer between two proper electrodes. PLEDs enable full-spectrum color displays and are relatively inexpensive compared to other display technologies such as LCD or OLED, and require little power to emit a substantial amount of light.

Polymer light-emitting diode was reported in 1990 by Jeremy Burroughs and his colleagues at the University of Cambridge, this device incorporates polymers — made from unions, often chains, of smaller organic molecules — to form the layers. Polymer LEDs are formed by spin coating: Applying a thin layer of polymer to a flat substrate and then spinning the substrate at a high speed (typically 1200 ～ 1500 revolutions per minute) to spread the polymer puddle by centrifugal force. The spin coating is followed by a baking step to remove the solvent, and in some cases, to complete the ***polymerization***. This film formation process is generally more economical than the thermal evaporation method discussed above. Polymers have had an edge over Kodak-type small-molecule devices in power efficiencies because the greater electronic conductivity of the polymer layers allows lower operating voltages.

The original PLED consisted of a single active layer of a polymer called poly phenyl inline (PPV) between dissimilar metal contacts such as indium tin oxide and calcium, as in an OLED, to provide injection of both holes and electrons. Indium tin oxide is a metal that tends to inject holes, and calcium is a metal that tends to inject electrons. Current PLEDs use a second polymer layer for hole injection and transport. The polymer PPV produces yellow light, with an excellent efficiency and lifetime; at computer-monitor luminance levels, such a PLED can last more than 10000 hours — about 10 years of regular use. Full color has been demonstrated, but the only commercial product to date uses yellow. Other polymers and mixed polymers (two different polymers in solution) have been developed, notably by Dow Chemical, based on the organic molecule poly fluorine. These configurations can be modified to produce a full range of colors, from red to green, by varying the lengths of the segments of the co-polymers. Unfortunately, the display lifetimes of these colors have not been comparable to that of PPV, and blue is not yet ready for prime time.

A PLED is constructed of several layers (see Fig. 2-24):

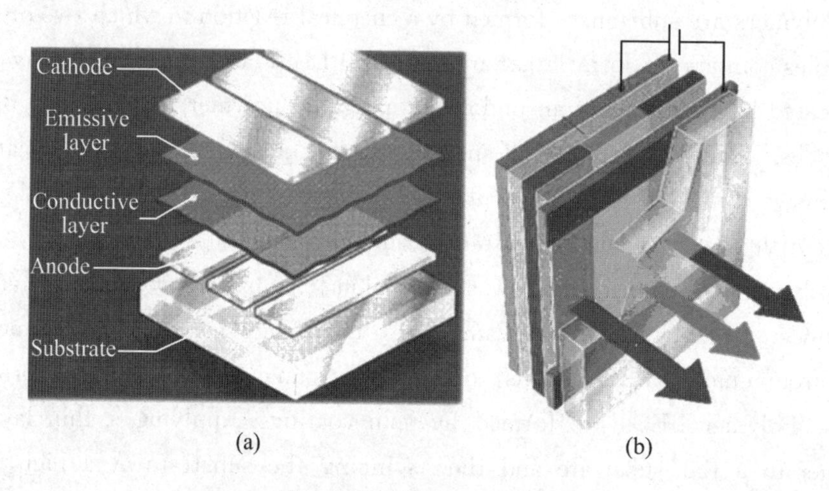

Fig. 2-24 PLED structure

(1) An engineer begins with a glass or plastic substrate — for PLED fabric displays, plastic tends to be a better choice because it's less fragile but more flexible than glass.

(2) Next comes a transparent electrode coating, which an engineer applies to

one side of the substrate.

(3) Then the engineer coats, the same side of the substrate with the light emitting polymer film.

(4) The last layer is an evaporated metal electrode, which the engineer applies to the other side of the polymer film.

The solution-processed PLED is typically prepared with a thin layer of semiconducting polymer film sandwiched between two charge injection contact electrodes, as shown in Fig. 2-25. The device is generally made onto a glass substrate or a thin plastic film with partially coated ***transparent electrode*** (such as ITO). A thin, semiconducting, luminescent polymer film with thickness typically in 50~200 nm range is then coated. Finally, the device is completed by depositing a low-work-function metal (such as calcium), as the cathode electrode.

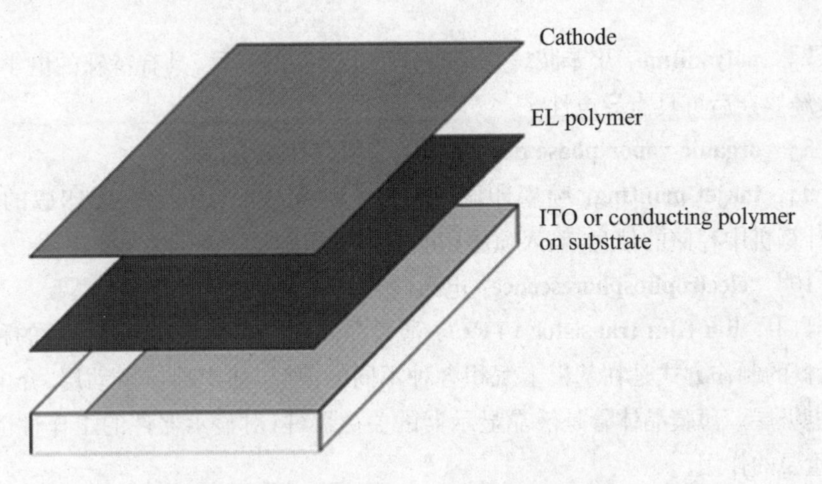

Fig. 2-25 PLED in sandwich configuration

When the engineer applies an electric field between the two electrodes, the polymer emits light, much like an LED. Because the polymers in PLED are made of organic molecules, they are also known as OLEDs — organic light-emitting diodes.

Using a PLED screen, it would be possible to create a fabric television. PLED displays are very thin and relatively light compared to other display technologies. Of course, the screen is just one important element in the overall fabric display — you would also need a power source, such as a lithium-ion battery, and a signal source. The signal source could be a small computer containing preloaded video clips or even a WiFi-enabled device that could stream audio and video directly to your clothes.

New Words and Expressions

[1]　**organic light-emitting diodes (OLEDs)**：有机发光二极管。

[2]　**electroluminescent**：电致发光。它是指电流通过物质时或物质处于强电场下发光的现象。

[3]　**passive-matrix**：无源矩阵。

[4]　**active-matrix**：有源矩阵。它是通过存储器元件来创建各个有源像素的一种 LCD 技术。

[5]　**substrate**：基底。

[6]　**anode**：阳极。它指的是失去电子的极，阳极总是与阴极相对应而存在的。

[7]　**polyaniline**：聚苯胺。它是高分子化合物的一种，具有特殊的电学、光学性质，经掺杂后可具有导电性。

[8]　**organic vapor phase deposition**：有机气相沉淀。

[9]　**inkjet printing**：喷墨印刷。它是一种无接触、无压力、无印版的印刷。电子计算机中存储的信息，输入喷墨印刷机即可印刷。

[10]　**electrophosphorescence**：电致磷光。

[11]　**thin film transistor (TFT)**：薄膜晶体管。它是场效应晶体管的种类之一，大概的制作方式是在基板上沉积各种不同的薄膜，如半导体主动层、介电层和金属电极层。薄膜晶体管是液晶显示器的关键器件，对显示器件的工作性能具有十分重要的作用。

[12]　**incandescent lighting**：白炽灯照明。

[13]　**dynamic range**：动态范围。

[14]　**flicker**：闪烁、闪光。

[15]　**amorphous silicon**：非晶硅，又称无定形硅。它是单质硅的一种形态，为棕黑色或灰黑色的微晶体。硅不具有完整的金刚石晶胞，纯度不高。熔点、密度和硬度也明显低于晶体硅。

[16]　**polysilicon**：多晶硅。它是单质硅的一种形态。熔融的单质硅在过冷条件下凝固时，硅原子以金刚石晶格形态排列成许多晶核，如这些晶核长成晶面取向不同的晶粒，这些晶粒结合起来，就结晶成多晶硅。

[17]　**subpixels**：亚像素。

[18]　**polymerization**：聚合作用。

［19］ **thermal evaporation**：热蒸发。把待镀膜的基片或工件置于真空室内，通过对镀膜材料加热使其蒸发气化而沉积于基体或工件表面，并形成薄膜或涂层的工艺过程，称为真空蒸发镀膜，简称蒸发镀膜或蒸镀。

［20］ **transparent electrode**：透明电极，如 ITO。

References

［1］ Dakin J P, Brown R. Handbook of Optoelectronics: Volume 1［M］. Boca Raton: CRC Press, 2010.

［2］ Mynbaev D K, Scheiner L L. Fiber-Optic Communications Technology［M］. New Jersey: Prentice Hall, Inc., 2000.

［3］ Kasap S O. Optoelectronics and Photonics: Principles & Practices［M］. 2nd ed. New Jersey: Prentice Hall, Inc., 2011.

Questions

1. List some application areas of LEDs.
2. List some advantages and disadvantages of LEDs.
3. Please simply illustrate the basic working principle of LD and especially point out three key steps of this working process.
4. Do you know the characteristics of laser light?
5. What is the basic working principle of LCD?
6. What is the advantage of OLEDs compares to LEDs?

Chapter 3 Optoelectronic Information Transformation

Optoelectronic information transformation is the core content of optoelectronic information technology. Thanks to the invention of all kinds of optoelectronic information conversion devices and the performance of continuous improvement, which led to the rapid development of modern optoelectronic information technology.

The principles of optoelectronic information conversion are mainly listed as follows:

(1) External photoelectric effect. Under the effect of the incident light energy, the electrons escape from the surfaces in some objects, optoelectronic information conversion devices which are made with this principle, for example, photomultiplier tubes, vacuum phototube, gas cell, etc.

(2) Photoconductive effect. Under the action of the incoming light, electrons in a semiconductor material are excited by the photon which is greater than or equal to the energy of forbidden band width. The electrons will from the valence band to the conduction band by crossing the forbidden band. So that the concentration of electrons in the conduction band increases, and the resistivity of materials decreases. Optoelectronic information conversion devices which are made with this principle are photosensitive resistance, etc.

(3) Photovoltaic effect. Under the effect of the incident light energy, which can make the object produces a certain direction of electromotive force. Optoelectronic information conversion devices which are made with this principle, for example, photovoltaic cells, photosensitive diode, light activated triode, etc.

(4) Thermo-effect of radiation. The illumination causes the temperature change of the material and generates an electric current, such as pyroelectric detector.

3.1 Photomultiplier Tubes (PMT)

Photomultiplier tubes (photomultipliers or PMTs for short), members of the class of vacuum tubes, and more specifically vacuum phototubes, are extremely sensitive detectors of light in the ultraviolet, visible, and near-infrared ranges of the electromagnetic spectrum as shown in Fig. 3-1. A photomultiplier tube is a vacuum tube consisting of an input window, a photocathode, focusing electrodes, an electron multiplier and an anode which is usually sealed into an evacuated glass tube.

Fig. 3-1 Photomultiplier tubes

These detectors multiply the current produced by incident light by as much as 100 million times (i. e., 160 dB), in multiple ***dynode*** stages, enabling individual photons to be detected when the ***incident flux*** of light is very low. Unlike most vacuum tubes, they are not obsolete.

The combination of high gain, low noise, high frequency response, equivalently, ultra-fast response, and large area of collection has earned photomultipliers an essential place in nuclear and particle physics, astronomy, medical diagnostics including blood tests, medical imaging, motion picture film scanning (telecine), radar jamming, and high-end image scanners known as drum scanners. Elements of photomultiplier technology, when integrated differently, are the basis of night vision devices.

Semiconductor devices, particularly ***avalanche photodiodes***, are alternatives to photomultipliers; however, photomultipliers are uniquely well-suited for applications requiring low-noise, high-sensitivity detection of light that is imperfectly collimated. While photomultipliers are extraordinarily sensitive and moderately efficient, as of 2012 research was still underway to create a photon-counting light detection device that is much more than 99% efficient. Such a detector is of interest for applications related to ***quantum information*** and ***quantum cryptography***.

In this section, we will introduce basic knowledge of PMT, and before that, some useful information about photoelectric effect and ***secondary emission*** is introduced, because it is necessary for us to better understand PMT.

3.1.1 Photoelectric Effect and Secondary Emission

In the photoelectric effect, electrons are emitted from solids, liquids or gases when they absorb energy from light. Electrons that emitted in this manner may be called ***photoelectrons***.

In 1887, Heinrich Hertz discovered that electrodes illuminated with ultraviolet light create ***electric sparks*** more easily. In 1905, Albert Einstein published a paper that explained experimental data from the photoelectric effect as being the result of light energy being carried in discrete quantized packets. This discovery led to the quantum revolution. Einstein was awarded the Nobel Prize in 1921 for "his discovery of the law of the photoelectric effect".

The photoelectric effect requires photons with energies from a few electron-volts to over 1 MeV in high atomic number elements. Study of the photoelectric effect led to important steps in understanding the quantum nature of light and electrons, and influenced the formation of the concept of ***wave-particle duality***. Other phenomena where light affects the movement of ***electric charges*** include the photoconductive effect (also known as ***photoconductivity*** or ***photoresistivity***), the ***photovoltaic effect***, and the photoelectrochemical effect. It also shed light on Max Planck's previous discovery of the Planck relation ($E=hf$) linking energy (E) and frequency (f) as arising from quantization of energy. The factor h is known as the Planck constant.

The photons of a light beam have a characteristic energy proportional to the frequency of the light. In the photoemission process, if an electron within some material absorbs the energy of one photon and acquires more energy than the work function (the ***electron binding energy***) of the material, it is ejected. If the photon energy is too low, the electron is unable to escape the material. Increasing the intensity of the light beam increases the number of photons in the light beam, and thus increases the number of electrons excited, but does not increase the energy that each electron possesses. The energy of the emitted electrons does not depend on the intensity of the incoming light, but only on the energy or frequency of the individual photons. It is an interaction between the incident photon and the outermost electrons.

Electrons can absorb energy from photons when irradiated, but they usually follow an "all or nothing" principle. All of the energy from one photon must be absorbed and used to liberate one electron from atomic binding, or else the energy is re-emitted. If the photon energy is absorbed, some of the energy liberates the electron from the atom, and the rest contributes to the electron's ***kinetic energy*** as a free particle. The process of light stimulating electronic is shown in Fig. 3-2.

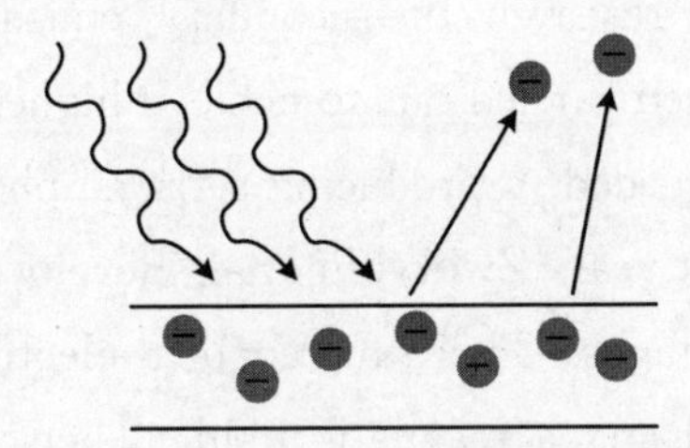

Fig. 3-2 The process of light stimulating electronic

Secondary emission in physics is a phenomenon where primary incident particles of sufficient energy, when hitting a surface or passing through some materials, induce the emission of ***secondary particles***. The primary particles are often charged particles like electrons or ions. If the secondary particles are electrons, the effect is termed secondary electron emission. In this case, the number of secondary electrons emitted per incident particle is called secondary emission yield. If the secondary particles are ions, the effect is termed ***secondary ion*** emission.

Photomultiplier tubes are vacuum tubes in which the first major component is a ***photocathode***. A light photon may interact in the photocathode to eject a low-energy electron into the vacuum. This process can be thought to occur in three steps.

(1) Absorption of the photon and energy transfers to the electron in the

photocathode material.

(2) The migration of the photoelectron to the surface of the photocathode.

(3) Escape of the electron from the photocathode surface.

In a photomultiplier tube, one or more electrons are emitted from a photocathode and accelerated towards a polished metal electrode (called dynode). They hit the electrode surface with sufficient energy to release a number of electrons through secondary emission. These new electrons are then accelerated towards another dynode, and the process is repeated several times, resulting in an overall gain ("electron multiplication") in the order of typically one million and thus generating an electronically detectable current pulse at the last dynodes.

Photomultiplier tubes have been making rapid progress since the development of photocathodes and secondary emission multipliers (dynodes). The first report on a secondary emissive surface was made by Austin et al. in 1902. Since that time, research of secondary emissive surfaces (secondary electron emission) has been carried out to achieve higher electron multiplication. In 1935, Iams et al. succeeded in producing a triode photomultiplier tube with a sound pickup. In the next year, Zworykin et al. developed a photomultiplier tube having multiple dynode stages. This tube enabled electrons to travel in the tube by using an electric field and a magnetic field. Then, in 1939, Zworykin and Rajchman developed an electrostatic-focusing type photomultiplier tube (this is the basic structure of photomultiplier tubes currently used). In this photomultiplier tube, an Ag-O-Cs photocathode was first used and later an Sb-Cs photocathode was employed. An improved photomultiplier tube structure was developed and announced by Morton in 1949 and 1956. Since then the dynode structure has been intensively studied, leading to the development of a variety of dynode structures including circular-cage, linear-focused and box-and-grid types. In addition, photomultiplier tubes using magnetic-focusing type multipliers, transmission-mode secondary-emissive surfaces and channel type multipliers have been developed.

3.1.2 Structure and Operating Principles of PMT

The invention of the photomultiplier is predicated upon two prior achievements, the discoveries of the photoelectric effect and secondary emission.

Photomultipliers are constructed from a ***glass envelope*** with a high vacuum inside, which houses a photocathode, several dynodes, and an anode. Incident photons strike the photocathode material, which is present as a thin deposit on the entry window of the device, with electrons being produced as a consequence of the photoelectric effect. These electrons are directed by the ***focusing electrode*** towards the electron multiplier, where electrons are multiplied by the process of secondary emission. The structure of PMT is shown in Fig. 3-3.

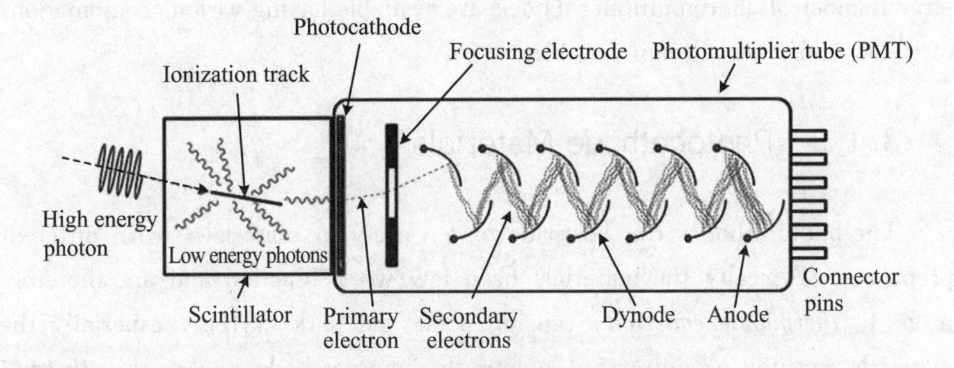

Fig. 3-3 The structure and principles of PMT

Light which enters a photomultiplier tube is detected and produces an output signal through the following processes: (1) Light passes through the input window. (2) Light excites the electrons in the photocathode so that photoelectrons are emitted into the vacuum (external photoelectric effect). (3) Photoelectrons are accelerated and focused by the focusing electrode onto the first dynode where they are multiplied by means of secondary electron emission. This secondary emission is repeated at each of the successive dynodes. (4) The multiplied secondary electrons emitted from the last dynode are finally collected by the anode.

The electron multiplier consists of a number of electrodes called dynodes. Each dynode is held at a more positive voltage than the previous one. The electrons leave the photocathode, having the energy of the incoming photon (minus the work function of the photocathode). As the electrons move towards the first dynode, they are accelerated by the electric field and arrive with much greater energy. Upon striking the first dynode, more low energy electrons are emitted, and these electrons in turn are accelerated toward the second dynode. The geometry of the dynode chain is such that a cascade occurs with an ever-increasing number of electrons being produced at each stage. Finally, the electrons reach the anode, where the accumulation of charge results in a sharp current pulse indicating the arrival of a photon at the photocathode.

There are two common photomultiplier orientations, the head-on or end-on (transmission mode) design. As shown above, where light enters the flat, circular top of the tube and passes the photocathode, and the side-on design (reflection mode), where light enters at a particular spot on the side of the tube, and impacts on an opaque photocathode. Besides, the different photocathode materials, performance is also affected by the transmission of the window material that the light passes through, and by the arrangement of the dynodes. A large number of photomultiplier models are available having various combinations of these, and other, design variables.

3.1.3 Photocathode Materials

The photocathodes can be made of a variety of materials, with different properties. Typically the materials have low work function and are therefore prone to ***thermionic emission***, causing noise and dark current, especially the materials sensitive in infrared; cooling the photocathode lowers this ***thermal noise***. The most common photocathode materials are as follows:

(1) Ag-O-Cs: also called S1. Transmission-mode, sensitive from 300 ~ 1200 nm; high ***dark current***; used mainly in near-infrared, with the photocathode cooled.

(2) GaAs-Cs: ***caesium-activated gallium arsenide***. Flat response from 300 ~ 850 nm, fading towards ultraviolet and to 930 nm.

(3) InGaAs-Cs: ***caesium-activated indium gallium arsenide***. Higher infrared sensitivity than GaAs-Cs. Between 900 ~ 1000 nm much higher signal-to-noise ratio than Ag-O-Cs.

(4) Sb-Cs: ***caesium-activated antimony***. Used for reflective mode photocathodes; response range from ultraviolet to visible. Widely used.

(5) Bialkali (Sb-K-Cs, Sb-Rb-Cs): caesium-activated antimony-rubidium or antimony-potassium alloy. Similar to Sb-Cs, with higher sensitivity and lower noise. Can be used for transmission-mode; favorable response to a NaI-Tl scintillator flashes makes them widely used in gamma spectroscopy and radiation detection.

(6) High-temperature bialkali (Na-K-Sb): less sensitive than the other bialkali cathodes; for prolonged use at temperatures above 60 ℃; can operate up to 175 ℃, used in well logging. Low dark current at room temperature.

(7) Multialkali (Na-K-Sb-Cs): more sensitive than the bialkali types in the range 600 ～ 850 nm, but with correspondingly higher noise; wide spectral response from ultraviolet to near-infrared; special cathode processing can extend range to 930 nm. Used in broadband ***spectrophotometers***.

(8) Solar-blind (Cs-Te, Cs-I): sensitive to vacuum-UV and ultraviolet; insensitive to visible light and infrared (Cs-Te has cutoff at 320 nm, Cs-I at 200 nm).

The response of the multialkali cathodes can be tailored to extend as required in the green and red regions. Other rarely used photocathode types are as follows:

(1) The monoalkali cathode (SbCs3).

(2) The AgOCs cathode which is sensitive from the visible region to the infrared region, but has a very low QE.

(3) Alkali tellurides on fused silica ("solar blind").

3.1.4 Window Materials

The windows of the photomultipliers act as wavelength filters; this may be irrelevant if the ***cut-off wavelengths*** are outside of the application range or outside of the photocathode sensitivity range, but special care has to be taken for uncommon wavelengths.

(1) Borosilicate glass (hard glasses), e.g., Corning Pyrex, is commonly used for near-infrared to about 300 nm. Glass with very low content of potassium can be used with bialkali photocathodes to lower the ***background radiation*** from the potassium-40 isotope.

(2) Ultraviolet glass, e.g., Schott 8337, transmits visible and ultraviolet down to 185 nm; used in spectroscopy.

(3) Synthetic silica transmits down to 160 nm, absorbs less UV than fused silica. Different thermal expansion than kovar (and than borosilicate glass that's expansion-matched to kovar), a graded seal needed between the window and the rest of the tube. The seal is vulnerable to mechanical shocks. The most commonly used materials are fused silica, e.g., Spectrosil, which is very transparent to UV radiation down to about 160 nm.

(4) Magnesium fluoride transmits ultraviolet down to 115 nm. ***Hygroscopic***,

though less than other alkali halides usable for UV windows.

Within each group of glass, there are many variants having different transmissions.

It is important to know beforehand the conditions of the incident light to be measured. Then, choose a photomultiplier tube that is best suited to detect the incident light and also select the optimum circuit conditions that match the application. Referring to the Table 3-1, select the optimum photomultiplier tubes, operating conditions and circuit configurations according to the incident light wavelength, intensity, beam size and the speed of optical phenomenon.

Table 3-1　Conditions of incident light

Conditions of incident light	Selection reference	
	Photomultiplier tubes	Circuit conditions
Light wavelength	Window material Photocathode spectral response	
Light intensity	Number of dynodes Dynode type Voltage applied to dynodes	Signal processing method (analog or digital method)
Light beam size	Effective diameter (size) Viewing configuration (side-on or head-on)	
Speed of optical phenomenon	Time response	Bandwidth of associated circuit

Fig. 3-4 shows an application example in which a photomultiplier tube is used in absorbtion spectroscopy.

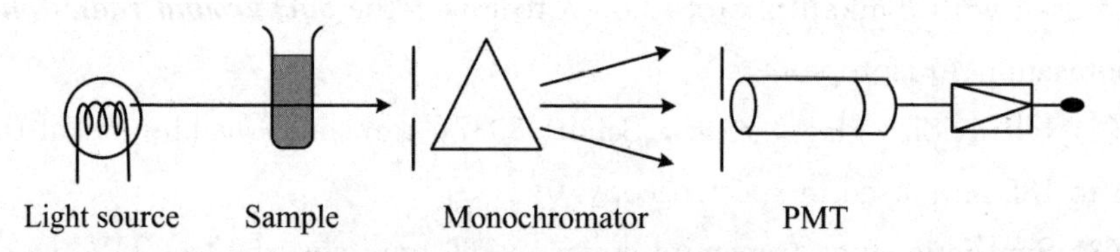

Fig. 3-4　An application example

3.1.5　Application of PMT

Photomultiplier tubes typically utilize 1000 ～ 2000 volts to accelerate electrons within the chain of dynodes. The most negative voltage is connected to the cathode, and the most positive voltage is connected to the anode. Negative high-voltage supplies (with the ***positive terminal*** grounded) are preferred, because this configuration enables the photocurrent to be measured at the low voltage side of the circuit for amplification by subsequent electronic circuits operating at low voltage. Voltages are distributed to the dynodes by a resistive voltage divider, although variations such as active designs (with transistors or diodes) are possible. The divider design, which influences frequency response or rise time, can be selected to suit varying applications. Some instruments that use photomultipliers have provisions to vary the anode voltage to control the gain of the system.

While powered (energized), photomultipliers must be shielded from ambient light to prevent their destruction through ***overexcitation***. In some applications this protection is accomplished mechanically by electrical interlocks or shutters that protect the tube when the photomultiplier compartment is opened. Another option is to add overcurrent protection in the external circuit, so that when the measured anode current exceeds a safe limit, the high voltage is reduced.

If used in a location with strong magnetic fields, which can curve electron paths, steer the electrons away from the dynodes and cause loss of gain, photomultipliers are usually magnetically shielded by a layer of mu-metal. This ***magnetic shield*** is often maintained at cathode potential. When this is the case, the external shield must also be electrically insulated because of the high voltage on it. Photomultipliers with large distances between the photocathode and the first dynode are especially sensitive to magnetic fields.

3.1.5.1 Typical Applications

(1) Photomultipliers are the first electric eye devices, being used to measure interruptions in beams of light.

(2) Photomultipliers are used in conjunction with scintillators to detect ionizing radiation by means of hand held and fixed radiation protection instruments, and particle radiation in physics experiments.

(3) Photomultipliers are used in research laboratories to measure the intensity and spectrum of light-emitting materials such as compound semiconductors and quantum dots.

(4) Photomultipliers are used as the detector in many spectrophotometers.

This allows an instrument design that escapes the thermal noises limit on sensitivity, and which can therefore substantially increase the dynamic range of the instrument.

(5) Photomultipliers are used in numerous medical equipment designs. For example, blood analysis devices used by clinical medical laboratories, such as flow cytometers, utilize photomultipliers to determine the relative concentration of various components in blood samples, in combination with optical filters and incandescent lamps.

(6) An array of photomultipliers is used in a Gamma camera.

(7) Photomultipliers are typically used as the detectors in Flying-spot scanners.

3.1.5.2 High Sensitivity Applications

After fifty years, during which solid-state electronic components have largely displaced the vacuum tube, the photomultiplier remains a unique and important optoelectronic component. Perhaps the most useful quality is that it acts, electronically, as a nearly perfect current source owing to the high voltage utilized in extracting the tiny currents associated with weak light signals. There is no Johnson noise associated with photomultiplier signal currents even though they are greatly amplified, e. g. , by 100 thousand times (i. e. , 100 dB) or more. The photocurrent still contains shot noise.

Photomultiplier-amplified photocurrents can be electronically amplified by a high-input-***impedance*** electronic amplifier (in the signal path, subsequent to the photomultiplier), thus producing appreciable voltages even for nearly infinitesimally small photon fluxes. Photomultipliers offer the best possible opportunity to exceed the Johnson noise for many configurations. The aforementioned refers to measurement of light fluxes, while small, nonetheless amount to a continuous stream of multiple photons.

For smaller photon fluxes, the photomultiplier can be operated in photon counting or Geiger mode. In Geiger mode, the photomultiplier gain is set so high (using high voltage) that a single photo-electron resulting from a single photon incident on the primary surface generates a very large current at the output circuit. However, owing to the avalanche of current, a reset of the photomultiplier is required. In either case, the photomultiplier can detect individual photons. The drawback, however, is that not every photon incident on the primary surface is counted either because of less-than-perfect efficiency of the photomultiplier, or because a second photon can arrive at the photomultiplier during the "dead time" associated with a first photon and never be noticed.

A photomultiplier will produce a small current even without incident photons; this is called the dark current. Photon counting applications generally demand photomultipliers designed to minimise dark current.

Nonetheless, the ability to detect single photons striking the primary ***photosensitive*** surface itself reveals the quantization principle that Einstein put forth. Photon-counting (as it is called) reveals that light, not only being a wave, consists of discrete particles (i. e., photons).

New Words and Expressions

[1] **photomultiplier tubes**：光电倍增管。它是将微弱的光信号转换成电信号的真空电子器件。

[2] **dynode**：电子倍增器电极。

[3] **incident flux**：入射光通量。光通量就是用来表示辐射功率经过人眼的视见函数影响后的光谱辐射功率大小的物理量。

[4] **avalanche photodiodes**：雪崩光电二极管。激光通信中使用的光敏元件。

[5] **quantum information**：量子信息。

[6] **quantum cryptography**：量子密码学，又称量子密钥分发。它利用量子力学特性来保证通信安全。它使通信的双方能够产生并分享一个随机的、安全的密钥，来加密和解密信息。

[7] **secondary emission**：二次发射。

[8] **photoelectrons**：光电子。它是指光波波段，即红外线、可见光、紫外线和软 X 射线(频率范围为 3×1011～3×1016 Hz 或波长范围为 1 mm～10 nm)波段的电子学。

[9] **electric sparks**：电火花。它是一种加工工艺，主要利用具有特定几何形状的放电电极(EDM 电极)在金属(导电)部件上烧灼出电极的几何形状。

[10] **wave-particle duality**：波粒二象性。它是指某物质同时具备波的特质及粒子的特质。

[11] **electric charges**：电荷。它是指带正负电的基本粒子。

[12] **photoconductivity**：光电导性。

[13] **photoresistivity**：光敏电阻率。

[14] **photovoltaic effect**：光伏效应。它是指光照使不均匀半导体或半导体与金属结合的不同部位之间产生电位差的现象。

[15] **electron binding energy**：电子结合能。

[16] **kinetic energy**：动能。它是指物质运动时所得到的能量。它通常被定义成使某物体从静止状态到运动状态所做的功。

[17] **secondary particles**：次级粒子。

[18] **secondary ion**：二次离子。

[19] **photocathode**：光电阴极。它是光电管(基于外光电效应的基本光电转换器件)发射光电子的一极(阴极)。

[20] **glass envelope**：玻壳。玻璃封装。

[21] **focusing electrode**：聚焦电极。

[22] **thermionic emission**：热离子发射。它是指金属或半导体表面的电子具有热运动的动能足以克服表面势垒而产生的电子发射现象。

[23] **thermal noise**：热噪声,又称白噪声。它是由导体中电子的热震动引起的,存在于所有电子器件和传输介质中。

[24] **dark current**：暗电流。光伏电池在无光照时,在外电压作用下 pn 结内流过的单向电流。

[25] **caesium-activated gallium arsenide**：铯激活的砷化镓。

[26] **caesium-activated indium gallium arsenide**：铯激活的铟砷化镓。

[27] **caesium-activated antimony**：铯激活的锑。

[28] **spectrophotometer**：分光光度计。它是利用分光光度法对物质进行定量定性分析的仪器。

[29] **cut-off wavelength**：截止波长。当所传输的光波长超过该波长时,光纤只能传播一种模式(基模)的光。

[30] **background radiation**：背景辐射。来自宇宙空间背景上的各向同性的微波辐射,也称为微波背景辐射。

[31] **hygroscopic**：吸湿的、湿度计的。

[32] **positive terminal**：正极端子。

[33] **overexcitation**：过激励,指过度激发。

[34] **magnetic shield**：磁屏蔽。

[35] **impedance**：阻抗。对交流电所起的阻碍作用叫作阻抗。

[36] **photosensitive**：光敏的、感性的。

3.2 Photodiode

An alternative way to detect the scintillation light is the use of a silicon photodiode. A photodiode, as shown in Fig. 3-5, is a type of ***photodetector*** capable of converting light into either current or voltage, depending upon the mode of operation. The common, traditional ***solar cell*** used to generate electric solar power is a large area photodiode.

Fig. 3-5 Photodiode

Photodiodes are similar to regular ***semiconductor diodes*** except that they may be either exposed (to detect vacuum UV or X-rays) or packaged with a window or optical fiber connection to allow light to reach the sensitive part of the device. Many diodes designed for use specifically as a photodiode use a PIN junction rather than a p-n junction, to increase the speed of response. A photodiode is designed to operate in ***reverse bias***.

3.2.1 Principle of Operation

A photodiode is a p-n junction or PIN structure. It is a form of light sensor that converts light energy into electrical energy (voltage or current). The photodiode accepts light energy as input to generate electric current. When a photon of sufficient energy strikes the diode, it excites an electron, thereby creating a free electron (and a positively charged electron hole). This mechanism is also known as the inner photoelectric effect. If the absorption occurs in the junction's ***depletion region***, or one ***diffusion length*** away from it, these carriers are swept from the junction by the built-in electric field of the depletion region. Thus holes move towards the anode, and electrons move towards the cathode, and a ***photocurrent*** is produced. The total current through the photodiode is the sum of the dark current (current that flows with or without light) and the photocurrent, so the dark current must be minimized to maximize the sensitivity of the device. The principle is shown in Fig. 3-6.

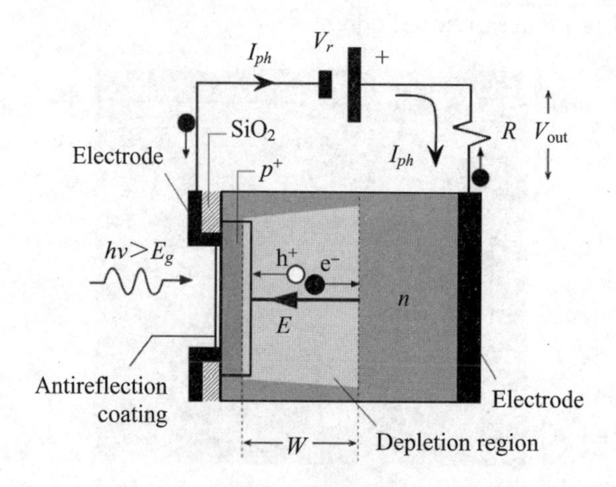

Fig. 3-6 The principle of photodiode①

Photodiode operates in three different modes. They are:

(1) Photovoltaic mode.

(2) Photoconductive mode.

(3) Avalanche diode mode.

Let us take a brief look at these mode.

3.2.1.1 Photovoltaic Mode

This is otherwise called as zero bias mode. When a photodiode operates in low frequency applications and ultra-level light applications, this mode is preferred. When used in ***zero bias*** or photovoltaic mode, the flow of photocurrent

① Kasap S O. Optoelectronics and Photonics: Principles & Practices[M]. 2nd ed. New Jersey: Prentice Hall, Inc., 2012.

out of the device is restricted and a voltage builds up. The voltage produced will have a very small dynamic range and it has a non-linear characteristic. When photodiode is configured with OP-AMP in this mode, there will be a very less variation with temperature. This mode exploits the photovoltaic effect, which is the basis for solar cells — a traditional solar cell is just a large area photodiode. Photovoltaic mode see Fig. 3-7.

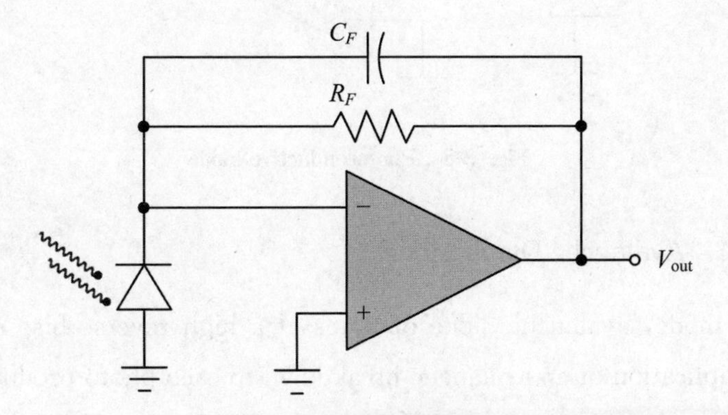

Fig. 3-7 Photovoltaic mode

3.2.1.2 Photoconductive Mode

In this mode, the diode is often reverse biased (with the cathode driven positive with respect to the anode). Compared to the forward bias, this dramatically reduces the response time at the expense of increased noise, because it increases the width of the depletion layer, which decreases the junction's ***capacitance***. The reverse bias induces only a small amount of current (known as saturation or dark current) along its direction while the photocurrent remains virtually the same. For a given spectral distribution, the photocurrent is linearly proportional to the ***illuminance*** (and to the ***irradiance***).

Although this mode is faster, the photoconductive mode (see Fig. 3-8) tends to exhibit more electronic noise. The ***leakage current*** of a good PIN diode is so low ($<$1 nA) that the Johnson-Nyquist noise of the ***load resistance*** in a typical circuit often dominates.

Transimpedance amplifiers are used as preamplifiers for photodiodes. Modes of such amplifiers keep the voltage maintains to be constant to make photodiode operate in the photoconductive mode.

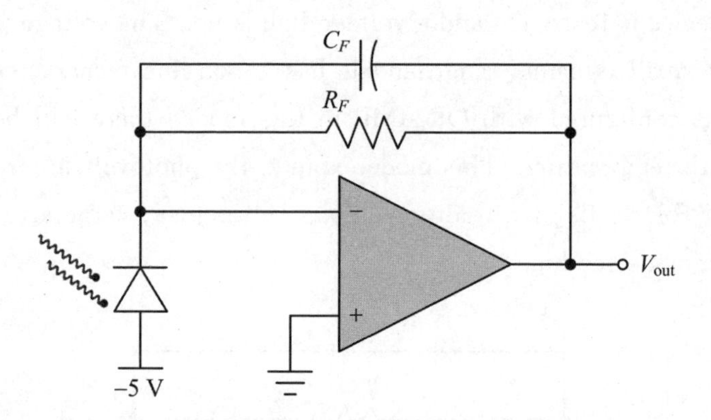

Fig. 3-8 Photoconductive mode

3.2.1.3 Avalanche Diode Mode

In this mode, avalanche diode operates at a high reverse bias condition. It allows multiplication of an avalanche breakdown to each photo-produced electron-hole pair. Hence, this produces internal gain within photodiode. The internal gain increases the device response.

3.2.1.4 Other Modes of Operation

Avalanche photodiodes have a similar structure to regular photodiodes, but they are operated with much higher reverse bias. This allows each photo-generated carrier to be multiplied by ***avalanche breakdown***, resulting in internal gain within the photodiode, which increases the effective responsivity of the device.

A phototransistor is in essence a ***bipolar transistor*** encased in a transparent case so that light can reach the base-***collector junction***. It was invented by Dr. John N. Shive (more famous for his wave machine) at Bell Labs in 1948, but it wasn't announced until 1950. The electrons that are generated by photons in the base-collector junction are injected into the base, and this photodiode current is amplified by the transistor's current gain β (or h_{fe}). If the ***emitter*** is left unconnected, the phototransistor becomes a photodiode. While phototransistors have a higher responsivity for light, they are not able to detect low levels of light any better than photodiodes. Phototransistors also have significantly longer response times.

3.2.2 Avalanche Photodiode

An avalanche photodiode (APD) is a semiconductor device in which charge carriers are generated and multiplied when exposed to light. Avalanche photodiodes are incorporated in many high performance ***optical communications***, imaging and sensing applications because they enable high ***signal to noise ratio*** (***SNR***) and high-speed operation. Modern optical acquisition and tracking applications often require the use of a plurality of photodetectors in order to steer the ***boresight*** of the optical system towards the target. Of various semiconductor photo detectors, an avalanche photo diode (APD) is a photo detector having characteristics of high sensitivity and wide band, and is widely employed in optical communications. In a typical fiber optic network, optical signals need to be converted to electronic signals, and electrical signals need to be converted to optical signals. Avalanche photodiodes are used in optical receivers for converting an optical signal into an electrical signal. The electrical signal output from the APD is coupled to an amplifier for amplification. The avalanche photodiode performs amplification of a photocurrent by applying a reverse bias voltage to its p-n junction to cause an ***avalanche multiplication*** under a high electric field as shown in Fig. 3-9. This develops a defect such that before the avalanche multiplication of the photocurrent occurs, electric fields center on the peripheral portion of the photo detecting region to cause a breakdown there. Avalanche photodiodes typically include an island or body of layers of appropriately doped ***crystalline silicon*** disposed between two electrodes electrically coupled to opposite surfaces of the APD body. Relatively high voltages are applied across the APD to generate the electric field necessary to cause the avalanche effect. Avalanche occurs when carriers generated by photon induced ***free carrier*** generation in a light receiving area of the device are introduced into a high electric field area formed in a highly reverse biased (near ***breakdown voltage***) semiconductor p-n junction. The introduced carriers collide with ***neutral atoms*** to release other carriers by ***impact ionization***. This collision process is then repeated in an avalanche fashion to effectively amplify the limited number of carriers. Typically, an APD has an absorption layer where an optical signal is absorbed. The optical signal includes a number of photons. Each photon impinging the absorption layer generates an electron-hole pair or a carrier as shown in Fig. 3-10.

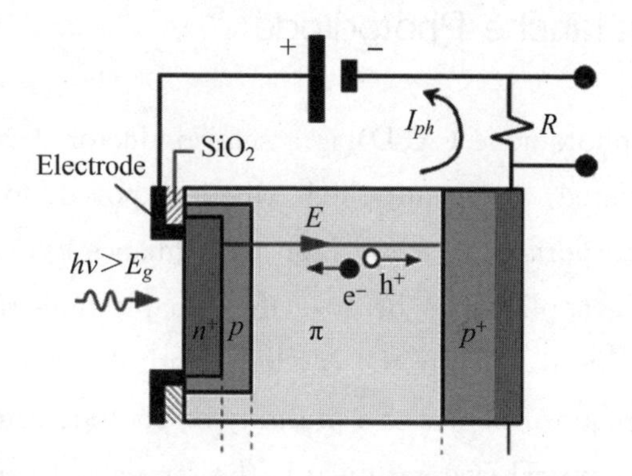

Fig. 3-9 The principle of APD①

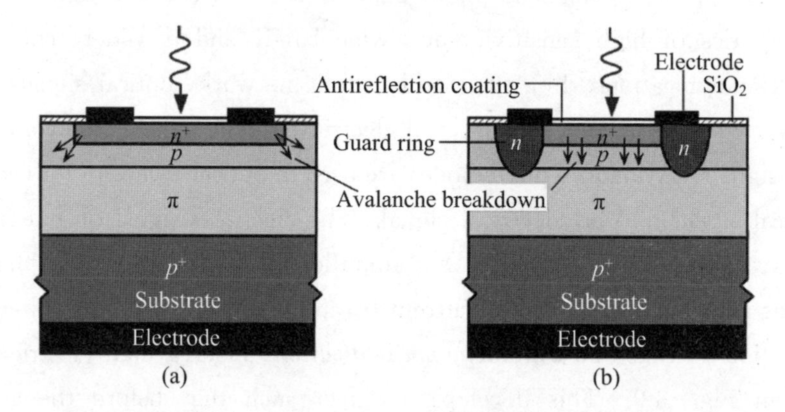

Fig. 3-10 APD structure with an absorption layer

(a) A Si APD structure without a guard ring. (b) A schematic illustration of the structure of a more practical Si APD①.

3.2.3 Phototransistor

A phototransistor is an ***electronic switching*** and current amplification component which relies on exposure to light to operate as shown in Fig. 3-11. A phototransistor has exposed-base sections, when exposed to light, which activate the component rather than the electric current used in conventional examples. As with most regular transistors, the phototransistor's operating range is also base-

① Kasap S O. Optoelectronics and Photonics: Principles & Practices[M]. 2nd ed. New Jersey: Prentice Hall, Inc., 2012:402,405.

input dependent. This means that the transistors range of operation may be controlled by the intensity of the applied light. The component is commonly used in devices such as optical remote controls, light pulse counters, and light measurement meters.

Fig. 3-11 Phototransistor

A phototransistor is a bipolar device that is completely made of silicon or another semi-conductive material, and is dependent on light energy. A typical transistor consists of a collector, emitter, and base sections. The collector is biased positively with respect to the emitter and the base-collector junction is reverse biased. A phototransistor remains inactive until light falls onto the base. Light activates the phototransistor, allowing the formation of hole-electron pairs and the flow of current across the collector or emitter. As the current spreads, it is concentrated and converted into voltage. Phototransistors are generally encased in an opaque or clear container in order to enhance light as it travels through it and allow the light to reach the phototransistor's sensitive parts. A phototransistor generally has an exposed base that amplifies the light that it comes in contact with. This causes a relatively high current to pass through the phototransistor. As the current spreads from the base to the emitter, the current is concentrated and converted into voltage. The principle of a phototransistor is shown in Fig. 3-12.

All transistors and most semiconductor components in fact are light sensitive. The phototransistor has been optimized to harness this characteristic. These components feature transparent base sections which allow unimpeded light gathering, and in most cases, do not have a base lead at all. Those that do have a base lead use it to bias or control the way the current flows rather than for activation. Apart from these differences, it is identical in construction and application to its conventional siblings.

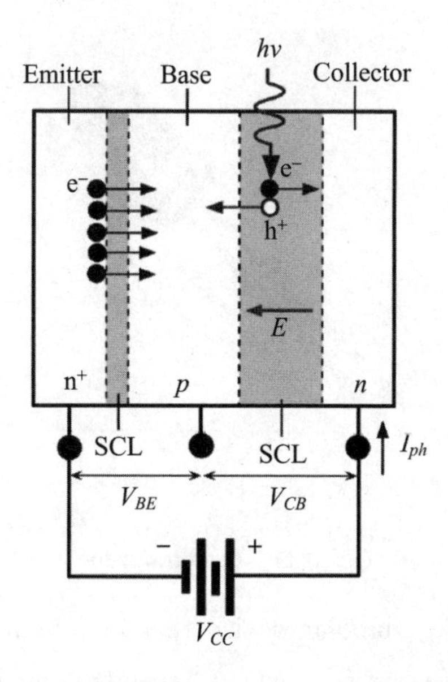

Fig. 3-12　The principle of a phototransistor[①]

The first phototransistors used single semiconductor materials such as germanium and silicone in their construction. Modern components use several differing material junctions including gallium and arsenide which lend the components far higher efficiency levels. The physical structure of the transistor is also optimized to allow for maximum light exposure. This usually entails placing the component contacts in an off-set configuration so as to avoid impeding light falling on the base.

The phototransistor's operational range is also base input dependent, i. e. , the extent to which the component conducts can be controlled by varying the intensity of light it is exposed to. This makes a phototransistor ideal for light measuring instruments such as photographers' light meters. Many optical remote controls also use this characteristic to allow the system to transmit a range of instructions. Counters which use light pulses also utilize phototransistors in their

① Kasap S O. Optoelectronics and Photonics: Principles & Practices[M]. 2nd ed. New Jersey: Prentice Hall, Inc. , 2012:417.

circuitry as do several types of day/night switches. The infrared phototransistor is also frequently used in light-dependent ***proximity switches*** such as door closure sensors and security motion detectors.

Phototransistors are used for a wide variety of applications. In fact, phototransistors can be used in any electronic device that senses light. For example, phototransistors are often used in smoke detectors, infrared receivers, and CD players. Phototransistors can also be used in astronomy, night vision, and laser range-finding.

Phototransistors have several important advantages that separate them from other optical sensors. They produce a higher current than photodiodes and also produce a voltage, something that photoresistors cannot do. Phototransistors are very fast and their output is practically instantaneous. They are relatively inexpensive, simple, and so small that several of them can fit onto a single integrated computer chip. While phototransistors can be advantageous, they also have several disadvantages. Phototransistors made of silicon cannot handle voltages over 1000 volts. They do not allow electrons to move as freely as other devices, such as electron tubes. Also, phototransistors are also more vulnerable to electrical surges/spikes and electromagnetic energy.

Both the photodiode and phototransistor are used for converting the energy of light to electrical current. However, the phototransistor is more responsive as contrasted to the photodiode due to the utilization of the transistor.

The transistor changes the base current which causes light absorption, therefore the huge output current can be gained throughout the collector terminal of the transistor. The photodiodes time response is very fast as compared with the phototransistor. So it is applicable where fluctuation in the circuit occurs. For better understating, here we have listed out some points of photodiode and phototransistor, as shown in Table 3-2.

Table 3-2 Some points of photodiode and phototransistor

Photodiode	Phototransistor
The semiconductor device that converts the energy of light to electrical current is known as a photodiode	The phototransistor is used to change the energy of light into electrical current by using the transistor
It generates both the current and voltage	It generates current

continued

Photodiode	Phototransistor
The response time is speed	The response time is slow
It is less responsive as compared with a phototransistor	It is responsive and generates a huge o/p current
This diode works in both the biasing conditions	This diode works in forward biasing only
It is used in a light meter, solar power plant, etc.	It is used to detect the light

3.2.4 Features of Photodiode

Critical performance parameters of a photodiode include the following features.

3.2.4.1 Responsivity

The spectral responsivity is a ratio of the generated photocurrent to incident light power, expressed in A/W when used in photoconductive mode. The wavelength-dependence may also be expressed as quantum efficiency, or the ratio of the number of ***photogenerated carriers*** to incident photons, a unitless quantity. In other words, it is a measure of the effectiveness of the conversion of the light power into electrical current. It varies with the wavelength of the incident light (see Fig. 3-13) as well as applied reverse bias and temperature.

3.2.4.2 Dark Current

The current through the photodiode in the absence of light, when it is operated in photoconductive mode. The dark current includes photocurrent generated by background radiation and the ***saturation current*** of the semiconductor junction. Dark current must be accounted for by calibration if a photodiode is used to make an accurate optical power measurement, and it is also a source of noise when a photodiode is used in an optical communication system.

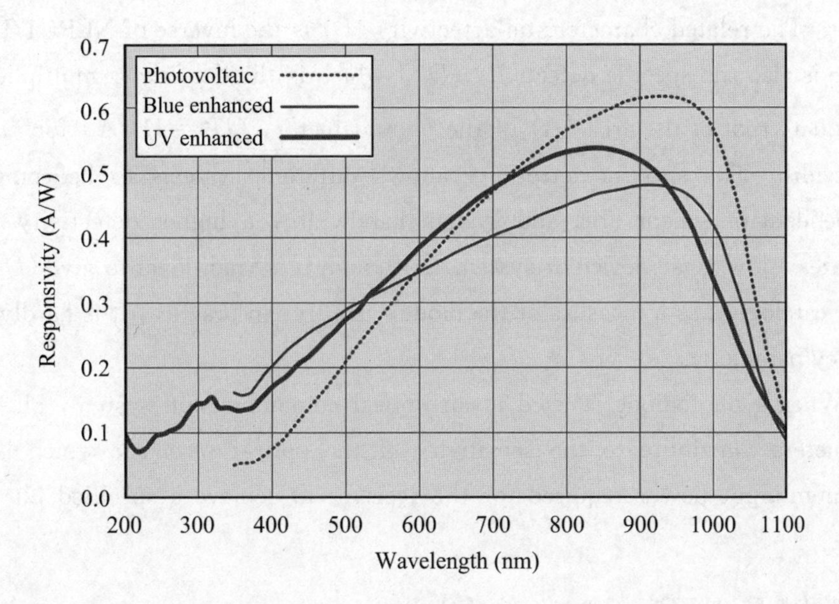

Fig. 3-13 Typical spectral responsivity of several different types of planar diffused photodiodes

3.2.4.3 Response Time

A photon absorbed by the semiconducting material will generate an electron-hole pair which will in turn start moving in the material under the effect of the electric field and thus generate a current. The finite duration of this current is known as the transit-time spread and can be evaluated by using Ramo's theorem. One can also show with this theorem that the total charge generated in the external circuit is well e and not 2e as might seem by the presence of the two carriers. Indeed the integral of the current due to both electron and hole over time must be equal to e. The resistance and capacitance of the photodiode and the external circuitry give rise to another response time known as RC time constant. This combination of R and C integrates the ***photoresponse*** over time and thus lengthens the impulse response of the photodiode. When used in an optical communication system, the response time determines the bandwidth available for signal modulation and thus data transmission.

3.2.4.4 Noise-equivalent Power (NEP)

The minimum input optical power to generate photocurrent, equal to the rms noise current in a 1 hertz bandwidth. NEP is essentially the minimum detectable power. The related characteristic detectivity (D) is the inverse of NEP, 1/NEP. There is also the specific detectivity (D^*) which is the detectivity multiplied by the square root of the area (A) of the photodetector, ($D^* = D\sqrt{A}$) for a 1 Hz bandwidth. The specific detectivity allows different systems to be compared independent of sensor area and system bandwidth; a higher detectivity value indicates a low-noise device or system. Although it is traditional to give (D^*) in many catalogues as a measure of the diode's quality, in practice, it is hardly ever the key parameter.

When a photodiode is used in an optical communication system, all these parameters contribute to the sensitivity of the optical receiver, which is the minimum input power required for the receiver to achieve a specified bit error rate.

3.2.4.5 V-I Characteristics of Photodiode

Photodiode operates in reverse bias condition. Reverse voltages are plotted along X-axis in volts and reverse current are plotted along Y-axis in microampere. Reverse current does not depend on reverse voltage. When there is no light illumination, reverse current will be almost zero. The minimum amount of current present is called as dark current. Once when the light illumination increases, reverse current also increases linearly (see Fig. 3-14).

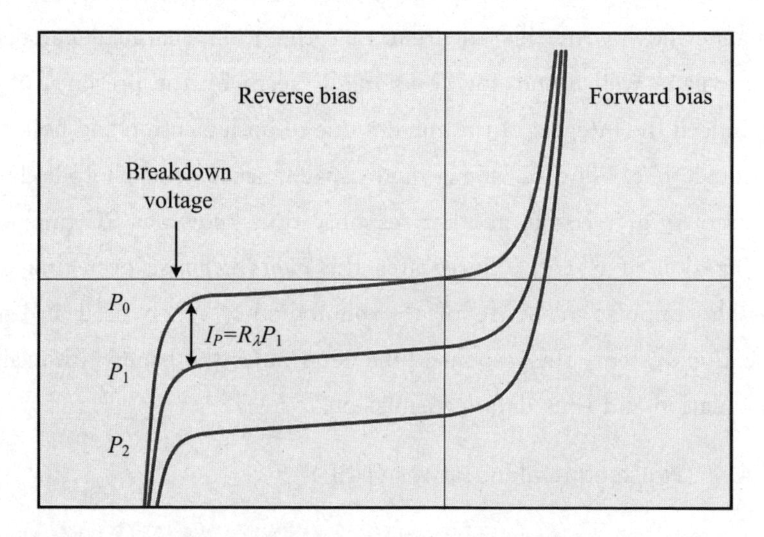

Fig. 3-14　V-I characteristics of photodiode

3.2.5 Applications of Photodiode

P-n photodiodes are used in similar applications to other photodetectors, such as photoconductors, charge-coupled devices, and photomultiplier tubes. They may be used to generate an output which is dependent upon the illumination (analog; for measurement and the like), or to change the state of circuitry (digital; either for control and switching, or digital signal processing).

Photodiodes are used in consumer electronics devices such as compact disc players, smoke detectors, and the receivers for infrared remote control devices used to control equipment from televisions to air conditioners. For many applications, either photodiodes or photoconductors may be used. Either type of photosensor may be used for light measurement, as in camera light meters, or to respond to light levels, as in switching on street lighting after dark.

Photosensors of all types may be used to respond to incident light, or to a source of light which is part of the same circuit or system. A photodiode is often combined into a single component with an emitter of light, usually a light-emitting diode (LED), either to detect the presence of a mechanical obstruction to the beam (slotted optical switch), or to couple two digital or analog circuits while maintaining extremely high electrical isolation between them, often for safety (optocoupler).

Photodiodes are often used for accurate measurement of light intensity in science and industry. They generally have a more linear response than photoconductors.

They are also widely used in various medical applications, such as detectors for computed tomography (coupled with scintillators), instruments to analyze samples (immunoassay), and pulse oximeters.

PIN diodes are much faster and more sensitive than p-n junction diodes, and hence are often used for optical communications and in lighting regulation.

P-n photodiodes are not used to measure extremely low light intensities. Instead, if high sensitivity is needed, avalanche photodiodes, intensified charge-coupled devices or photomultiplier tubes are used for applications, such as astronomy, spectroscopy, night vision equipment and laser-range finding.

3.2.6 Summary

Photodiodes are semiconductor devices responsive to high-energy particles and photons. Photodiodes are operated by absorption of photons or charged particles and generate a flow of current in an external circuit. Photodiodes can be used to detect the presence or absence of minute quantities of light and can be calibrated for extremely accurate measurements from intensities below 1 pW/cm^2 to intensities above 100 mW/cm^2. Silicon photodiodes are utilized in such diverse applications as spectroscopy, photography, analytical instrumentation, optical position sensors, beam alignment, surface characterization, laser range finders, optical communications, and medical imaging instruments.

New Words and Expressions

[1] **photodetector**：光电探测器。它用来把光信号转换为电信号。

[2] **solar cell**：太阳能电池。它是通过光电效应或者光化学效应直接把光能转化成电能的装置。

[3] **semiconductor diodes**：半导体二极管。它主要用于检波、混频、参量放大、开关、稳压、整流等。

[4] **reverse bias**：反向偏压。

[5] **depletion region**：耗尽层。它是指 pn 结中在漂移运动和扩散作用的双重影响下载流子数量非常少的一个高电阻区域。

[6] **diffusion length**：扩散长度。

[7] **photocurrent**：光电流。很多光电子形成的电流叫光电流。

[8] **zero bias**：零偏差。当一个测量器在无任何测量物件时(测量器的读数应当为零),测量器的读数是一个非零数字,此现象被称为零偏差。

[9] **capacitance**：电容。它是在给定电位差下的电荷储藏量。

[10] **illuminance**：照度。它是单位面积上所接受可见光的能量。

[11] **irradiance**：辐照度。它是照射在面元上的辐射通量与该面元的面积之比。

[12] **leakage current**：泄漏电流。电器在正常工作时,其火线与零线之间产生的极为微小的电流。

[13] **load resistance**：负载电阻。这类电阻功率大,一般为无感的功率电阻。

［14］ **avalanche breakdown**：雪崩击穿。

［15］ **bipolar transistor**：双极型晶体管。它是由两个背靠背的 pn 结构成的具有电流放大作用的晶体三极管。

［16］ **collector junction**：集电极。它是集电区和基区之间的 pn 结。

［17］ **emitter**：发射器。

［18］ **optical communications**：光通信。它是以光波为载波的通信方式。

［19］ **signal to noise ratio(SNR)**：信噪比。它是放大器的输出信号的功率与同时输出的噪声功率的比。

［20］ **boresight**：瞄准线，孔径。

［21］ **avalanche multiplication**：雪崩倍增。它是在强电场区内因碰撞离化而引起的自由载流子数目的增加。

［22］ **crystalline silicon**：晶体硅。其性质为带有金属光泽的灰黑色固体、熔点高(1410 ℃)、硬度大、有脆性、常温下化学性质不活泼。

［23］ **free carrier**：自由载流子。它是可以自由移动的带有电荷的物质微粒，如电子和离子。

［24］ **breakdown voltage**：击穿电压。它是使电介质击穿的电压。

［25］ **neutral atoms**：中性原子。对原子来说，核外电子等于核内质子数，即正负电量相等，原子不显电性。

［26］ **impact ionization**：碰撞电离。气体原子、分子的电离可以通过光、X 射线、伽马射线照射，即电磁波的吸收，加速电子、离子或高能中性粒子的碰撞等方式发生。

［27］ **electronic switching**：电子开关。它是利用电力电子器件实现电路通断的运行单元。

［28］ **photogenerated carriers**：光生载流子。用光照射半导体时，若光子的能量等于或大于半导体的禁带宽度，则价带中的电子吸收光子后进入导带，产生电子-空穴对。这种类型的载流子称为光生载流子。

［29］ **proximity switches**：引发开关。

［30］ **saturation current**：饱和电流。

［31］ **photoresponse**：光响应。

3.3 Photoresistor

A ***photoresistor***, as shown in Fig. 3-15, often referred to as a ***light dependent resistor*** (***LDR***) or photo-conductive cell, is a resistor that reacts to increased ***exposure*** to light by decreasing its resistance in a circuit. They are used in a variety of devices that require ***sensitivity to light*** to operate, such as glow-in-the-dark watches and street lamps that turn on when the sun goes down. Photoresistors are part of a larger group of sensors known as photodetectors, which are devices that react to light.

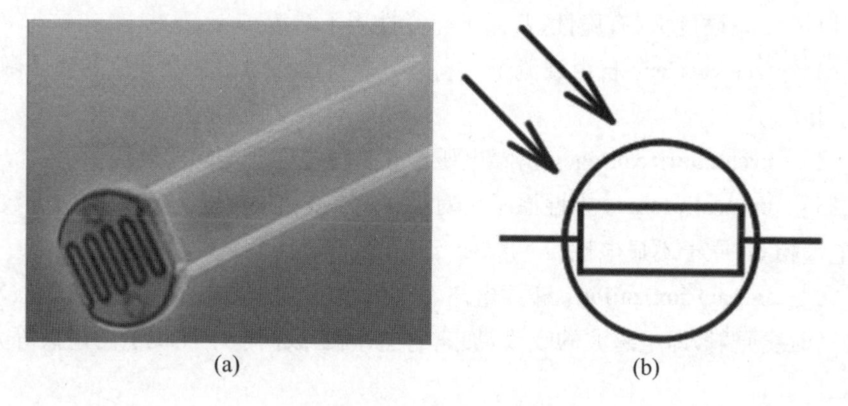

Fig. 3-15 Photoresistor

Resistors are present in almost all types of electric circuits. They function to block the flow of electricity through a circuit so that it stays within a safe range. In the case of a photoresistor, they also serve as a switch, regulating the flow of electricity based on the amount of light that they are exposed to.

Photoresistors are essentially semiconductors, meaning they conduct electricity by means of the flow of electrons. They typically have two prongs connected to a photosensitive plate. When light shining on the plate reaches a sufficiently high level frequency, this stimulates the electrons in the device and gives them enough energy to break free from their bonds. These freed electrons allow electricity to flow through the photoresistor.

Since the discovery of photoconductivity in selenium, many other materials have been found that are light dependent. In the 1930s and 1940s, PbS, PbSe and PbTe were studied following the development of photoconductors made of silicon and germanium. Modern light dependent resistors are made of lead sulfide, lead selenide, indium antimonide, and the most commonly cadmium sulfide and cadmium selenide. The popular cadmium sulfide types are often indicated as CdS photoresistors.

The uses of a photoresistor are widespread. They are commonly seen in smaller devices in the form of the cadmium sulfide (CdS) cell as shown in Fig. 3-16. The CdS cell, a term that is largely considered synonymous with the term photoresistor, is found in many forms of clocks and watches, light meters in cameras, and street lamps.

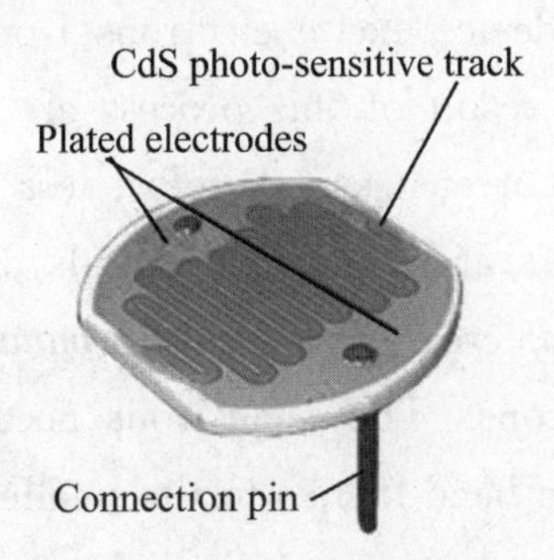

Fig. 3-16 The structure of photoresistor

To manufacture a cadmium sulfide LDR, highly purified cadmium sulfide powder and inert binding materials are mixed. This mixture is then pressed and sintered. Electrodes are vacuum evaporated onto the surface of one side to form interleaving combs and connection leads are connected. The disc is then mounted in a glass envelope or encapsulated in transparent plastic to prevent surface contamination. The spectral response curve of cadmium sulfide matches that of the human eye. The peak sensitivity wavelength is about 560～600 nm which is in the visible part of the spectrum. It should be noted that devices containing lead or cadmium are not RoHS compliant and are banned for use in countries that adhere to RoHS laws.

3.3.1 Working Principle of Photoresistor

The photoresistor doesn't have a p-n junction like photodiodes. It is a passive component. Photoresistors are made of highly resistant semiconductors that are sensitive to high photonic frequencies. When light is incident on the photoresistor, photons get absorbed by the semiconductor material. The energy from the photon gets absorbed by the electrons. When these electrons acquire sufficient energy to break the bond, they jump into the conduction band. As more photons hit the semiconductor, more electrons are knocked loose. This creates a very effective conductive flow of electricity that only travels through the semiconductor when in the presence of light.

Based on the materials used, photoresistors can be divided into two types: ***intrinsic*** and ***extrinsic***.

Intrinsic photoresistors use undoped materials such as silicon or germanium. Photons that fall on the device excite electrons from the valence band to the conduction band, and the result of this process are more free electrons in the material, which can carry current, and therefore less resistance.

Extrinsic photoresistors are made of materials doped with impurities, also called dopants. The dopants create a new ***energy band*** above the existing valence band, populated by electrons. These electrons need less energy to make the transition to the conduction band thanks to the smaller energy gap. The result is a device sensitive to different wavelengths of light.

Extrinsic photoresistors can react to infrared waves. Intrinsic photoresistors can detect higher frequency light waves. Regardless, both types will exhibit a decrease in resistance when illuminated. The higher the light intensity, the larger the ***resistance drop*** is. Therefore, the resistance of LDRs is an inverse, nonlinear function of light intensity. Depending upon the type of semiconductor material used for photoresistor, their resistance range and sensitivity differs. In the absence of light, the photoresistor can have resistance values in megaohms. And during the presence of light, its resistance can decrease to a few hundred ohms.

3.3.2 Characteristics of Photoresistor

3.3.2.1 Wavelength Dependency

The sensitivity of a photoresistor varies with the light wavelength. If the wavelength is outside a certain range, it will not affect the resistance of the device at all. It can be said that the LDR is not sensitive in that light wavelength range. Different materials have different unique spectral ***response curves*** of wavelength versus sensitivity. Extrinsic light dependent resistors are generally designed for longer wavelengths of light, with a tendency towards the infrared (IR). When working in the IR range, care must be taken to avoid heat buildup, which could affect measurements by changing the resistance of the device due to ***thermal effects***. Fig. 3-17 shown here represents the spectral response of photoconductive detectors made of different materials, with the operating temperature expressed in K and written in the parentheses.

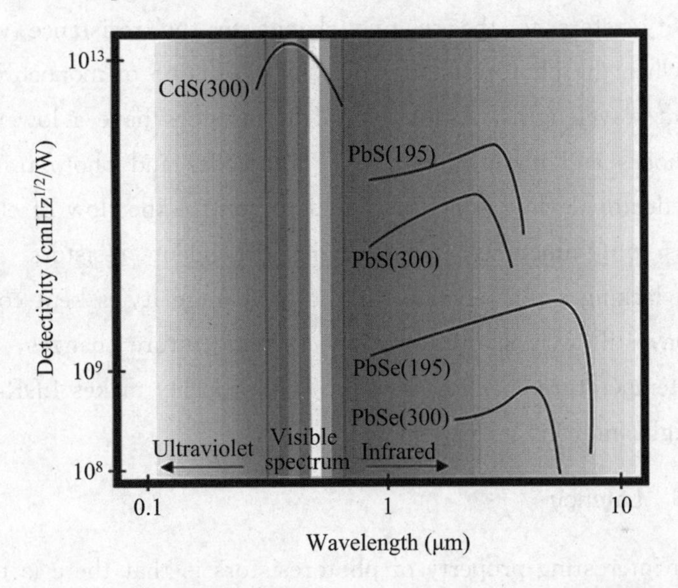

Fig. 3-17 Spectral response of photoconductive detectors made of different materials

According to the spectral characteristics of the photoresistor, it can be divided into three types of photoresistors:

(1) Ultraviolet photoresistor: more sensitive to ultraviolet light, including cadmium sulfide, cadmium selenide photoresistors, etc.

(2) Infrared photoresistor: mainly lead sulfide, lead telluride, lead selenide. Photosensitive resistors such as indium antimonide are widely used in missile guidance, astronomical detection, non-contact measurement, human lesion detection, infrared spectroscopy, infrared communications and other defense, scientific research, and industrial and agricultural production.

(3) Visible light photoresistor: including selenium, cadmium sulfide, cadmium selenide, cadmium telluride, gallium arsenide, silicon, germanium, and

zinc sulfide photoresistors. It is mainly used in various photoelectric control systems, such as photoelectric automatic opening and closing of portals, automatic turning on and off of navigation lights, street lights and other lighting systems, automatic water supply and automatic water stopping devices, mechanical automatic protection devices, and "position detectors", camera automatic exposure device, photo electric counter, smoke alarm, photoelectric tracking system, etc.

3.3.2.2 Sensitivity

Sensitivity refers to the relative change in the resistance value (dark resistance) when the photoresistor under the irradiation of monochromatic light with different wavelengths. Light dependent resistors have a lower sensitivity than photodiodes and phototransistors. Photodiodes and photo transistors are true semiconductor devices which use light to control the flow of electrons and holes across p-n junctions, while light dependent resistors are ***passive components***, lacking a p-n junction. If the light intensity is kept constant, the resistance may still vary significantly due to temperature changes, so they are sensitive to temperature changes as well. This property makes LDRs unsuitable for precise light intensity measurements.

3.3.2.3 Latency

Another interesting property of photoresistors is that there is time latency between changes in illumination and changes in resistance. The rate at which the resistance changes is called the resistance recovery rate. It takes usually about 10 ms for the resistance to drop completely when light is applied after total darkness, while it can take up to 1 second for the resistance to rise back to the starting value after the complete removal of light. For this reason, the LDR cannot be used where rapid fluctuations of light are to be recorded or used to actuate ***control equipment***, but this same property is exploited in some other devices, such as audio ***compressors***, where the function of the light dependent resistor is to smooth the response.

There are several specifications that are important for light dependent resistors and photoresistors when considering their use in any electronic circuit design. These photoresistor specifications include in Table 3-3.

Table 3-3 Photoresistor specifications

Parameters	Details
Maximum power dissipation	This is the maximum power that the device is able to dissipate within a given temperature range. Derating may be applicable above a certain temperature
Maximum operating voltage	Particularly as the device is semiconductor based, the maximum operating voltage must be observed. This is typically specified at 0 lux, i. e. , darkness
Peak wavelength	This photoresistor specification details the wavelength of maximum sensitivity. Curves may be provided for the overall response in some instances. The wavelength is specified in nm
Resistance (when illuminated)	The resistance under illumination is a key specification and a key parameter for any photoresistor. Often a minimum and maximum resistance is given under certain light conditions, often 10 lux. A minimum and maximum vale may be given because of the spreads that are likely to be encountered. A "fully on" condition may also be given under extreme lighting, e. g. , 100 lux
Dark resistance	Dark resistance values will be given for the photoresistor. These may be specified after a given time because it takes a while for the resistance to fall as the charge carrier recombine-photoresistors are noted for their slow response times

3.3.3 Typical Applications for Photoresistors

Photoresistors are found in many different applications and can be seen in many different electronic circuit designs. They have a very simple structure and they are low cost and rugged devices. They are widely used in many different items of electronic equipment and circuit designs.

Photoresistors are most often used as ***light sensors***. They are often utilized when it is required to detect the presence and absence of light or measure the light intensity. Examples are night lights and ***photography light meters***. An interesting hobbyist application for light dependent resistors is the line following robot, which uses a light source and two or more LDRs to determine the needed change of course. Sometimes, they are used outside sensing applications, for example, in audio compressors, because their reaction to light is not instantaneous, and so the function of LDR is to introduce a ***delayed response***.

3.3.3.1 Light Sensor

If a basic light sensor is needed, an LDR circuit such as the one in the figure can be used. The LED lights up when the intensity of the light reaching the LDR resistor is sufficient. A circuit example of light sensor is shown in Fig. 3-18. The 10 K variable resistor is used to set the threshold at which the LED will turn on. If the LDR light is below the ***threshold intensity***, the LED will remain in the off state. In real-world applications, the LED would be replaced with a relay or the output could be wired to a ***microcontroller*** or some other device. If a darkness sensor was needed, where the LED would light in the absence of light, the LDR and the two 10 K resistors should be swapped.

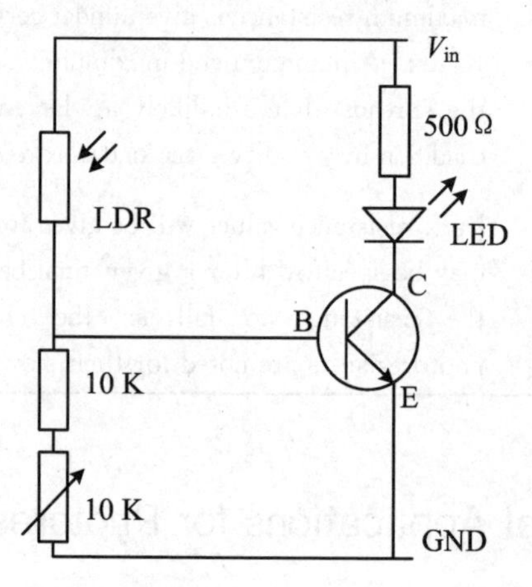

Fig. 3-18 Light sensor circuit example

3.3.3.2 Audio Compressors

Audio compressors are devices which reduce the gain of the ***audio amplifier*** when the ***amplitude*** of the signal is above a set value. This is done to amplify soft sounds while preventing the loud sounds from clipping. Some compressors use an LDR and a small lamp (LED or electroluminescent panel) connected to the signal source to create changes in ***signal gain***. This technique is believed by some to add smoother characteristics to the signal because the response times of the light and the resistor soften the attack and release. The delay in the response time in these applications is on the order of 0.1 s.

3.3.4 Advantages and Disadvantages of Photoresistor

3.3.4.1 Advantages

Photoresistors have several important advantages that separate them from other devices. They are completely dependent on how much light they receive. This means that external forces will not interfere with the devices that they are connected to. A photoresistor is also very simple because it is merely a semiconductor with a conductive pathway connected to one end in order to transfer a current from the semiconductor to the external device that it is powering.

They are small enough to fit into virtually any electronic device and are used all around the world as a basic component in many electrical systems. Also, photoresistors are simply designed and are made from materials that are widely available, allowing hundreds of thousands of units to be produced each year.

3.3.4.2 Disadvantages

There are different types of photoresistors and each has its own disadvantages, but they have several disadvantages in common. Most photoresistors cannot detect low light levels and may not work in certain conditions or circumstances. Photoresistors are also slow to respond to new levels of light and may take up to several seconds to recognize the change. This is because electrons are still moving through the semiconductor and take a few seconds to slow down or speed up.

3.3.5 Conclusion

The photoresistor is an important photoelectric conversion element. With the rapid development of electronic information technology and the continuous enhancement of the performance requirements of electronic components, the automation of photoresistor production will greatly improve the development of industrialization.

New Words and Expressions

［1］ **photoresistor**：光敏电阻。其特性是在特定波长的光的照射下，其阻值迅速减小。

［2］ **light dependent resistor（LDR）**：光敏电阻器，又称光导管。其特性是在特定光的照射下，其阻值迅速减小，可用于检测可见光。

［3］ **exposure**：曝光。

［4］ **sensitivity to light**：感光灵敏度，又称光敏度。视觉的敏感度可从绝对阈强度和辨别阈两个方面来测量。绝对阈强度指在生理条件下能引起感觉的最小光量，而其倒数即光敏度。

［5］ **intrinsic**：本质的、固有的。

［6］ **extrinsic**：外在的、外来的、非固有的。

［7］ **energy band**：能带。在形成分子时，原子轨道构成具有分立能级的分子轨道，可以将所形成的分子轨道的能级看成是准连续的，即形成了能带。

［8］ **resistance drop**：电阻电压降。

［9］ **response curves**：响应曲线。

［10］ **thermal effects**：热效应。物质系统在物理的或化学的等温过程中只做膨胀功时所吸收或放出的热量。

［11］ **passive components**：无源元件。它是指在不需要外加电源的条件下，就可以显示其特性的电子元件。

［12］ **control equipment**：控制设备、控制仪器。

［13］ **compressors**：压缩机。

［14］ **photography**：摄影。它是指使用某种专门的设备进行影像记录的过程，一般我们使用机械照相机或者数码照相机进行摄影。

［15］ **light meters**：照度计。它是一种专门测量光度、亮度的仪器仪表。测量光照强度（照度）是物体被照明的程度，即物体表面所得到的光通量与被照面积之比。

［16］ **light sensors**：光敏元件。基于半导体光电效应的光电转换传感器。

［17］ **delayed response**：延迟响应。

［18］ **threshold intensity**：临界强度。

［19］ **microcontroller**：微控制器。它诞生于 20 世纪 70 年代中期，经过 20 多年的发展，其成本越来越低，而性能越来越强大，这使其应用已经无处不在，遍及各个领域。

[20]　**audio amplifier**：音频放大器。

[21]　**amplitude**：振幅。振动物体离开平衡位置的最大距离叫振动的振幅。

[22]　**signal gain**：信号增益。在输入功率相等的条件下，实际天线与理想的辐射单元在空间同一点处所产生的信号的功率密度之比。

3.4　Solar Cell

A solar cell, also known as a ***photovoltaic cell***, is the name given to an energy capturing device. It absorbs the energy of light and transforms it into electricity by way of the photovoltaic effect. These cells have evolved dramatically since their creation, and in the past few years, particularly large strides have been made in this technology. The application of solar cell is shown in Fig. 3-19.

(a)　　　　　　　　　　(b)

Fig. 3-19　The application of solar cells

Most simply, a solar cell works by absorbing sunlight. The photons from the light run into the ***panel*** and are absorbed by some sort of semiconducting materials. Most contemporary ones are made out of silicon, although other substances are being experimented with as semiconductors to make them more cost effective and environmentally friendly. Electrons are then freed from their host atom, and they move freely as electricity. From the solar cell, this electricity then passes through a larger array, where it is turned into ***direct current*** (***DC***) electricity, which may then later be converted to ***alternating current*** (***AC***).

3.4.1 History of Solar Cell

The photovoltaic effect was first presented in the early 19th century. In the 1880s, the idea was put to practical use in the creation of the first solar cell, made with selenium as the semiconductor. The first one was around 1% efficient, meaning that it managed to capture 1% of the total solar energy that hit the cell.

In 1954, Bell Labs discovered that silicon could be modified slightly to make it incredibly ***photo-sensitive***. This led to the modern revolution in photovoltaic cells, with the early ***silicon cells*** operating at around 6% efficiency.

In 1958, a satellite, the Vanguard 1, was launched with them as a source of energy. This allowed the satellite to remain in ***geosynchronous orbit*** indefinitely, since it didn't rely on a finite amount of fuel.

Through the 1970s and 1980s, solar technology continued to improve. By 1988, ones were being mass produced that were capable of 17% efficiency, and by the end of the decade, those made from both gallium arsenide and silicon had surpassed 20% efficiency.

In the late 1980s, a new type of technology also appeared, using lenses to concentrate sunlight on to a single cell. This high ***energy density*** allowed for efficiencies of up to 37% at the time.

There are three main classifications of solar cells, referred to as "generations" because of when the technologies first appeared.

A first generation cell is what most people think of when they think of this technology. They account for around 90% of the solar cells in the world, and have a theoretical maximum efficiency of around 33%. The first generation cells — also called conventional, traditional or wafer-based cells — are made of crystalline silicon, the commercially predominant PV technology, that includes materials such as polysilicon and monocrystalline silicon.

A second generation solar cell is designed to be substantially cheaper and easier to produce. Using technologies such as ***electroplating*** and ***vapor deposition***, second generation ones can be mass produced relatively cheaply. They are usually just a thin film of some sort of materials, such as amorphous silicon or cadmium telluride (CdTe), applied in a very thin sheet to a material like ceramic or glass. They are commercially significant in utility-scale photovoltaic power stations, building integrated photovoltaics or in small stand-alone power system.

A third generation solar cell takes the second generation technology and tries to greatly improve their efficiency. These are the cutting edge technologies, trying new concentration methods, using extra heat to increase the voltage generated, and other technologies to work towards target efficiencies in the range of 30% to 60%. Most of them have not yet been commercially applied and are still in the research or development phase. Most of them use organic materials, organometallic compounds and inorganic substances. Despite the fact that their efficiencies had been low and the stability of the absorber material was often too short for commercial applications, there is a lot of research invested into these technologies as they promise to achieve the goal of producing low-cost, high-efficiency solar cells.

3.4.2 Working Principle of Solar Cell

Solar cells are typically named after the semiconducting material that they are made of. These materials must have certain characteristics in order to absorb sunlight. Some cells are designed to handle sunlight that reaches the Earth's surface, while others are optimized for use in space. Solar cells can be made of only one single layer of light-absorbing material (single-junction) or use multiple physical configurations (multi-junctions) to take advantage of various absorption and charge separation mechanisms. Basically, when light strikes the cell, a certain portion of it is absorbed within the semiconductor material. This means that the energy of the absorbed light is transferred to the semiconductor. The energy knocks electrons loose, allowing them to flow freely.

PV cells also all have one or more electric field that acts to force electrons freed by light absorption to flow in a certain direction. This flow of electrons is a current, and by placing metal contacts on the top of and at the bottom of the PV cell, we can draw that current off for external use, say, to power a calculator. This current, together with the cell's voltage (which is a result of its built-in electric field or fields), defines the power (or ***wattage***) that the solar cell can produce.

Fig. 3-20 shows a schematic diagram of a typical solar cell and the basic processes that occur during the photovoltaic effect. The solar cell device comprises two major regions, which are specially tailored to conduct negative and positive charges. An n-type material conducts electrons well, while the p-type material conducts a positive charge to a high degree. To understand how these materials differ, one must examine the materials themselves. The atoms in the solar cell are bonded to adjacent atoms in each material by their shared electrons. If an impurity, or dopant, such as a phosphorous atom is introduced into Si, it contributes an extra electron when it is incorporated into the structure, creating an n-type material. Likewise, if a boron atom is introduced, it is missing a ***bonding electron*** compared to Si and will create a p-type material. If a slab of p-type material is subsequently doped on one side with phosphorous, a so-called p-n junction is formed at the interface near the surface. Electrons from the ***donor atoms*** in the n region diffuse into the p region and combine with positive "holes" from the ***acceptor atoms***, producing a layer of negatively charged ***impurity atoms***.

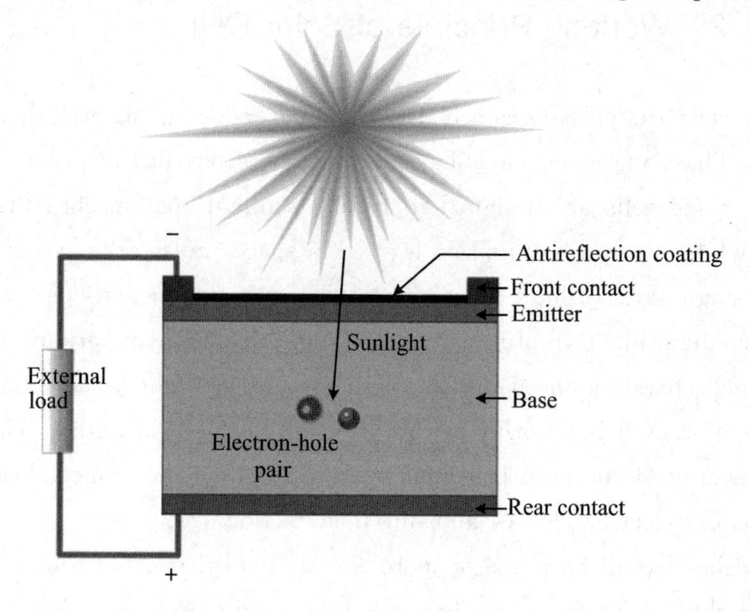

Fig. 3-20　The principle of solar cells

When light enters the device and a photon is absorbed, an additional free electron is produced within these two layers. By "free" it is meant that the electron is no longer tightly bound to its host atom — it is mobile. What is left is a place in the light-absorbing material where an electron once resided. If ***charge neutrality*** was present before the light was turned on, the result is a positive charge at the site, called a hole, where the electron was ejected. Far from being fixed, this hole can also have "mobility". Electrons and holes are not only produced by light, but also constantly produced (and destroyed) from ***thermal excitation***. If a free electron within the device encounters the hole, it will fill it, and the hole will then reside on the site from where the electron came. It is not enough to have ***mobile electrons*** and holes. Ultimately, to produce work these charges must be separated and collected at the external contacts in the front and at the back of the device.

In conclusion, the solar cell works in the following three steps:

(1) Photons in sunlight hit the solar panel and are absorbed by semiconducting materials, such as silicon.

(2) Electrons (negatively charged) are knocked loose from their atoms, causing an ***electric potential difference***. Current starts flowing through the material to cancel the potential and this electricity is captured. Due to the special composition of solar cells, the electrons are only allowed to move in a single direction.

(3) An array of solar cells converts solar energy into a usable amount of direct current (DC) electricity.

3.4.3 Efficiency of Solar Cell

The efficiency of a solar cell may be broken down into ***reflectance*** efficiency, thermodynamic efficiency, charge carrier separation efficiency and ***conductive efficiency***. The overall efficiency is the product of each of these individual efficiencies.

A solar cell usually has a ***voltage dependent*** efficiency curve, temperature coefficients, and shadow angles.

Due to the difficulty in measuring these parameters directly, other parameters are measured instead: ***thermodynamic efficiency***, quantum efficiency, integrated quantum efficiency, VOC ratio, and fill factor. Reflectance losses are a portion of the quantum efficiency under "external quantum efficiency". ***Recombination losses*** make up a portion of the quantum efficiency, VOC ratio, and fill factor. Resistive losses are predominantly categorized under fill factor, but also make up minor portions of the quantum efficiency, VOC ratio.

The ***fill factor*** is defined as the ratio of the actual maximum obtainable power to the product of the ***open circuit voltage*** and ***short circuit current***. This is a key parameter in evaluating the performance of solar cells. Typical commercial solar cells have a fill factor > 0. 7. Grade B cells have a fill factor usually between 0. 4 and 0. 7. Cells with a high fill factor have a low ***equivalent series resistance*** and a high equivalent ***shunt resistance***, so less of the current produced by the cell is dissipated in ***internal losses***.

Single p-n junction crystalline silicon devices are now approaching the theoretical limiting power efficiency of 33. 7%, noted as the Shockley-Queisser limit in 1961. In the extreme, with an infinite number of layers, the corresponding limit is 86% using concentrated sunlight.

3.4.4 Construction of Solar Cell

Solar cells share some of the same processing and manufacturing techniques as other semiconductor devices. However, the strict requirements for cleanliness and quality control of semiconductor fabrication are more relaxed for solar cells.

Polycrystalline silicon wafers are made by wire-sawing block-cast silicon ingots into 180 to 350 micrometer wafers. The wafers are usually lightly p-type-doped. A surface diffusion of n-type dopants is performed on the front side of the wafer. This forms a p-n junction a few hundred nanometers below the surface.

Anti-reflection coatings are then typically applied to increase the amount of light coupled into the solar cell. Silicon nitride has gradually replaced titanium dioxide as the preferred material, because of its excellent surface passivation qualities. It prevents carrier recombination at the cell surface. A layer several hundred nanometers thick is applied using plasma-enhanced chemical vapor deposition. Some solar cells have textured front surfaces, that like anti-reflection coatings, increase the amount of light reaching the wafer. Such surfaces were first applied to single-crystal silicon, followed by multicrystalline silicon somewhat later.

A full area metal contact is made on the back surface, and a grid-like metal contact made up of fine "fingers" and larger "bus bars" are screen-printed onto the front surface using a silver paste. This is an evolution of the so-called "wet" process for applying electrodes, first described in a US patent filed in 1981 by Bayer AG. The rear contact is formed by screen-printing a metal paste, typically aluminium. Usually this contact covers the entire rear, though some designs employ a grid pattern. The paste is then fired at several hundred degrees Celsius to form metal electrodes in ohmic contact with the silicon. Some companies use an additional electroplating step to increase efficiency. After the metal contacts are made, the solar cells are interconnected by flat wires or metal ribbons, and assembled into modules or "solar panels". Solar panels have a sheet of tempered glass on the front, and a polymer encapsulation on the back.

Solar cells are manufactured in volume in Japan, Germany, China, Malaysia and the United States, whereas Europe, China, the United States and Japan have dominated (94% or more as of 2013) in installed systems. Other nations are acquiring significant solar cell production capacity.

3.4.5 Applications of Solar Cell

Solar cells are often electrically connected and encapsulated as a module. Photovoltaic modules often have a sheet of glass on the front (sun up) side, allowing light to pass while protecting the semiconductor wafers from abrasion and impact due to wind-driven debris, rain, hail, etc. Solar cells are also usually connected in series in modules, creating an additive voltage. Connecting cells in parallel will yield a higher current. However, very significant problems exist with parallel connections. For example, shadow effects can shut down the weaker (less illuminated) parallel string (a number of series connected cells) causing substantial power loss and even damaging the weaker string because of the excessive reverse bias applied to the shadowed cells by their illuminated partners. Strings of series cells are usually handled independently and not connected in parallel, special paralleling circuits are the exceptions. Although modules can be interconnected to create an array with the desired peak DC voltage and loading ***current capacity***, using independent MPPTs (maximum power point trackers) provides a better solution. In the absence of paralleling circuits, ***shunt diodes*** can be used to reduce the power loss due to shadowing in arrays with series/parallel connected cells.

To make practical use of the solar-generated energy, the electricity is most often fed into the electricity grid using inverters (grid-connected photovoltaic systems); in stand-alone systems, batteries are used to store the energy that is not needed immediately. Solar panels can be used to power or recharge portable devices.

3.4.6 Conclusion

Though solar cell has some disadvantage associated it (for example, high cost of installation, low efficiency, etc.), but the disadvantages are expected to overcome as the technology advances, since the technology is advancing, the cost of solar plates as well as the installation cost will decrease down, so that everybody can afford to install the system. Furthermore, the government is laying much emphasis on the solar energy so after some years we may expect that every household and also every electrical system is powered by solar or the renewable energy source.

New Words and Expressions

[1] **photovoltaic cell**：光电池。它是一种在光的照射下产生电动势的半导体元件。

[2] **panel**：仪表板、嵌板。

[3] **direct current (DC)**：直流电。它是指方向不随时间作周期性变化的电流。

[4] **alternating current (AC)**：交流电。它是指大小和方向随时间作周期性变化的电压或电流。

[5] **photo-sensitive**：感光的。

[6] **silicon cells**：硅光电池、硅电池。一种直接把光能转换成电能的半导体器件。

[7] **geosynchronous orbit**：同步轨道。地球同步轨道，俗称“24 小时轨道”，它是运行周期与地球自转周期(23 小时 56 分 4 秒)相同的顺行人造地球卫星轨道。

[8] **energy density**：能量密度。它是指在一定的空间或质量物质中储存能量的大小。

[9] **electroplating**：电镀。它是指利用电解原理在某些金属表面镀上一薄层其他金属或合金的过程。

[10] **vapor deposition**：蒸镀。在真空环境中，将材料加热并镀到基片上。

[11] **wattage**：瓦特数。它是功率单位，用时间内消耗的能量来表示。

[12] **bonding electron**：价电子。它是原子核外电子中能与其他原子相互作用形成化学键的电子。

[13] **donor atoms**：施主原子。在半导体层中掺入5价元素的杂质后，杂质元素会失去一个电子，或者说在里面提供了一个自由移动的电子。

[14] **acceptor atoms**：受主原子。

[15] **impurity atoms**：杂质原子。

[16] **charge neutrality**：电中性区。

[17] **thermal excitation**：热激励、热激发。

[18] **mobile electrons**：流动电子。

[19] **electric potential difference**：电势差。它是衡量单位电荷在静电场中由于电势不同所产生的能量差的物理量。

[20] **reflectance**：反射比。它是反射的能量与入射的能量之比。

[21] **conductive efficiency**：导电效率。它可反映导电性能。

[22] **voltage dependent**：电压依赖性、电位依从性。

[23] **thermodynamic efficiency**：热力效率、热力学效率。

[24] **recombination losses**：复合损耗。

[25] **fill factor**：填充因素。它是指像素上的光电二极管相对于像素表面的大小。

[26] **open circuit voltage**：开路电压。它是指在开路状态下的端电压。

[27] **short circuit current**：短路电流。它是指电力系统在运行中相与相之间或相与地(或中性线)之间发生非正常连接(短路)时流过的电流。

[28] **equivalent series resistance**：等效串联电阻。

[29] **shunt resistance**：并联电阻。几个电阻两端是用导线连通的，其主要的特征是几个电阻两端的电压相等。

[30] **current capacity**：电流容量。

[31] **internal losses**：内损失。

[32] **shunt diodes**：旁路二极管。为防止太阳能电池在强光下由于遮挡而造成其中一些因为得不到光照而成为负载，导致严重发热而受损，因此需要在太阳能电池组件输出端的两极并联旁路二极管。

3.5 Charge-coupled Device (CCD)

A ***charge-coupled device*** (***CCD***) is a device for the movement of electrical charge, usually from within the device to an area where the charge can be manipulated, for example, conversion into a ***digital value***. This is achieved by "shifting" the signals between stages within the device one at a time. CCDs move charge between capacitive bins in the device, with the shift allowing for the transfer of charge between bins. The application of CCD is shown in Fig. 3-21.

(a) (b)

Fig. 3-21 The application of CCD①

The CCD is a major piece of technology in digital imaging. In a CCD image sensor, pixels are represented by p-doped ***metal oxide semiconductor*** (MOS) capacitors. These capacitors are biased above the threshold for inversion when image acquisition begins, allowing the conversion of incoming photons into electron charges at the semiconductor-***oxide*** interface; the CCD is then used to read out these charges. Although CCDs are not the only technology to allow for light detection, CCD image sensors are widely used in professional, medical, and scientific applications where high-quality image data is required. In applications where a somewhat lower quality can be tolerated, such as webcams, cheaper active pixel sensors (CMOS) are generally used.

---

① Kasap S O. Optoelectronics and Photonics: Principles & Practices[M]. 2nd ed. New Jersey: Prentice Hall, Inc., 2012:434.

3.5.1 History and Development of CCD

The charge-coupled device was invented in 1969 at AT&T Bell Labs by Willard Boyle and George E. Smith. The lab was working on semiconductor ***bubble memory*** when Boyle and Smith conceived of the design of what they termed, in their notebook, "Charge 'Bubble' Devices", a description of how the device could be used as a ***shift register***. The essence of the design was the ability to transfer charge along the surface of a semiconductor from one ***storage capacitor*** to the next. The concept was similar in principle to the ***bucket-brigade device*** (BBD), which was developed at Philips Research Labs during the late 1960s. The first patent (4085456) on the application of CCDs to imaging was assigned to Michael Tompsett. Smith and Boyle were only thinking of making a memory device and did not conceive of imaging in their notebook entry or patent, or participate in the development of CCD imagers or cameras as incorrectly described in their Nobel citation. This was first done by Michael Tompsett.

The initial paper describing the concept listed possible uses as a memory, a delay line, and an imaging device. The first experimental device demonstrating the principle was a row of closely spaced metal squares on an oxidized silicon surface electrically accessed by wire bonds.

The first working CCD made with integrated circuit technology was a simple 8-bit shift register. This device had input and output circuits and was used to demonstrate its use as a shift register and as a crude eight pixel linear imaging device. Development of the device progressed at a rapid rate. By 1971, Bell researchers Michael F. Tompsett et al. were able to capture images with simple linear devices. Bell Laboratories led by Michael F. Tompsett were the first to demonstrate CCD imagers and CCD cameras leading to a 525 line camera. Several companies, including Fairchild Semiconductor, RCA and Texas Instruments, picked up on the invention and began development programs. Fairchild's effort, led by ex-Bell researcher Gil Amelio, was the first with commercial devices, and by 1974 had a linear 500-element device and 2-D 100×100 pixel devices. Steven Sasson, an electrical engineer working for Kodak, invented the first digital still camera using a Fairchild 100 × 100 CCD in 1975. The first KH-11 KENNAN ***reconnaissance satellite*** equipped with charge-coupled device array (800×800 pixels) technology for imaging was launched in December, 1976. Under the leadership of Kazuo Iwama, Sony also started a large development effort on CCDs involving a significant investment. Eventually, Sony managed to mass-produce CCDs for their ***camcorders***. Before this happened, Iwama died in August, 1982; subsequently, a CCD chip was placed on his tombstone to acknowledge his contribution.

In January, 2006, Boyle and Smith were awarded the National Academy of Engineering Charles Stark Draper Prize, and in 2009 they were awarded the Nobel Prize for physics, for their invention of the CCD concept. Michael F. Tompsett was awarded the 2010 National Medal of Technology and Innovation for pioneering work and electronic technologies including the design and development of the first charge coupled device (CCD) imagers. He was also awarded the 2012 IEEE Edison Medal "For pioneering contributions to imaging devices including CCD imagers, cameras and ***thermal imagers***".

3.5.2 Basic Operation of CCD

In a CCD for capturing images, there is a photoactive region (an epitaxial layer of silicon), and a transmission region made out of a shift register (the CCD, properly speaking). One moment of a CCD imaging sensor is shown in Fig. 3-22.

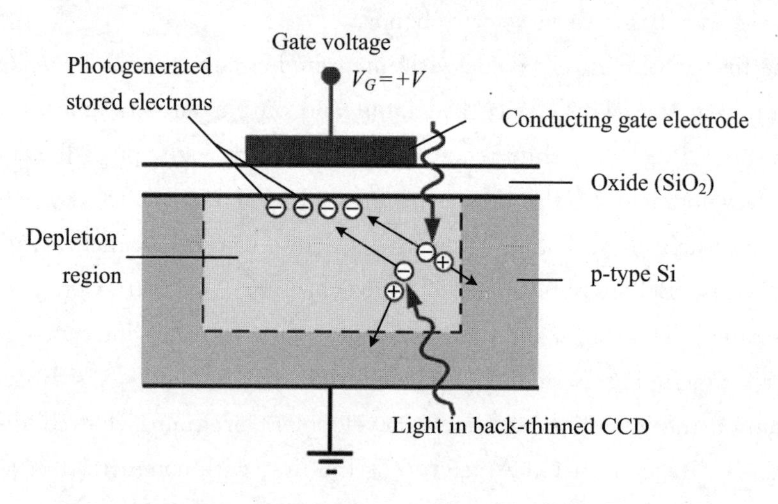

Fig. 3-22 One element of a CCD imaging sensor ①

① Kasap S O. Optoelectronics and Photonics: Principles & Practices[M]. 2nd ed. New Jersey: Prentice Hall, Inc., 2012:435,436.

The principle of CCD is shown in Fig. 3-23 and Fig. 3-24. An image is projected through a lens onto the capacitor array (the photoactive region), causing each capacitor to accumulate an electric charge proportional to the light intensity at that location. A one-dimensional array, used in line-scan cameras, captures a single slice of the image, while a two-dimensional array, used in video and still cameras, captures a two-dimensional picture corresponding to the scene projected onto the ***focal plane*** of the sensor. Once the array has been exposed to the image, a control circuit causes each capacitor to transfer its contents to its neighbor (operating as a shift register). The last capacitor in the array dumps its charge into a charge amplifier, which converts the charge into a voltage. By repeating this process, the controlling circuit converts the entire contents of the array in the semiconductor to a sequence of voltages. In a digital device, these voltages are then sampled, digitized, and usually stored in memory; in an analog device (such as an analog video camera), they are processed into a continuous analog signal (e. g. , by feeding the output of the charge amplifier into a low-pass filter) which is then processed and fed out to other circuits for transmission, recording, or other processing.

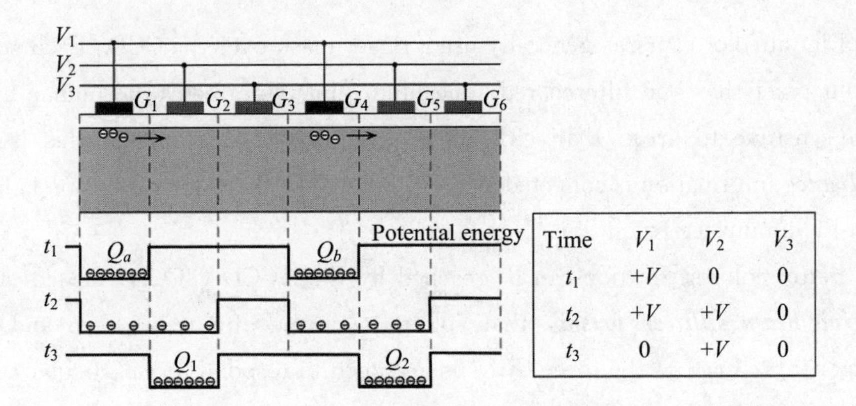

Time	V_1	V_2	V_3
t_1	$+V$	0	0
t_2	$+V$	$+V$	0
t_3	0	$+V$	0

Fig. 3-23 Transfer of charge from one well to another by clocking the gate voltages

The figure shows the gate voltage sequences in a three phase CCD (Schematic only). ①

① Kasap S O. Optoelectronics and Photonics: Principles & Practices[M]. 2nd ed. New Jersey: Prentice Hall, Inc. , 2012:435,436.

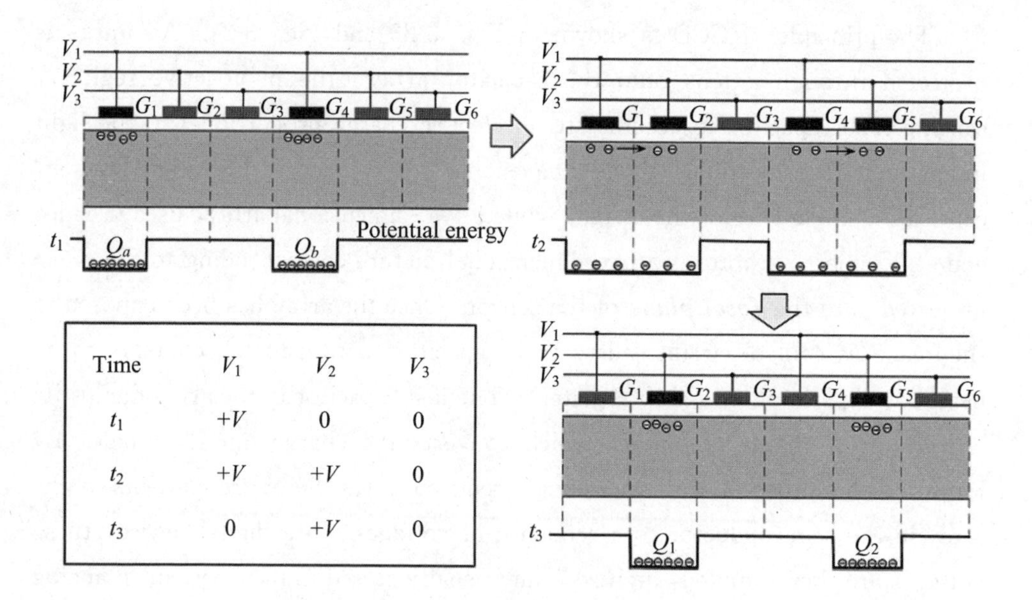

Time	V_1	V_2	V_3
t_1	$+V$	0	0
t_2	$+V$	$+V$	0
t_3	0	$+V$	0

Fig. 3-24 The process of the charge transfer①

3.5.3 Color Cameras

Digital color cameras generally use a Bayer mask over the CCD. Each square of four pixels has one filtered red, one blue, and two green (the human eye is more sensitive to green than either red or blue). The result of this is that ***luminance*** information is collected at every pixel, but the color resolution is lower than the luminance resolution.

Better color separation can be reached by three-CCD (3CCD) devices and a ***dichroic beam splitter prism***, that splits the image into red, green and blue components. Each of the three CCDs is arranged to respond to a particular color. Many professional video camcorders, and some semi-professional camcorders, use this technique, although developments in competing CMOS technology have made CMOS sensors, both with ***beam-splitters*** and bayer filters, increasingly popular in high-end video and ***digital cinema*** cameras. Another advantage of 3CCD over a Bayer mask device is higher quantum efficiency (and therefore higher light

① Kasap S O. Optoelectronics and Photonics: Principles & Practices[M]. 2nd ed. New Jersey: Prentice Hall, Inc., 2012:436.

sensitivity for a given aperture size). This is because in a 3CCD device most of the light entering the aperture is captured by a sensor, while a Bayer mask absorbs a high proportion (about 2/3) of the light falling on each CCD pixel.

For still scenes, for instance, in microscopy, the resolution of a Bayer mask device can be enhanced by ***microscanning*** technology. During the process of color co-site sampling, several frames of the scene are produced. Between acquisitions, the sensor is moved in pixel dimensions, so that each point in the visual field is acquired consecutively by elements of the mask that are sensitive to the red, green and blue components of its color. Eventually every pixel in the image has been scanned at least once in each color and the resolution of the three channels become equivalent (the resolutions of red and blue channels are quadrupled while the green channel is doubled).

3.5.4 Construction of CCD

The photoactive region of a CCD is, generally, an epitaxial layer of silicon. It is lightly p-doped (usually with boron) and is grown upon a substrate material, often p++. In buried-channel devices, the type of design utilized in most modern CCDs, certain areas of the surface of the silicon are ion implanted with phosphorus, giving them an n-doped designation. This region defines the channel in which the photogenerated charge packets will travel. Simon Sze details the advantages of a buried-channel device.

This thin layer (0.2～0.3 micron) is fully depleted and the accumulated photogenerated charge is kept away from the surface. This structure has the advantages of higher transfer efficiency and lower dark current, from reduced surface recombination. The penalty is smaller charge capacity, by a factor of 2～3 compared to the surface-channel CCD.

The gate oxide, i.e., the capacitor dielectric, is grown on top of the epitaxial layer and substrate. Later in the process, polysilicon gates are deposited by chemical vapor deposition, patterned with photolithography, and etched in such a way that the separately phased gates lie perpendicular to the channels. The channels are further defined by utilization of the LOCOS (Local Oxidation of Silicon) process to produce the channel stop region.

Channel stops are thermally grown oxides that serve to isolate the charge packets in one column from those in another. These channel stops are produced before the polysilicon gates, as the LOCOS process utilizes a high-temperature step that would destroy the gate material. The channel stops are parallel to, and exclusive of, the channel, or "charge carrying", regions.

Channel stops often have a p+ doped region underlying them, providing a further barrier to the electrons in the charge packets (this discussion of the physics of CCD devices assumes an electron transfer device, though hole transfer is possible).

The clocking of the gates, alternately high and low, will forward and reverse bias the diode that is provided by the buried channel (n-doped) and the epitaxial layer (p-doped). This will cause the CCD to deplete, near the p-n junction and will collect and move the charge packets beneath the gates — and within the channels — of the device.

CCD manufacturing and operation can be optimized for different uses. The above process describes a frame transfer CCD. While CCDs may be manufactured on a heavily doped p++ wafer, it is also possible to manufacture a device inside p-wells that have been placed on an n-wafer. This second method, reportedly, reduces smear, dark current, and infrared and red response. This method of manufacture is used in the construction of interline-transfer devices.

Another version of CCD is called a peristaltic CCD. In a peristaltic charge-coupled device, the charge-packet transfer operation is analogous to the peristaltic contraction and dilation of the digestive system. The peristaltic CCD has an additional implant that keeps the charge away from the silicon/silicon dioxide interface and generates a large lateral electric field from one gate to the next. This provides an additional driving force to aid in transfer of the charge packets.

3.5.5 Complementary Metal Oxide Semiconductor (CMOS)

A ***complementary metal oxide semiconductor*** (***CMOS***), as shown in Fig. 3-25, is an integrated circuit design on a ***printed circuit board*** (***PCB***) that uses semiconductor technology. The PCB has microchips and a layout of electric circuits that connect the chips. All circuit boards are typically either CMOS chips, n-type metal oxide semiconductor (NMOS) logic, or ***transistor-transistor logic*** (***TTL***) chips. The CMOS chip is most commonly used, as it produces less heat and requires less electricity than the others.

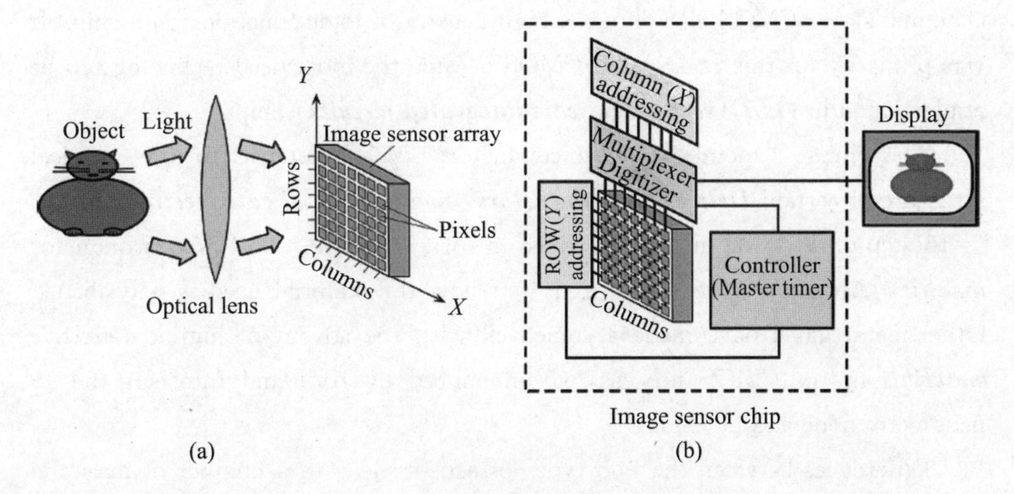

Fig. 3-25 Simple flowchart of CMOS technology

(a) The basic image sensing operation using an array of photosensitive pixels. (b) The image sensor chip that incorporates the auxiliary electronics that run the sensor array. ①

CMOS is used in static RAM, digital logic circuits, microprocessors, microcontrollers, image sensors, and the conversion of computer data from one file format to another. Most configuration information on newer CPUs is stored on one CMOS chip. The configuration information on a CMOS chip is called the real-time clock/nonvolatile RAM (RTC/NVRAM) chip, which works to retain data when the computer is shut off.

CMOS sometimes is also referred to as complementary-symmetry metal-oxide-semiconductor (or COS-MOS). The phrase "complementary-symmetry" refer to the fact that the typical digital design style with CMOS uses complementary and symmetrical pairs of p-type and n-type metal oxide semiconductor field effect transistors (MOSFETs) for logic functions.

Two important characteristics of CMOS devices are high ***noise immunity*** and low ***static power consumption***. Since one transistor of the pair is always off, the series combination draws significant power only momentarily during switching between on and off states. Consequently, CMOS devices do not produce as much waste heat as other forms of logic, for example transistor-transistor logic (TTL)

① Kasap S O. Optoelectronics and Photonics: Principles & Practices[M]. 2nd ed. New Jersey: Prentice Hall, Inc., 2012:432.

or NMOS logic, which normally have some standing current even when not changing state. CMOS also allows a high density of logic functions on a chip. It was primarily for this reason that CMOS became the most used technology to be implemented in ***VLSI*** (***very large scale integrated circuits***) chips.

The phrase "metal-oxide-semiconductor" is a reference to the physical structure of certain ***field-effect transistors***, having a ***metal gate electrode*** placed on the top of an oxide insulator, which in turn is on the top of a semiconductor material. Aluminium was once used, but now the material used is polysilicon. Other metal gates have made a comeback with the advent of high-k ***dielectric materials*** in the CMOS process, as announced by IBM and Intel for the 45 nanometre node and beyond.

Differences between the two types of sensors lead to a number of pros and cons:

(1) CCD sensors create high-quality, low-noise images. CMOS sensors are generally more susceptible to noise.

(2) Because each pixel on a CMOS sensor has several transistors located next to it, the ***light sensitivity*** of a CMOS chip is lower. Many of the photons hit the transistors instead of the photodiode.

(3) CMOS sensors traditionally consume little power. CCDs, on the other hand, use a process that consumes lots of power. CCDs consume as much as 100 times more power than an equivalent CMOS sensor.

(4) CCD sensors have been mass produced for a longer period of time, so they are more mature. They tend to have higher quality pixels, and more of them.

Although numerous differences exist between the two sensors, they both play the same role in the camera — they turn light into electricity. For the purpose of understanding how a digital camera works, you can think of them as nearly identical devices.

3.5.6 Conclusion

There are a relatively few manufacturers worldwide who maintain CCD fabrication facilities. CMOS imagers are now used for most imaging applications (cell phones, consumer electronics, etc.) which drive the imaging market much more than the scientific community. However, large pixel image sensors will likely remain CCDs as CMOS pixel sizes continue to shrink. Large pixels are required for most scientific and industrial imaging applications due to their larger dynamic range, although progress continues to improve the full-well capacity of smaller pixels for all image sensors. Very large area scientific CCDs are in demand and continue to grow in size. The simplicity of CCD devices (fewer transistors and other structures compared to CMOS sensors) tends to produce higher yield at very large areas.

New Words and Expressions

[1] **charge-coupled device (CCD)**：电荷耦合器件。

[2] **digital value**：数字量。它是物理量的一种。数字量是不随时间连续变化的量，抗干扰能力强，表达数据比模拟量精确。

[3] **metal oxide semiconductor**：金属氧化物半导体。

[4] **oxide**：氧化物。它是负价氧和另外一个化学元素组成的二元化合物。

[5] **bubble memory**：磁泡存储器。在某些磁性石榴石单晶薄膜中，其易磁化方向是与膜面垂直的。在一定条件下，这些薄膜形成磁化向量与膜面垂直的磁畴；因其形状似水泡而称为磁泡。

[6] **shift register**：移位寄存器。在数字电路中，用来存放二进制数据或代码的电路称为寄存器。

[7] **storage capacitor**：储存电容。

[8] **bucket-brigade device**：斗链器件。

[9] **reconnaissance satellite**：侦察卫星、勘测卫星。它是窃取军事情报的卫星，它既能监视，又能窃听，是个名副其实的“超级间谍”。

[10] **camcorder**：便携式摄像机。

[11] **thermal imagers**：热成像仪、热成像器。它是能反映出千分之一度的温差，并将其转换为电信号，进而在显示器上生成热图像和温度值，并可以对温度值进行计算的一种检测设备，使用时只需指向目标，对焦仪器，它就会自动调整温度范围来显示清晰鲜明的图像。

[12] **focal plane**：焦平面。过第一焦点（前焦点或物方焦点）且垂直于系统主光轴的平面称第一焦平面，又称前焦面或物方焦面。过第二焦点（后焦点或象方焦点）且垂直于系统主光轴的平面称第二焦平面，又称后焦面或象方焦面。

[13] **luminance**：亮度。

［14］ **dichroic beam splitter**：分色光束分离器。

［15］ **prism**：棱镜。用以分光或使光束发生色散。

［16］ **beam-splitters**：分光板。

［17］ **digital cinema**：数字电影，又称数码电影。它是使用数码技术制作、发行、传播的电影。其载体不再是以胶片为载体，发行方式也不再是拷贝，而代之以数字文件形式，通过网络、卫星直接传送到电影院以及家庭。

［18］ **microscanning**：微扫描。

［19］ **complementary metal oxide semiconductor（CMOS）**：互补金属氧化物半导体。

［20］ **printed circuit board（PCB）**：印刷电路板。

［21］ **transistor-transistor logic（TTL）**：晶体管。它是一种固体半导体器件，可以用于检波、整流、放大、开关、稳压、信号调制等。

［22］ **noise immunity**：抗干扰度。

［23］ **static power consumption**：静态功耗。它是指漏电流功耗，是电路状态稳定时的功耗。

［24］ **VLSI(very large scale integrated circuits)**：超大规模集成电路。

［25］ **field-effect transistors**：场效应晶体管。它由多数载流子参与导电。

［26］ **metal gate electrode**：金属栅电极。

［27］ **dielectric materials**：介质材料。

［28］ **light sensitivity**：光敏性。它是指电阻由于光照导致其自身电阻阻值改变的性质。

Reference

[1] Kasap S O. Optoelectronics and Photonics: Principles & Practices [M]. 2nd ed. New Jersey: Prentice Hall, Inc., 2012.

Questions

1. Please describe the basic working principle of PMT and especially

illustrate the concept of photoelectric effect and secondary emission.

2. What is the typically properties of the materials for making of photocathodes?

3. What is the structure of photodiode and how the photodiode works?

4. How does the dark current generate?

5. Which three steps does the solar cell work?

6. Please find some applications of solar cells in our daily lives.

7. What are the two most important characteristics of CMOS devices?

8. What is the difference between CCD and CMOS?

Chapter 4 Optical Fibers and Dielectric Waveguides

Optical fibers and optical waveguides are channels through which light waves propagate in optical devices, similar to wires in electronics. The basic principle is to confine light in the channel through total internal reflection, which occurs when light travels from a high-refractive-index material to a low-refractive-index material. Therefore, the waveguide usually has a sandwich structure, with a high refractive index of the core material in the middle and a low refractive index of the cladding materials above and below. This chapter mainly discusses the basic principle of symmetric planar dielectric slab waveguide and basic properties for step index fiber such as numerical aperture, dispersion, and attenuation. The manufacture process is also introduced.

Before we start introducing the content of this chapter, let's first introduce Charles Kuen Kao who is the "father of optical fiber".

Charles Kuen Kao (November 4, 1933 ~ September 23, 2018), born in Jinshan County, Jiangsu Province (now Jinshan District, Shanghai), is a Chinese physicist, educator, expert in optical fiber communication and electrical engineering, and former president of the Chinese University of Hong Kong. He is known as the "father of optical fiber", "father of optical fiber communication" and "godfather of broadband". Kuen Kao went to Britain to study electrical engineering in 1954. He received his bachelor's and doctor's degrees from University College London in 1957 and 1965. In 1970, he joined the Chinese University of Hong Kong to organize the Department of electronics and served as the dean of the Department. From 1987 to 1996, he was the third president of the Chinese University of Hong Kong. He was elected academician of the National Academy of Engineering in 1990. In 1992, he was elected academician of

Academia Sinica. In 1996, he was elected as a foreign academician of the Chinese Academy of Sciences. He was elected academician of the Royal Society in 1997. He won the Nobel Prize in physics in 2009 and awarded the grand Bauhinia medal in 2010. He was elected honorary academician of the Hong Kong Academy of Sciences in 2015.

Kao has long been engaged in the research on the application of optical fiber in the field of communication. Since 1957, Kao has been engaged in the research on the application of optical fiber in the field of communication. In 1964, he proposed to replace current with light and wire with glass fiber in telephone network. In 1965, Kao and Hockham jointly concluded that the basic limit of glass optical attenuation is less than 20 dB/km (dB/km, which is a method to measure signal attenuation over distance), which is the key threshold of optical communication. However, in this measurement, optical fibers usually exhibit optical losses of up to 1000 dB/km or more. This conclusion opens the mileage of finding low loss materials and suitable fibers to meet this standard.

In 1966, Kao published a paper entitled optical frequency dielectric fiber surface waveguide, which creatively put forward the basic principle of the application of optical fiber in communication, and described the structure and material characteristics of insulating fiber required for long-range and high information optical communication. In short, as long as the problems of glass purity and composition are solved, optical fibers can be made of glass, so as to transmit information efficiently. After this idea was put forward, some people called it incredible, and others praised it. However, in the debate, Kao's idea has gradually become a reality: Optical fibers made of quartz glass are more and more widely used, and a revolution in optical fiber communication has been set off all over the world.

Kao played a leading role in the early stage of optical communication engineering and commercial implementation. In 1969, Kao measured the inherent loss of fused silica at 4 dB/km, which is the first evidence of the effectiveness of ultra transparent glass in transmitting signals. Driven by his efforts, the world's first 1 km long optical fiber came out in 1971, and the first optical fiber communication system was put into operation in 1981.

In the mid-1970s, Kao made a pioneering study on the fatigue strength of glass fiber. When he was appointed as the first executive scientist of ITT, Kao launched the "terabit technology" program to solve the high-frequency limitation of signal processing. Therefore, Kao is also known as the "father of terabit technology concept"(see Fig. 4-1).

Fig. 4-1 Professor Charles Kuen Kao

Kao also developed the auxiliary subsystem needed to realize optical fiber communication. He has done a lot of research in many fields, such as the structure of single-mode fiber, the strength and durability of fiber, fiber connectors and couplers and diffusion equalization characteristics, and these research results are the key to the success of transmitting the signal in gigabits per second to the distance in 10000 meters without amplification.

4.1 Symmetric Planar Dielectric Slab Waveguide

4.1.1 Waveguide Condition

In order to understand the general properties of light wave propagation in ***optical waveguides***, we first consider the planar dielectric slab waveguide shown in Fig. 4-2, which is the simplest waveguide in terms of processable analysis. A dielectric plate with a thickness of $2a$ and a refractive index of n_1 is sandwiched between two semi-infinite regions with a ***refractive index*** of n_2 ($n_2 < n_1$). The region with higher refractive index (n_1) is called the core, and the region with lower refractive index (n_2) sandwiching the core is called the cladding.

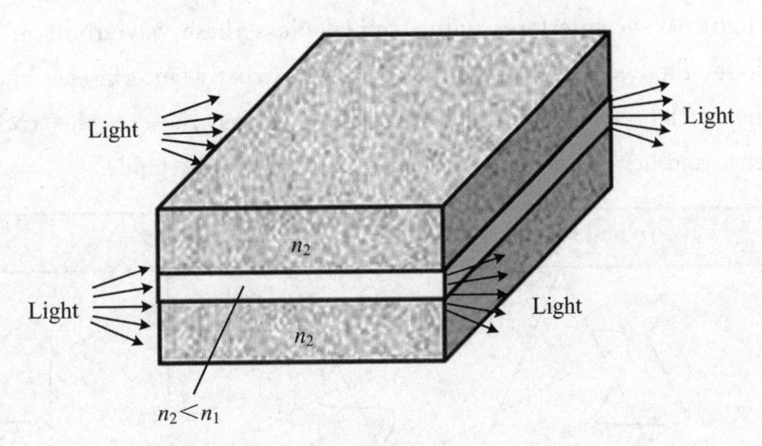

Fig. 4-2 Planar dielectric slab waveguide

The slab dielectric waveguide has a central rectangular region with a higher refractive index n_1 than the surrounding region with a refractive index n_2. It is assumed that the waveguide is infinitely wide and the thickness of the central region is $2a$. One end of it is illuminated by a monochrome light source.

A light ray can readily propagate in a zigzag manner along this waveguide, provided that it can experience ***total internal reflection*** (***TIR***) at the dielectric boundaries. It seems that any light wave that has an angle of incidence (θ) greater than the critical angle (θ_c) for TIR will be propagated. However, this applies only to very thin light beams with a diameter much smaller than the plate thickness $2a$. We consider the actual situation when the entire end of the waveguide is illuminated, as shown in Fig. 4-2. To simplify the analysis, we assume that light is emitted from a line source in a medium with a refractive index of n_1. Generally speaking, the refractive index of the transmitting medium will be different from n_1, but this will only affect the amount of light coupled into the waveguide.

Consider the propagation of plane wave type light ray in dielectric slab waveguide, as shown in Fig. 4-3. We assume that the electric field E is along x, parallel to the interface and perpendicular to z. The reflection of light passing through the boundary of the core cladding (n_1/n_2) is guided in a zigzag manner along the guide axis z. The result is the effective propagation of the electric field E along z. The figure also shows the constant phase wavefronts perpendicular to the propagation direction on the ray. This special light is reflected at B and then at C. After reflection at C, the wavefront at C overlaps the wavefront at A on the original light. Wave interferes with itself. Unless these wavefronts at A and C are in phase, the two wavefronts will produce destructive interference and destroy each other. Only certain reflection angle θ gives rise to the constructive interference and hence only certain waves can exist in the guide.

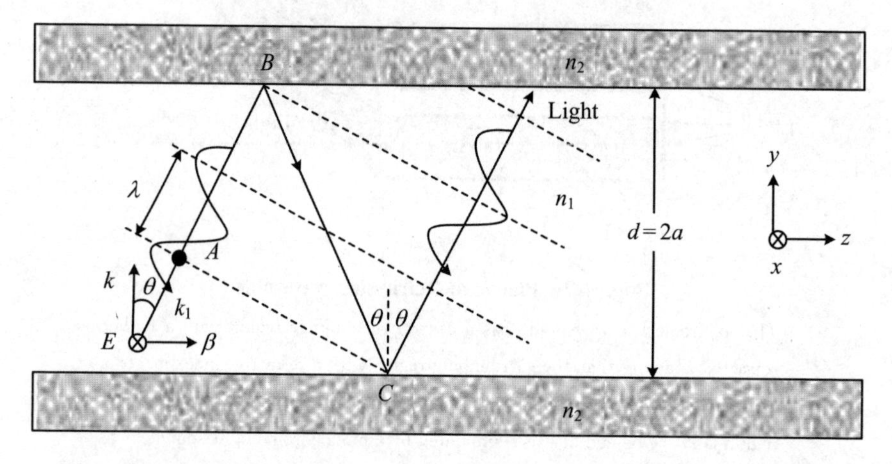

Fig. 4-3 Propagation of plane wave type light ray in dielectric slab waveguide

A light ray traveling in the guide must interfere constructively with itself to propagate successfully. Otherwise, destructive interference will destroy the wave. The x axis is into the paper.

The phase difference between points A and C corresponds to the optical path length $AB+BC$. In addition, there are two total internal reflections (TIRs) at B and C, each introducing a further phase change of ϕ. Suppose that k_1 is the wavevector in n_1, i. e., $k_1=kn_1=2\pi n_1/\lambda$, in which k and λ are the free space wavevector and wavelength. For constructive interference, the phase difference between A and C must be a multiple of 2π.

$$\Delta\phi(AC) = k_1(AB+BC) - 2\phi = m(2\pi) \quad (4\text{-}1)$$

in which $m=0,1,2,\cdots$ is an integer.

$AB+BC$ can be easily evaluated from geometrical considerations.

$BC=d/\cos\theta$ and $AB=BC\cos(2\theta)$. Therefore,

$$AB+BC = BC\cos(2\theta)+BC = BC[(2\cos^2\theta-1)+1] = 2d\cos\theta$$

Thus, for wave propagation along the guide we need:

$$k_1(2d\cos\theta) - 2\phi = m(2\pi) \quad (4\text{-}2)$$

Obviously, for a given integer m, there are only certain θ and ϕ values can satisfy this equation. But ϕ depends on θ and also depends on the polarization state of the wave (direction of the electric field). So for each m, there will be one allowed angle θ_m and one corresponding ϕ_m. Dividing equation (4-2) by 2 we can obtain the waveguide condition:

$$\left[\frac{2\pi n_1(2a)}{\lambda}\right]\cos\theta_m - \phi_m = m\pi \tag{4-3}$$

in which ϕ_m indicates that ϕ is a function of the incidence angle θ_m.

It might think that the above treatment is artificial because we took a narrow angle for θ. The results show that the general waveguide conditions of guided waves can be derived from equation (4-3) whether we use a narrow or wider angle, one or more rays. As shown in Fig. 4-4, if we take two arbitrary parallel rays into the waveguide, we can get the same condition. The rays 1 and 2 represent the same "planwave" when they are initially in phase. Ray 1 then undergoes two reflections at A and B, and then propagates parallel to ray 2 again. Unless the wavefront on ray 1 reflected at B is in phase with that on ray 2 at B', the two rays will destroy each other. Both rays are initially in phase. Ray 1 is at A before reflection and ray 2 is at A'. After two reflections, ray 1 at B has a phase $k_1 AB - 2\phi$. Ray 2 has a phase $k_1(A'B')$ at B. The difference between the

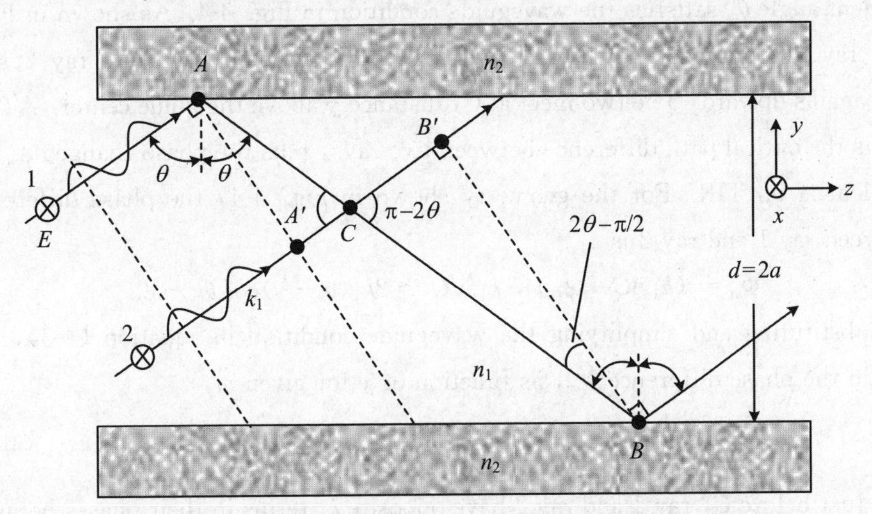

Fig. 4-4 Taking two arbitrary parallel rays into the waveguide

Two arbitrary waves 1 and 2 that are initially in phase must remain in phase after reflection. Otherwise, the two will interfere with each other and cancel each other.

two phases must be $m(2\pi)$ and leads to the waveguide condition in equation (4-3).

The wave vector k_1 can be decomposed into two propagation constants β and κ, along and perpendicular to the guide axis z as shown Fig. 4-2. For those satisfying waveguide conditions θ_m, we define:

$$\beta_m = k_1 \sin\theta_m = \left(\frac{2\pi n_1}{\lambda}\right)\sin\theta_m \tag{4-4}$$

$$\kappa_m = k_1 \cos\theta_m = \left(\frac{2\pi n_1}{\lambda}\right)\cos\theta_m \tag{4-5}$$

The simplified analysis of the embedded waveguide condition in equation (4-3) clearly shows that only some reflection angles are allowed in the waveguide corresponding to $m=0, 1, 2, \cdots$. We notice that the higher the value of m, θ_m is lower. Each m choice results in a different reflection angle θ_m. The simultaneous interpreting of the different propagation along the equation (4-4) is caused by each m value.

If the interference of many rays were taken into account, as shown in Fig. 4-3, a fixed electric field mode of the sheet wave in the y direction would be found, and this field pattern propagates along the z axis, with a propagation constant β_m. We can illustrate this by considering the synthesis of two parallel rays whose incident angle θ_m satisfies the waveguide condition in Fig. 4-4. As shown in Fig. 4-5, ray 1 propagates downward after being reflected at A, while ray 2 still propagates upward. The two meet at C, distance y above the guide center. $A'C-AC$ is the optical path difference between two rays, plus the phase changed ϕ_m for ray 1 at A on TIR. For the geometry shown in Fig. 4-4, the phase difference between ray 1 and ray 2 is

$$\Phi_m = (k_1 AC - \phi_m) - k_1 A'C = 2k_1(a-y)\cos\theta_m - \phi_m$$

by substituting and simplifying the waveguide condition in equation (4-3), we obtain the phase difference Φ_m, as function of y for given m,

$$\Phi_m = \Phi_m(y) = m\pi - \frac{y}{a}(m\pi + \phi_m) \tag{4-6}$$

Just before C, ray 1 and ray 2 have opposite k_y terms in their phases because they move in the opposite y direction. The electric fields of ray 1 and ray 2 at C are

$$E_1(y,z,t) = E_0\cos(\omega t - \beta_m z + \kappa_m y + \Phi_m)$$

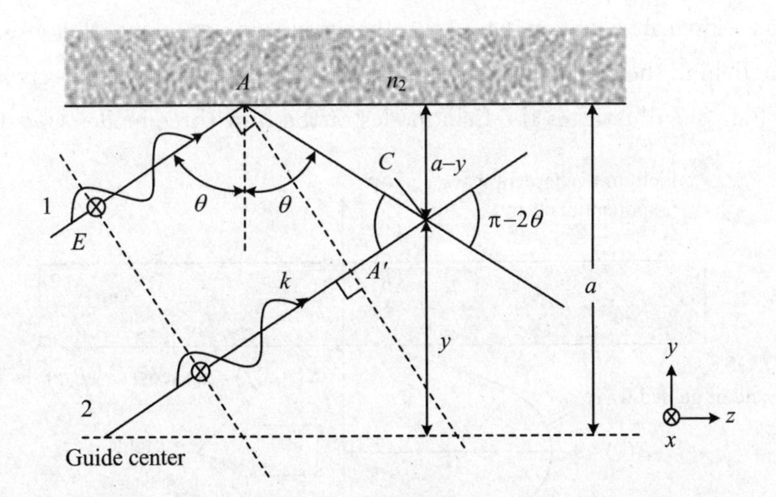

Fig. 4-5 Interference of waves

Interference of waves such as ray 1 and ray 2 leads to a standing wave pattern along the y direction which propagates along z.

and

$$E_2(y,z,t) = E_0 \cos(\omega t - \beta_m z - \kappa_m y)$$

These two waves interfere to give:

$$E(y,z,t) = 2E_0 \cos\left(\kappa_m y + \frac{1}{2}\Phi_m\right) \cos\left(\omega t - \beta_m z + \frac{1}{2}\Phi_m\right) \tag{4-7}$$

Equation (4-7) is a traveling wave along the z direction, and its amplitude along the y direction is modulated by the $\cos\left(\kappa_m y + \frac{1}{2}\Phi_m\right)$ term due to the $\cos(\omega t - \beta_m z)$ term. The $\cos\left(\kappa_m y + \frac{1}{2}\Phi_m\right)$ term is independent of time and corresponds to the standing wave mode along the y direction. Since each m value gives different κ_m and Φ_m, we obtain different field modes along y for each m. Thus a light wave propagating along the guide is of the form:

$$E(y,z,t) = 2E_m(y) \cos(\omega t - \beta_m z) \tag{4-8}$$

in which $E_m(y)$ is the field distribution of a given m along y. The distribution $E_m(y)$ through the guide rail moves downward along the guide rail along z.

Fig. 4-6 shows the field mode with the lowest mode $m=0$, which has the maximum intensity at the center. The whole field distribution moves along the z direction, and the propagation vector is β_0. Also notice that the field penetrates into the cladding, which is caused by evanescent waves propagating into the cladding near the boundary. The field mode in the core presents a harmonic change as shown in equation (4-7) on the waveguide or along the y direction, while the field in the cladding is an evanescent wave field and decays exponentially with y. Fig. 4-7 illustrates the field modes of the first three modes ($m=0,1,3$).

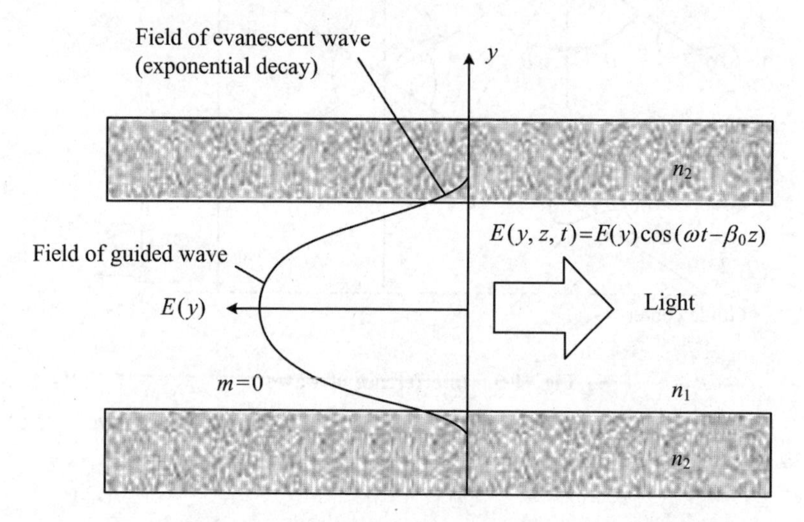

Fig. 4-6　The field modes with the lowest mode

The electric field distribution of the lowest mode traveling wave along the waveguide. This mode has $m=0$ and the lowest θ. It is often referred to as glass incident light. It has the highest phase velocity along the guide rail.

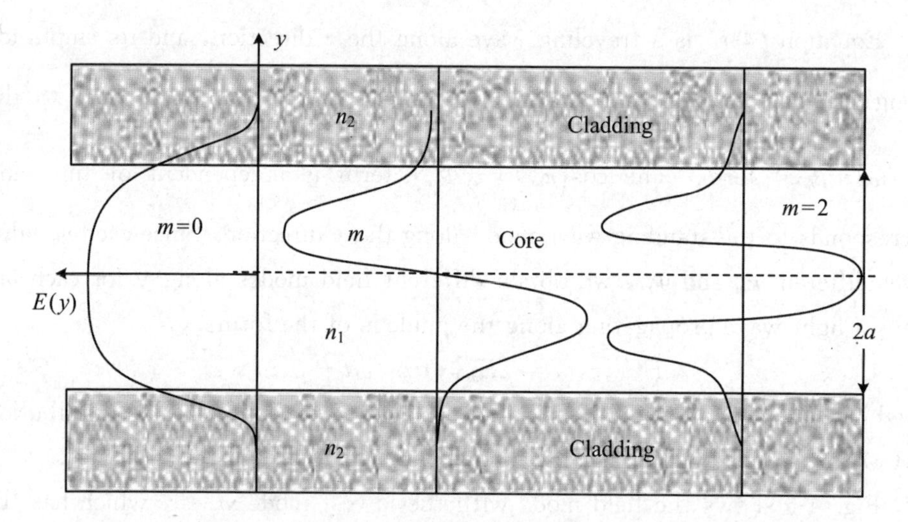

Fig. 4-7　The field modes of the first three modes

The electric field distribution of the first three modes ($m=0,1,2$) traveling waves along the waveguide. Note that the magnetic field penetration in the cladding is different.

We have seen that each m leads to an allowable θ_m value corresponding to the specific traveling wave in the z direction described in equation (4-8) and the specific wave vector m defined in equation (4-4). Each of these traveling waves has a different field mode $E_m(y)$, which constitutes a mode of propagation. The integer m identifies these patterns and is called the mode number. As shown in Fig. 4-8, the light energy can only be transmitted along the waveguide through one or more possible propagation modes. Notice that the rays have been shown to penetrate the cladding, as shown in Fig. 4-8, and reflected from the view plane in the cladding. Since θ_m is smaller for larger m, higher modes show more reflections, but they also penetrate the cladding more, as shown in Fig. 4-8. For the lowest mode $m=0$, which leads to θ_m being closest to 90° and the wave is said to travel axially. Light that is launched into the core of the waveguide can travel only along the guide in the allowed modes specified by equation (4-3). These modes will move down the guide rail at different group speeds. When they reach the end of the guide rail, they form an outgoing beam. If we emit a short-time optical pulse into the dielectric waveguide, the light emitted from the other end will be a widened optical pulse, because the light energy will propagate along the waveguide at different group velocities, as shown in Fig. 4-8. Therefore, the optical pulse propagates along the waveguide.

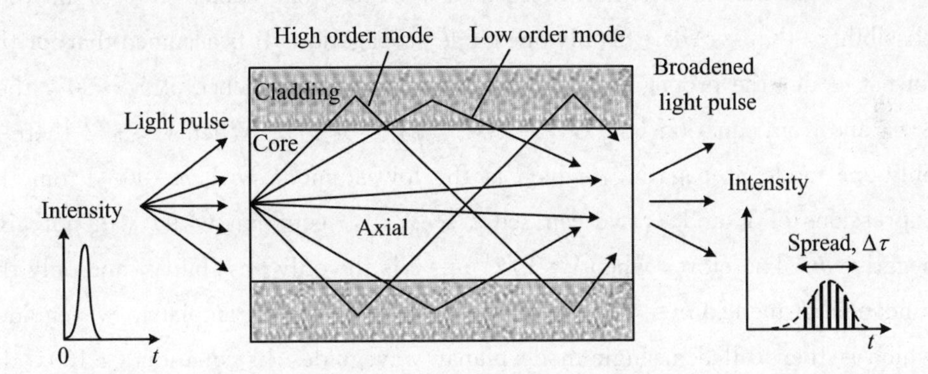

Fig. 4-8 Schematic diagram of light propagation in flat dielectric waveguide

The optical pulse entering the waveguide is decomposed into various modes and then propagated along the waveguide at different group velocities. At the end of the waveguide, these modes combine to form an output optical pulse wider than the input optical pulse.

4.1.2 Single-mode and Multi-mode Waveguides

Although the waveguide condition in equation (4-3) specifies the allowable θ_m values, θ_m must meet TIR, i. e., $\sin\theta_m > \sin\theta_c$. With the latter condition, we can only have a certain number of modes in the waveguide. From equation (4-3), we can get the expression of $\sin\theta_m$, and then apply TIR condition $\sin\theta_m > \sin\theta_c$ to indicate that the number of modes m must be satisfied:

$$m \leqslant (2V - \phi)/\pi \tag{4-9}$$

in which V, called the V-number, is a quantity defined by

$$V = \frac{2\pi a}{\lambda}(n_1^2 - n_2^2)^{1/2} \tag{4-10}$$

The V-number also has other names, V-parameter, normalized thickness, and normalized frequency. The term normalized thickness is more common in current planar waveguides, while the term V-number is more common in optical fiber (discussed later). For a given free space wavelength λ, the V-number depends on the waveguide geometry ($2a$) and waveguide properties (n_1 and n_2). Therefore, it is a characteristic parameter of the waveguide.

The question is whether there is a V value that makes $m = 0$ the only possibility, that is, there is only one mode propagation. It is assumed that for the lowest mode, the propagation is caused by the glass incidence of $\theta_m \to 90°$, then $\phi \to \pi$ and from equation (4-9), $V = (m\pi + \phi)/2$ or $\pi/2$. When $V < \pi/2$ there is only one mode propagating, which is the lowest mode with $m = 0$. From the expressions of V and ϕ, we can see $\phi \leqslant 2V$, so equation (4-9) will not give negative m. Therefore, when $V < \pi/2$, $m = 0$ is the only possibility, and only the fundamental mode ($m = 0$) propagates along the dielectric planar waveguide, which is then called a single-mode planar waveguide. In equation (4-10), the free-space wavelength λ_c leading to $V = \pi/2$ is the cut-off wavelength. Above this wavelength, there is only one-mode, that is, the basic mode, which will propagate.

4.1.3 TE and TM Modes

We have shown that for a particular mode, the variation of field strength along y, $E_m(y)$ is harmonic. Fig. 4-9 (a) and Fig. 4-9 (b) consider two possibilities of the electric field direction of the wave propagating to the core cladding boundary.

(1) The electric field is perpendicular to the incident plane (paper plane) as indicated by $E_{\perp}$, as shown in Fig. 4-9 (a). $E_{\perp}$ is along x, so that $E_{\perp}=E_x$.

(2) The magnetic field is perpendicular to the incident plane, as indicated by $B_{\perp}$, as shown in Fig. 4-9 (b). In this case, the electric field is parallel to the incident plane, represented by $E_{/\!/}$.

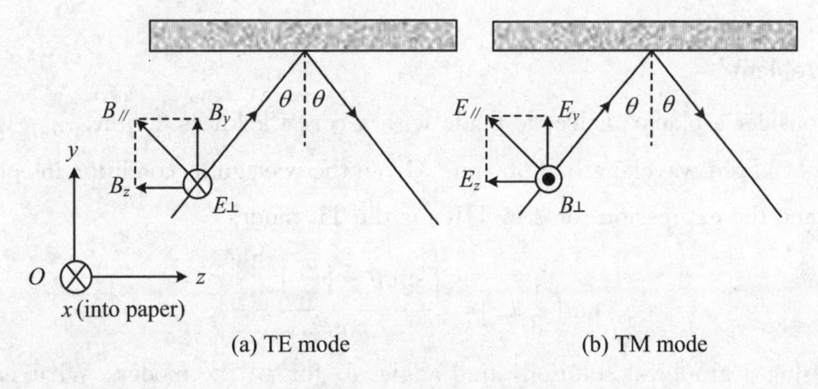

(a) TE mode (b) TM mode

Fig. 4-9 TE and TM modes

Possible modes can be classified according to (a) transverse electric field (TE) and (b) transverse magnetic field (TM). The incident plane is the focus of this paper.

Any other field direction (perpendicular to the ray path) can be decomposed into electric field components along $E_{/\!/}$ and $E_{\perp}$. The two fields undergo different phase transitions, $\phi_{/\!/}$ and $\phi_{\perp}$, so different angles θ_m are required to propagate along the waveguide. Therefore, we have a different set of patterns for $E_{/\!/}$ and $E_{\perp}$. The mode associated with $E_{/\!/}$ (or $E_{\perp}$) is called the ***transverse electric field*** mode and is represented by TE_m because $E_{\perp}$ is actually perpendicular to the propagation direction z.

The mode related to field $E_{/\!/}$ has a magnetic field $B_{\perp}$ perpendicular to the propagation direction, which is called ***transverse magnetic field*** mode, represented by TM_m. Interestingly, $E_{/\!/}$ has a field component parallel to the z axis, as shown by E_z, along the propagation direction. Obviously, E_z is a propagating longitudinal electric field. In free space, such a longitudinal electric field cannot exist, but in optical waveguides, due to interference, there may indeed be a longitudinal electric field. Similarly, TE mode, those with $B_{/\!/}$, have a magnetic field along the z direction, which propagates in the form of longitudinal waves along this direction.

The phase change ϕ that accompanies TIR depends on the polarization of the field and is different for $E_{/\!/}$ and $E_{\perp}$. The difference, however, is negligibly small for $n_1 - n_2 \ll 1$ and thus the waveguide condition and the cut-off condition can be taken to be identical for both TE and TM modes.

Example 4.1.1

Problem

Consider a planar dielectric guide with a core thickness 20 μm, $n_1 = 1.455$, $n_2 = 1.44$, light wavelength of 900 nm. Given the waveguide condition in equation (4-3) and the expression for ϕ in TIR for the TE mode:

$$\tan\left(\frac{1}{2}\phi_m\right) = \frac{\left[\sin^2\theta - \left(\frac{n_2}{n_1}\right)^2\right]^{1/2}}{\cos\theta_m}$$

Using a graphical solution, find angles θ_m for all the modes. What is your conclusion?

Solution

Consider the waveguide condition in equation (4-3), using $k_1 \cos\theta_m = \kappa$, can be written as:

$$(2a)k_1\cos\theta_m - m\pi = \phi_m$$

For example,

$$\tan\left(ak_y\cos\theta_m - m\frac{\pi}{2}\right) = \frac{\left[\sin^2\theta_m - \left(\frac{n_2}{n_1}\right)^2\right]^{1/2}}{\cos\theta_m} = f(\theta_m) \qquad (4\text{-}11)$$

The left-hand side (LHS) even simply reproduces itself whenever $m = 0, 2, 4, \cdots$. It becomes a cotangent function whenever $m = 1, 3, \cdots$ odd integer. The solutions therefore fall into odd and even m categories. Fig. 4-10 shows the right-hand side (RHS), $f(\theta_m)$ vs. θ_m as well as the LHS, tan $(ak_1\cos\theta_m - m\pi/2)$. Since the critical angle, $\theta_c = \arcsin(n_2/n_1) = 81.77°$, we can only find solutions in the range $\theta_m = 81.77° \sim 90°$. For example, the intersections for $m=0$ and $m=1$ are at 89.17°, 88.34°. We can also calculate the penetration δ_m of the field into the cladding using,

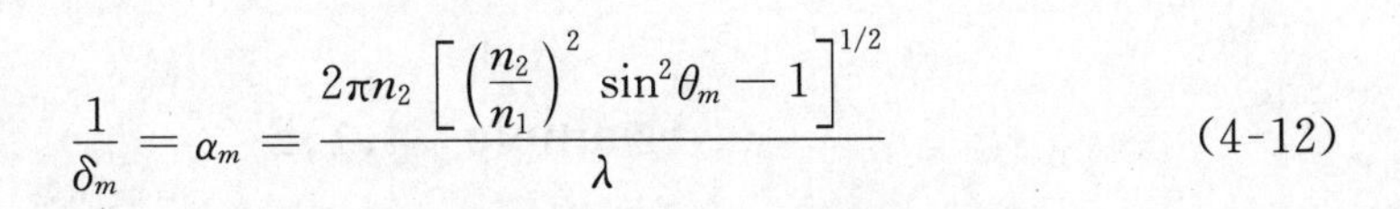

$$\frac{1}{\delta_m} = \alpha_m = \frac{2\pi n_2\left[\left(\frac{n_2}{n_1}\right)^2 \sin^2\theta_m - 1\right]^{1/2}}{\lambda} \tag{4-12}$$

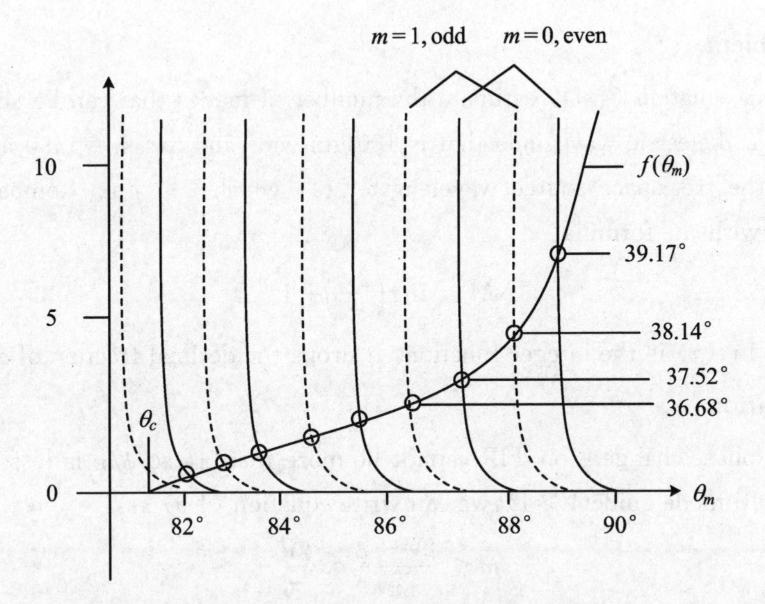

Fig. 4-10 Modes in a planar dielectric waveguide

It can be determined by plotting the LHS and the RHS of equation (4-11).

Using equation (4-11) and equation (4-12), we can find the following mode angles θ_m and corresponding penetrations δ_m of the field into the cladding. The highest mode (highest m) has substantial field penetration into the cladding. See Table 4-1.

Table 4-1 θ_m and δ_m

m	0	1	2	3	4	5	6	7	8	9
θ_m	89.2°	88.3°	87.5°	86.7°	85.9°	85°	84.2°	83.4°	82.6°	81.9°
δ_m (μm)	0.691	0.702	0.722	0.751	0.793	0.886	0.97	1.15	1.57	3.83

It is apparent that $f(\theta_m)$ intersects the LHS tangent function once for each choice of m until $\theta_m \leqslant \theta_c$. There are 10 modes.

The exact solution of equation (4-11) will show that the basic mode angle of TE mode is actually 89.172°. If we use the phase transition ϕ_m of TM mode, we will find that 89.172°, which is almost the same as ϕ_m for the TE mode.

Example 4.1.2

Problem

Using equation (4-9), estimate the number of modes that can be supported in a planar dielectric waveguide that is 100 μm wide and has $n_1=1490$ and $n_2=1470$ at the free-space source wavelength (λ), which is 1 μm. Compare your estimate with the formula:

$$M=\mathrm{In}\,t\left(\frac{2V}{\pi}\right)+1 \qquad (4\text{-}13)$$

in which $\mathrm{In}\,t(x)$ is the integer function, it drops the decimal fraction of x.

Solution

The phase change ϕ on TIR cannot be more than π, so ϕ/π is less than 1. For a multi-mode guide($V\gg 1$), we can write equation (4-9) as:

$$m\leqslant\frac{2V-\phi}{\pi}\approx\frac{2V}{\pi}$$

We can calculate V since we are given $a=50$ μm, $n_1=1490$, $n_2=1470$, and $\lambda=1\times 10^{-6}$ m or 1 μm:

$$V=(2\pi a/\lambda)(n_1^2-n_2^2)^{1/2}=76.44$$

Then, $m\leqslant 2\times 76.44/\pi=48.7$, or $m\leqslant 48$. There are about 49 modes, because we must also include $m=0$ as a mode. Using equation (4-13):

$$M=\mathrm{In}\,t\left(\frac{2\times 76.44}{\pi}\right)+1=49$$

Example 4.1.3

Problem

As shown in Fig. 4-6, the field distribution along the y direction penetrates into the cladding. Therefore, the range of electric field passing through the waveguide exceeds $2a$. In the core, the electric field is harmonic, and from the boundary to the cladding, the electric field is generated by evanescent wave and it decays exponentially according to

$$E_{\text{cladding}}(y')=L_{\text{cladding}}(0)\exp(-\alpha_{\text{cladding}}y')$$

in which $E_{\text{cladding}}(y')$ is the field in the cladding at a position y' measured from the boundary and α is the decay constant (or attenuation) for the evanescent wave in medium 2, which is given by

$$\alpha_{\text{cladding}} = \frac{2\pi n_2}{\lambda}\left[\left(\frac{n_1}{n_2}\right)\sin^2\theta_1{}^2 - 1\right]^{1/2}$$

in which λ is the wavelength in free space. For the axial mode we can take the approximation $\theta_i \rightarrow 90^\circ$, then:

$$\alpha_{\text{cladding}} = \frac{2\pi n_2}{\lambda}\left[\left(\frac{n_1}{n_2}\right)^2 \sin^2\theta_1 - 1\right]^{1/2} \approx \frac{2\pi}{\lambda}(n_1^2 - n_2^2)^{1/2} = \frac{V}{a}$$

The field in the cladding decays by a factor of e^{-1} when $y' = \delta = 1/\alpha_{\text{cladding}} = a/V$.

The extent of the field in the cladding is about δ. The total extent of the field across the whole guide is therefore $2a + 2\delta$ which is called the mode field distance (MFD) and denoted by 2ω. Thus:

$$2\omega_\nu \approx 2a + 2\frac{a}{\nu}$$

For example,

$$2\omega_\nu \approx 2a\frac{(\nu + 1)}{\nu} \tag{4-14}$$

We note that as V increases, MFD becomes the same as the core thickness, $2a$. In single-mode operation $\nu < \pi/2$ and MFD is considerably larger than $2a$. In fact at $V = \pi/2$, MFD is 1.6 times $2a$. In the case of cylindrical dielectric waveguide, e. g. , optical fibers, MFD is called the mode field diameter.

New Words and Expressions

[1] **optical waveguide**：光波导。它是引导光波在其中传播的介质装置，又称介质光波导。

[2] **refractive index**：折射率。光在真空中的传播速度与光在该介质中的传播速度之比。

[3] **total internal reflection(TIR)**：全反射。当光线从较高折射率的介质进入较低折射率的介质时，当入射角大于某一临界角时，折射光线将会消失，所有的入射光线将被反射而不进入低折射率的介质。

[4] **transverse electric field**：横电模。它是指电场完全分布在与电磁波传播方向垂直的横截面内，磁场具有传播方向分量的波形。

[5] **transverse magnetic field**：横磁模。它是指磁场完全分布在与电磁波传播方向垂直的横截面内，电场具有传播方向分量的波形。

4.2 Step Index Fiber

The general ideas for guided wave propagation in a planar dielectric waveguide can be readily extended to the step index optical fiber shown in Fig. 4-11, with only certain modifications. This is essentially a cylindrical dielectric waveguide, the refractive index n_1 of the inner core medium is greater than the refractive index n_2 of the outer dielectric cladding. The normalized index difference Δ is defined by

$$\Delta = (n_1 - n_2)/n_1 \tag{4-15}$$

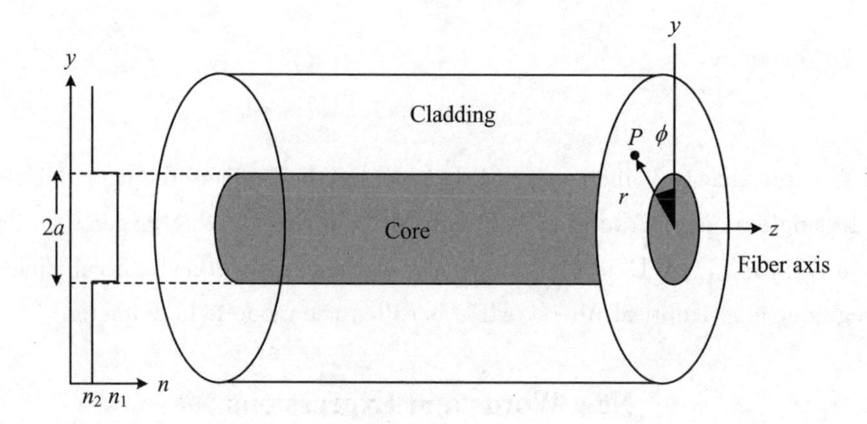

Fig. 4-11 The step index optical fiber (schematic)

The central region, the core, has greater refractive index than the outer region, the cladding. Thc fiber has cylindrical symmetry. The coordinators r, ϕ, z are used to represent any point P in the fiber. Cladding is normally much thicker than shown.

For all practical fibers used in optical communications, the difference between n_1 and n_2 is small, typically less than a few percent, so $\Delta \ll 1$.

The planar waveguide is bounded only in one dimension so that reflections occur only in the y direction. The requirement of constructive interference of waves results in the existence of distinct modes each labeled by m. As shown in Fig. 4-11, the cylindrical guide is bounded in two dimensions and the reflections occur from all the surfaces, i. e. , from a surface encountered along any radial direction r, along a radial direction at any angle ϕ to the y axis in Fig. 4-11. Since any radial direction can be represented by x and y, the reflection participates in the constructive interference of the waves in both x and y directions, so two integers, l and m, are required to exist in the guide for all possible travel, wave or guidance modes.

In the case of a planar waveguide, the guided propagating wave is imagined as rays that were zigzagging down the guide and all these rays necessarily passed through the axial plane of the guide. In addition, all waves were either TE (transverse electric) or TM (transverse magnetic). A distinctly different feature of the step index fiber from the planar waveguide is existence of rays that zigzag down the fiber without necessarily crossing the fiber axis, called skew rays. A ***meridional ray*** enters the fiber through the fiber axis and therefore also crosses the fiber axis on each reflection as it zigzags down the fiber. It travels in a plane containing the fiber axis as shown in Fig. 4-12(a). On the other hand, a ***skew ray*** enters the fiber off the fiber axis and zigzags down the fiber without crossing the axis. When viewed looking down the fiber (its projection in a plane normal to the fiber axis), it traces out a polygon around the fiber axis as illustrated in Fig. 4-12(b). Therefore, a skew ray has a helical path around the fiber axis. In a step index fiber, both meridional and skew rays produce guided modes (propagating waves) along the fiber, and each mode has a propagation constant β along z. In this planar waveguide, the guided mode generated by the meridian can be TE type or TM type.

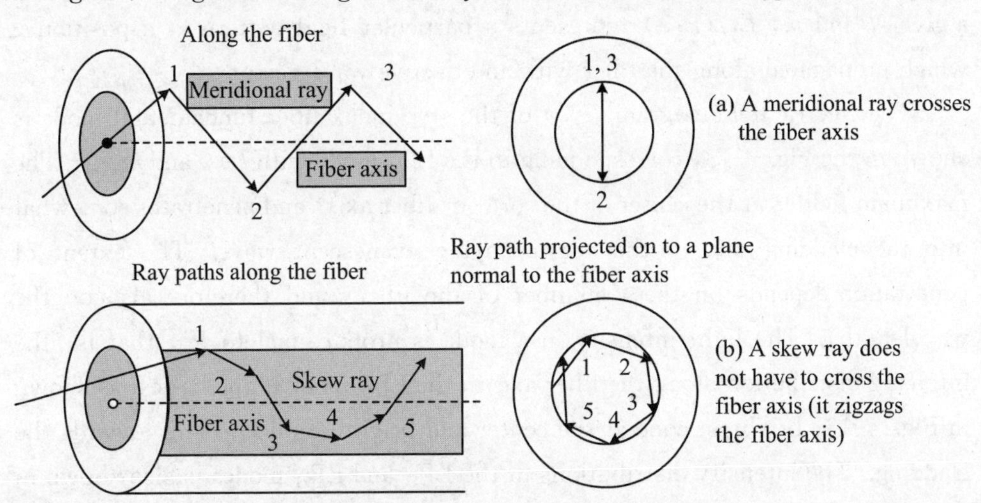

Fig. 4-12 Difference between a meridional ray and a skew ray

Numbers represent reflections of the ray.

On the other hand, the modes generated by skew rays have both Ez and Bz (or Hz) components, so they are not TE or TM waves. They are called HE or EH modes because both electric and magnetic fields have components along the z direction, which called hybrid modes. Obviously, the guided mode in a step index fiber cannot be described as easily as the mode in a planar waveguide.

Guided modes in a step index fiber with $\Delta \ll 1$ are called a ***weakly-guiding fiber***, which usually displayed by a traveling wave that is almost plane polarized. Transverse electric and magnetic fields (E and B are perpendicular to each other and also to z) have properties similar to the field direction of a plane wave, but the field magnitudes are not constant in the plane. These waves are called ***linearly polarized*** (***LP***) and have transverse electric and magnetic field characteristics. The LP mode guided along the fiber can be expressed as the propagation of an electric field distribution $E(r, \phi)$ along z. This field distribution, or pattern, is in the plane normal to the fiber axis and hence depends on rand p but not on z. Further, because of the presence of two boundaries, it is characterized by two integers, l and m. So the propagating field distribution in an LP mode is given by $E_{lm}(r, \phi)$, which is denoted as LP_{lm}. Thus, an LP_{lm} mode can be described by a traveling wave along z of the form:

$$E_{LP} = E_{lm}(r, \phi) \exp j(\omega t - \beta_{lm} z) \tag{4-16}$$

in which E_{LP} is the LP mode field and β_{lm} is its propagation constant along z. For a given l and m, $E_{lm}(r, \phi)$ represents a particular field pattern at a position z which propagated along the fiber with an effective wave vector β_{lm}.

The electric field diagram (E_{01}) of the step index fiber fundamental mode is shown in the Fig. 4-13, corresponding to the LP_{01} mode with $l=0$ and $m=1$. The maximum field is at the center of the core (or fiber axis) and penetrates somewhat into the cladding due to the accompanying evanescent wave. The extent of penetration depends on the V-number of the fiber (and therefore also on the wavelength). The light intensity in a mode is proportional to E^2, that is, the intensity distribution along the fiber axis in the LP_{01} mode is the largest as shown in Fig. 4. 13; brightest zone at the center and brightness decreasing towards the cladding. The intensity distributions in the LP_{11} and LP_{21} modes is also shown in Fig. 4-13. The integers l and m are related to the intensity pattern in a LP_{lm} mode. The m maximums are along r, starting from the center and $2l$ maximums are around a circumference, as shown in Fig. 4-13. Moreover, in the ray diagram, l represents the extent of helical propagation, or the contribution of skew rays to the mode. In the fundamental mode this is zero. In a planar waveguide, m is directly related to the reflection angle θ of the rays.

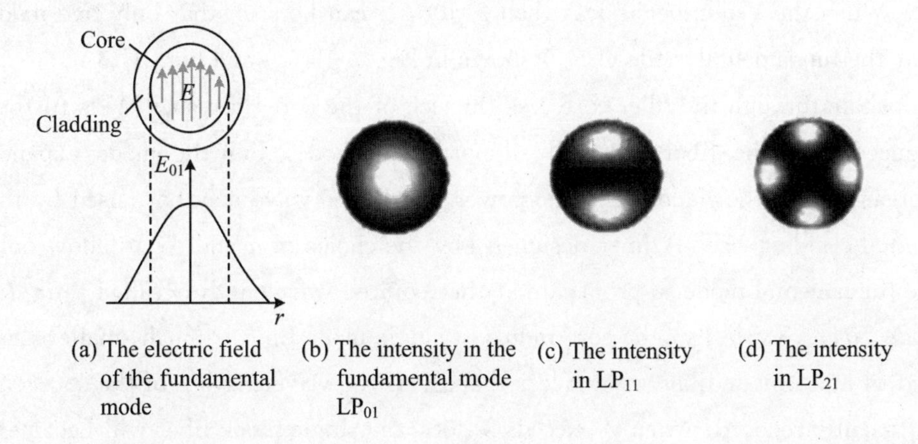

(a) The electric field of the fundamental mode (b) The intensity in the fundamental mode LP_{01} (c) The intensity in LP_{11} (d) The intensity in LP_{21}

Fig. 4-13 The intensity distributions in the LP_{01}, LP_{11}, LP_{21}

The electric field distribution of the fundamental mode in the transverse plane to the fiber axis z. The light intensity is greatest at the center of the fiber. Intensity patterns in LP_{11} and LP_{01}, LP_{11} and LP_{21} modes.

It can be seen from the above discussion that light propagates through various modes in the optical fiber, and each mode has its own propagation vector β_{lm} and its own electric field mode $E_{lm}(r,\phi)$. The group velocity $V_g(l,m)$ exists in each mode, which depends on the dispersion behavior of the ω vs. β_{lm}. When a light pulse is input into the fiber, it will propagate in the fiber through various modes. These modes propagate with different group velocities and therefore emerge at the end of the fiber with a spread of arrival times, which means that the output pulse is a broadened version of the input pulse. In the case of a planar waveguide, the broadening of the optical pulse is an ***intermodal dispersion*** phenomenon. However, a suitable fiber can be designed that only allows the fundamental mode to propagate, so there is no ***modal dispersion***.

For a step index fiber, the V-number or normalized frequency is defined in a manner similar to that of a planar waveguide:

$$V = \frac{2\pi a}{\lambda}(n_1^2 - n_2^2)^{1/2} = \frac{2\pi a}{\lambda}(2n_1 n\Delta)^{1/2} \tag{4-17}$$

in which a is the radius of the fiber core, λ is the free space wavelength, n is the average refractive index of the core and cladding, i. e., $n=(n_1+n_2)/2$, and Δ is the normalized index difference, that is,

$$\Delta = \frac{n_1 - n_2}{n_1} \approx \frac{n_1^2 - n_2^2}{2n_1^2} \tag{4-18}$$

When the V-number is less than 2.405, it can be seen that only one mode that the fundamental mode (LP_{01}) shown in Fig. 4-13 (a) and Fig. 4-13 (b), can propagate through the fiber core. As the size of the core decreases, V is further reduced, and the fiber can still support LP_{01} mode, but the mode expands increasingly to the cladding. Some power loss in the wave may be caused by the limited cladding size. A fiber designed (by the choice of a and Δ) to allow only the fundamental mode to propagate at the required wavelength is called a ***single-mode fiber***. Typically, the core radius of single-mode fiber is much smaller than that of multi-mode fiber, which less than Δ. If the wavelength λ of the source is sufficiently reduced, when V exceeds 2.405, the single-mode fiber will become a ***multi-mode fiber***; higher modes also contribute to propagation. The cut-off wavelength λ_c above which the fiber becomes single mode is given by

$$V_{\text{cut-off}} = \frac{2\pi a}{\lambda_c}(n_1^2 - n_2^2)^{1/2} = 2.405 \tag{4-19}$$

The number of modes rises sharply as the V-parameter increases above 2.405. In a ***step index multi-mode fiber***, a good approximation to the number of modes M is given by

$$M = \frac{V^2}{2} \tag{4-20}$$

From the V-number in equation (4-17), it is easy to deduce the influence of various physical parameters of the step index fiber on the propagation modulus. For example, increasing the core radius (a) or its refractive index (n_1) increases the number of modes. On the other hand, the increase in wavelength or cladding refractive index (n_2) will cause the modulus to decrease. Since the cladding diameter does not enter the V-number equation, it has no significant effect on wave propagation. In multi-mode fibers, light propagates through many modes, which mainly all confined to the core. The fundamental mode field penetrates into the cladding as an evanescent wave propagating along the boundary in the step index fiber. If the cladding is not thick enough, this field will reach the end of the cladding and escape, resulting in a loss of intensity. For example, the cladding diameter for a single-mode step index fiber is at least 10 times that of the core.

Since the waveguide properties and the wavelength of the light source are dependent on the propagation constant β_{lm} of the LP mode, it is convenient to describe light propagation in terms of a "normalized" propagation constant that only depends on the V-number. Given $k=2\pi/\lambda$ and guide indices n_1 and n_2, the normalized propagation constant b is related to $\beta=\beta_{lm}$ by the definition:

$$b=\frac{(\beta/k)^2-n_2^2}{n_1^2-n_2^2} \tag{4-21}$$

According to the above definition, the lower limit $b=0$ corresponds to the propagation of $\beta=kn_2$ in the cladding material, and the upper limit $b=1$ corresponds to the propagation of $\beta=kn_1$ in the core material. As shown in Fig. 4-14, the dependence of b on the V-number for various modes for a few lower-order LP modes has been calculated in the literature. It is worth noticing that for all V-numbers, the fundamental mode (LP_{01}) exists, and LP_{11} is cut-off at $V=2.405$. For each specific LP mode higher than the fundamental mode, there is a cut-off V and the corresponding cut-off wavelength. Depending on the V parameters of the fiber, b, and hence β, can be easily found for the allowed LP modes from Fig. 4-14.

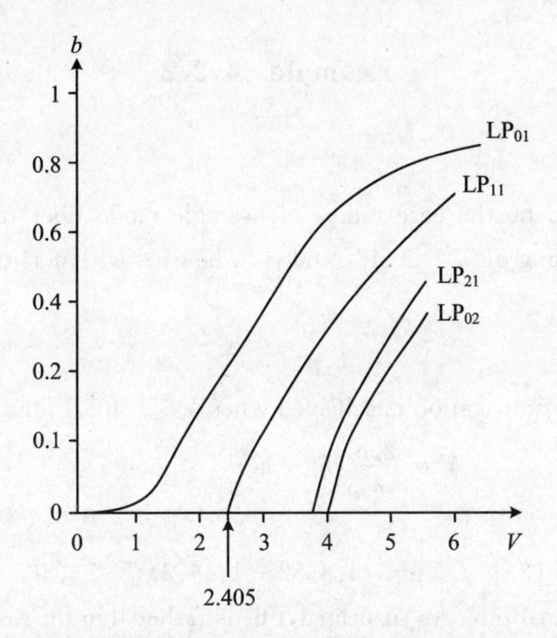

Fig. 4-14 Normalized propagation constant b vs. V-number for a step index fiber various LP modes

Example 4.2.1

Problem

The number of allowable modes for the multi-mode step refractive index fiber is calculated, with a core refractive index of 1.468 and diameter of 100 μm. When the source wavelength is 850 nm, the cladding refractive index is 1.447.

Solution

The V-number for this fiber can be calculated by substituting $a=50\ \mu$m, $\lambda=0.850\ \mu$m, $n_1=1.468$, and $n_2=1.447$ into the expression for the V-number:

$$V=\frac{2\pi a}{\lambda}(n_1^2-n_2^2)^{1/2}=(2\pi 50/0.85)(1.4682-1.4472)^{1/2}=91.44$$

Since $V\gg 2.405$, the number of modes is

$$M\approx\frac{V^2}{2}=91.44^2/2=4181$$

which is large. There are numerous modes.

Example 4.2.2

Problem

What should be the core radius of a single mode fiber that has a core of $n_1=1.468$, cladding of $n_2=1.447$, and is to be used with a source wavelength of 1.3 μm?

Solution

Single mode propagation is achieved when $V\leqslant 2.405$. Thus we need:

$$V=\frac{2\pi a}{\lambda}(n_1^2-n_2^2)^{1/2}\leqslant 2.405$$

or

$$(2\pi a/1.3\ \mu\text{m})(1.4682-1.4472)^{1/2}\leqslant 2.405$$

which gives $a\leqslant 2.01\ \mu$m. As suspected, this is rather thin for easy coupling of the fiber to a light source or to another fiber and special coupling techniques must be used. The size of a is comparable in magnitude to the wavelength, which means that the geometric ray picture, strictly, cannot be used to describe light propagation.

Example 4.2.3

Problem

Calculate the cut-off wavelength for single mode operation for a fiber that has a core with diameter of 7 μm, a refractive index of 1. 458, and a cladding of refractive index of 1. 452. What is the V-number and the mode field diameter (MFD) when operating at λ=1. 3 μm?

Solution

For single-mode operation:

$$V = \frac{2\pi a}{\lambda}(n_1^2 - n_2^2)^{1/2} \leqslant 2.405$$

Substituting for a, n_1 and n_2, and rearranging we get:

$$\lambda > 2\pi(3.5\ \mu\text{m})(1.458^2 - 1.452^2)^{1/2}/2.405 = 1.208\ \mu\text{m}$$

Wavelengths shorter than 1. 028 μm will result in a multi-mode propagation.

At λ=1. 3 μm,

$$V = 2\pi[(3.5\ \mu\text{m})/(1.3\ \mu\text{m})](1.458^2 - 1.452^2)^{1/2} = 2.235$$

The mode field diameter MFD is then,

$$2\omega_0 \approx 2a(V+1)/V = (7\ \mu\text{m})(2.235+1)/2.235 = 10.13\ \mu\text{m}$$

Example 4.2.4

Problem

Consider a single-mode fiber with core and cladding indices of 1. 448 and 1. 44, core radius of 3 μm, operating at 1. 5 μm. Given that we can approximate the fundamental mode normalized propagation constant b by

$$b \approx \left(1.1428 - \frac{0.996}{V}\right)^2 \quad (1.5 < V < 2.5) \tag{4-22}$$

calculate the propagation constant β. Change the operating wavelength to λ' by a small amount, say 0. 1%, and then recalculate the new propagation constant. Then determine the group velocity V_g of the fundamental mode at 1. 5 μm, and the group delay τ_g over 1 km of fiber.

Solution

Equation (4-21) for a weakly guiding fiber can also be written as:

$$b = \frac{(\beta/k)^2 - n_2^2}{n_1^2 - n_2^2}, \quad \beta = n_2 k(1 + b\Delta) \tag{4-23}$$

V can be calculated from the fiber properties that are given, and then use equation (4-22) to calculate b. From b, $k=2\pi/\lambda$ is used in equation (4-23), β can be calculated. In addition, $\omega=2\pi c/\lambda$. Thus, $V=(2\pi a/\lambda)(n_1^2-n_2^2)^{1/2}=1.910088$. From equation (4-22), $b=0.3860859$, and from equation (4-23), $\beta=6.044796\times 10^6\ \mathrm{m}^{-1}$. Calculations are summarized in Table 4-2. The group velocity is

$$V_g = \frac{\omega' - \omega}{\beta' - \beta} = \frac{(1.256511 - 1.256624) \times 10^{15}}{(6.044189 - 6.044796) \times 10^6} = 2.0714 \times 10^8\ \mathrm{m \cdot s^{-1}}$$

The group delay τ_g over 1 km is 4.83 μs.

Table 4-2　Calculations

Calculations	V	$K(\mathrm{m}^{-1})$	$\omega\,(\mathrm{rad \cdot s^{-1}})$	b	$\beta(\mathrm{m}^{-1})$
$\lambda=1.5\ \mu$m	1.910088	4188790	1.256624×10^{15}	0.3860859	6.044796×10^6
$\lambda'=1.50015\ \mu$m	1.909897	4188371	1.256511×10^{15}	0.3860211	6.044189×10^6

New Words and Expressions

［1］ **skew ray**：斜射光线。它是指不经过光纤轴线传输的光线。

［2］ **meridional ray**：子午光线。它是指入射光线通过光纤轴线，且入射角大于界面临界角的光线。

［3］ **weakly-guiding fiber**：弱导光纤。它是指纤芯中最大的折射率和均匀包层最小折射率之差很小的光纤，通常其折射率差小于1%。

［4］ **linearly polarized (LP)**：线偏振光。在光的传播方向上，光矢量只沿一个固定的方向振动，光矢量端点的轨迹为一条直线。

［5］ **intermodal dispersion**：模内色散。它是指在光纤中由光源有限宽度产生的色散，包括材料色散和波导色散。

［6］ **modal dispersion**：模间色散。一种频率的光波以不同的角度入射到光纤中，形成不同的模式，每种模式都具有不同的轴向速度，因而同时发出的不同模式到达输出端的时间是不相同的，从而导致输出端信号的畸变。

［7］ **single-mode fiber**：单模光纤。在给定的工作波长上只能传输一种模式的光纤。

［8］ **multi-mode fiber**：多模光纤。在给定的工作波长上可以传输多种模式的光纤。

[9] **step index multi-mode fiber**：阶跃折射率多模光纤。它是指具有阶跃型折射率分布的一类多模光纤。

4.3 Numerical Aperture

Not all source radiation can be guided along an optical fiber. Only rays falling within a certain cone at the input of the fiber can normally be propagated through the fiber. In Fig. 4-15, the path of a light ray launched from the outside medium of refractive index n_0 (not necessarily air) into the fiber core. Assume that the incident angle at the end of the fiber core is a and the angle between the light and the normal of the fiber axis is θ inside the waveguide. Then unless the angle is greater than the critical angle for TIR, the ray will escape into the cladding. Therefore, for the propagation of light, the launching angle α has to be such that TIR is supported within the fiber. From Fig. 4-15, it should be apparent that the maximum value a that which results in $\theta = \theta_c$. At the n_0/n_1 interface, Snell's law gives:

$$\sin\alpha_{\max}/\sin(90° - \theta_c) = n_1/n_0$$

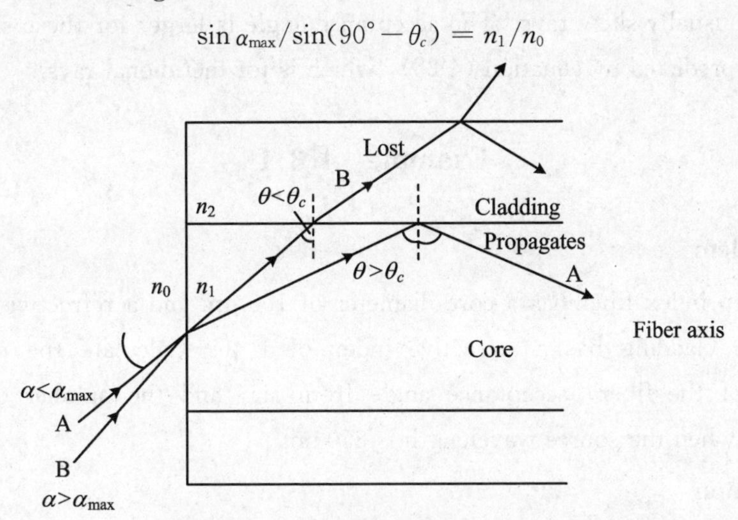

Fig. 4-15 Maximum acceptance angle $\alpha_{\max}$

Maximum acceptance angle $\alpha_{\max}$ is that which just gives total internal reflection at the core-cladding interface, i. e., when $\alpha = \alpha_{\max}$ then $\theta = \theta_c$. Rays with $\alpha > \alpha_{\max}$ (e. g., ray B) become refracted and penetrate the cladding and are eventually lost.

in which θ_c is determined by the onset of TIR, that is $\sin\theta_c = n_2/n_1$. We can now eliminate θ_c to obtain:

$$\sin\alpha_{max} = \frac{(n_1^2 - n_2^2)^{1/2}}{n_0}$$

The ***numerical aperture*** (***NA***) is a characteristic parameter of an optical fiber defined by

$$NA = (n_1^2 - n_2^2)^{1/2} \tag{4-24}$$

so that in terms of NA, the maximum ***acceptance angle***, α_{max}, becomes:

$$\sin\alpha_{max} = \frac{NA}{n_0} \tag{4-25}$$

The angle $2\alpha_{max}$ is called the total acceptance angle and depends on the NA of the fiber and the refractive index of the launching medium. NA is an important factor in light launching designs into the optical fiber. We should note that equation (4-25) is strictly applicable for meridional rays. Skew rays have a wider acceptance angle. Since NA is defined in terms of the refractive indices, we can obtain a relationship between the V-number and NA:

$$V = \frac{2\pi a}{\lambda} NA \tag{4-26}$$

Multi-mode propagation in a fiber involves many modes, the majority of which are usually skew rays. The acceptance angle is larger for these skew rays than that predicted by equation (4-25), which is for meridional rays.

Example 4.3.1

Problem

A step index fiber has a core diameter of 100 μm and a refractive index of 1.48. The cladding has a refractive index of 1.46. Calculate the numerical aperture of the fiber, acceptance angle from air, and the number of modes sustained when the source wavelength is 850 nm.

Solution

The numerical aperture is

$$NA = (n_1^2 - n_2^2)^{1/2} = (1.48^2 - 1.46^2)^{1/2} = 0.2425$$

From $\sin\alpha_{max} = NA/n_0 = 0.2425/1$, the acceptance angle is $\alpha_{max} = 14°$ and the total acceptance angle is 28°.

The V-number in terms of the numerical aperture can be written as:

$$V = (2\pi a/\lambda)NA = (2\pi 50/0.85)0.2425 = 89.62$$

The number of modes, $M \approx V^2/2 = 4016$.

Example 4.3.2

Problem

A typical single-mode optical fiber has a core of diameter 8 μm and a refractive index of 1.46. The normalized index difference is 0.3%. The cladding diameter is 125 μm. Calculate the numerical aperture and the acceptance angle of the fiber. What is the single mode cut-off wavelength λ_c of the fiber?

Solution

The numerical aperture is

$$NA = (n_1^2 - n_2^2)^{1/2} = [(n_1 + n_2)(n_1 - n_2)]^{1/2}$$

Substituting $(n_1 - n_2) = n_1\Delta$ and $(n_1 + n_2) \approx 2n_1$, we get:

$$NA \approx (2n_1^2\Delta)^{1/2} = n_1(2\Delta)^{1/2} = 1.46(2 \times 0.003)^{1/2} = 0.113$$

The acceptance angle is given by

$$\sin\alpha_{\max} = NA/n_0 = 0.113/1 \quad \text{or} \quad \alpha_{\max} = 6.5°$$

The condition for single-mode propagation is $V \leqslant 2.405$, which corresponds to a minimum wavelength λ_c given by

$$\lambda_c = (2\pi aNA)/2.405 = [(2\pi 4\ \mu\text{m})0.113]/2.405 = 1.18\ \mu\text{m}$$

Illumination wavelengths shorter than 1.18 μm will result in a multimode operation.

New Words and Expressions

[1] **numerical aperture**（**NA**）：数值孔径。它是一个无量纲的数，用以衡量系统能够收集的光的角度范围。

[2] **acceptance angle**：接收角。它是将光射入光纤纤芯或者波导能允许的光线最大的入射角。

4.4 Dispersion in Single-mode Fibers

4.4.1 Material Dispersion

The advantage of single-mode fiber is that only one mode propagates in the light without the multi-mode dispersion of the input light pulse. However, even if light propagates in a single mode, there will still be dispersion due to changes in the refractive index of the core glass and the wavelength of the light coupled into the optical fiber. The propagation velocity of the guided wave along the fiber core depends on the refractive index, which in turn depends on the wavelength. This dispersion caused by the wavelength dependence of the material properties of the waveguide is called ***material dispersion***. The characteristic of the waveguide is material dispersion. It is perfectly monochromatic when there is no actual light source input, so there are various free-space wavelength waves in the waveguide, that is, the range of wavelengths. For example, the excitation from the laser causes the spectrum of the source wavelength to be fed into the optical fiber, as shown in Fig. 4-16. The wavelength width ($\Delta\lambda$) of the spectrum depends on the nature of the light source, but it will never be zero. Since the group index N_g (related to n_1) of the medium depends on the wavelength, each radiation with a different wavelength will propagate through the fundamental mode with a different ***group velocity*** (V_g). The wave will arrive the end of the fiber at different times, so a broadened output light pulse will be generated as schematically illustrated in Fig. 4-16. The group index of silica glass is almost constant at a wavelength of about 1.3 μm, which means that the dispersion of the material near this wavelength is zero.

Material dispersion can be evaluated by considering the dependence of the group velocity V_g on the wavelength through the N_g vs. λ behavior and the spectrum of the light source that defines the range $\Delta\lambda$ of emitted wavelengths from the light source. Here, a very short-duration light pulse is regarded as the input signal with a spectrum of wavelengths over $\Delta\lambda$, as shown in Fig. 4-16. Due

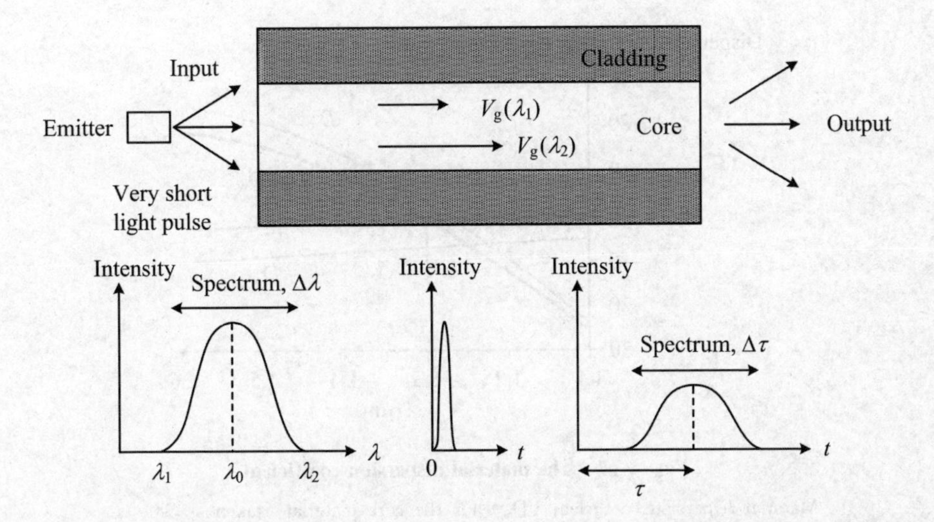

Fig. 4-16 The excitation from the laser causes the spectrum of the source wavelength

All excitation sources are inherently non-monochromatic and emit within a spectrum ($\Delta\lambda$) of wavelengths. Waves of different free-space wavelengths in the waveguide propagate at different group velocities due to the dependence of wavelength of n_1. These waves arrive at the end of the fiber at different times, thus producing a broadened output pulse.

to the expansion of the wave's arrival time τ, the output pulse is widened by $\Delta\tau$. $\Delta\tau$ increases as the fiber length L increases, because slower waves fall further behind faster waves over a longer distance. Therefore, dispersion is usually expressed as the propagation per unit length, which can be expressed as:

$$\frac{\Delta\tau}{L} = |D_m| \Delta\lambda \tag{4-27}$$

where D_m is the dispersion coefficient of the material. It can be calculated approximately from the second derivative of the refractive index:

$$D_m \approx -\frac{\lambda}{c}\left(\frac{d^2 n}{d\lambda^2}\right) \tag{4-28}$$

It should be noted that although D_m can be negative or positive, $\Delta\tau$ and $\Delta\lambda$ are defined as positive numbers, which is why the amplitude sign appears in equation (4-27). In order to find $\Delta\tau/L$ in equation (4-27), D_m must be calculated at the center wavelength λ_0 of the spectrum. Fig. 4-17 shows the dependence of silicon as a typical fiber core glass material D_m on λ. Notice that the curve passes through zero at $\lambda \approx 1.27$ μm. When silica (SiO_2) is doped with germania (GeO_2) to increase the refractive index for the core, the D_m vs. λ curve shifts slightly to higher wavelengths.

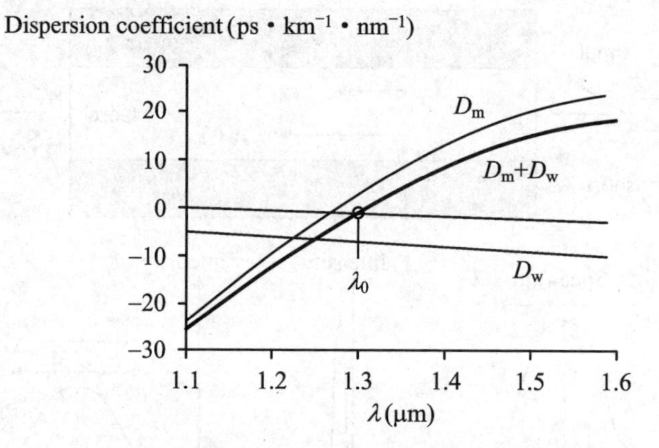

Fig. 4-17 The material dispersion coefficient

Material dispersion coefficient (D_m) for the core material (taken as SiO_2), waveguide dispersion coefficient (D_m) (a=4.2 μm) and the total or ***chromatic dispersion*** coefficient D_{ch} (=D_m+D_w) as a function of free space wavelength, λ.

The transit time τ of a light pulse represents a delay of information between the output and the input. The signal delay time per unit distance, τ/L, is called the group delay (τ_g) and is determined by the group velocity V_g as it refers to the transit time of signals (energy). Therefore, $\Delta\tau/L$ in equation (4-27) is the expansion of the group delay time due to the limited input spectrum. The group delay as a function of wavelength can be expressed as:

$$\tau_g = \frac{1}{V_g} = \frac{d\beta_{01}}{d\omega} \tag{4-29}$$

where β_{01} is the propagation constant of the fundamental mode. Material dispersion in equation (4-27) is the spread in τ_g in equation (4-29) due to the dependence of β_{01} on wavelength through N_g.

4.4.2 Waveguide Dispersion

Waveguide dispersion is another dispersion mechanism. Even if n_1 and n_2 are constant, the group velocity V_g(01) of the fundamental mode depends on the V-number, and the V-number depends on the source wavelength λ. It should be emphasized that waveguide dispersion is quite separate from material dispersion. Even if n_1 and n_2 were wavelength-independent (no material dispersion), light propagation still suffers from waveguide dispersion, because V_g(01) depends on

V, and V depends on λ instead. Waveguide dispersion arises as a result of the guiding properties of the waveguide, which imposes a nonlinear ω-β_{lm} relationship, as discussed in Section 4. 2. Therefore, the spectrum of source wavelengths will produce a different number of V for each source wavelength, resulting in different propagation velocities. As shown in Fig. 4-16, the group delay times of fundamental mode waves with different λ will expand, so the output light pulse will be broadened.

If we use a light pulse of very short duration as input, as in Fig. 4-16, with a wavelength spectrum of $\Delta\lambda$, then the broadening or dispersion per unit length, $\Delta\tau/L$, in the output light pulse due to waveguide dispersion can be found from:

$$\frac{\Delta\tau}{L} = |D_{\mathrm{w}}|\Delta\lambda \tag{4-30}$$

where D_{w} is a coefficient of waveguide dispersion. It depends on the waveguide properties (in a non-trivial way) and, over the range $1.5<V<2.4$, it is written approximately by

$$D_{\mathrm{w}} \approx \frac{1.984N_{\mathrm{g}2}}{(2\pi a)^2 2cn_2^2} \tag{4-31}$$

where $N_{\mathrm{g}2}$ and n_2 are the group and refractive indices of the cladding, respectively. Obviously, D_{w} depends on the waveguide geometry of the core radius a. Fig. 4-17 also shows the dependence of D_{w} on the wavelength for a core radius of $a=4.2$ μm. In addition, D_{m} and D_{w} have opposite trends.

4.4.3 Chromatic Dispersion or Total Dispersion

In a single-mode fiber, the dispersion of the propagating pulse is due to the limited width of the source spectrum $\Delta\lambda$, it is not completely monochromatic. Dispersion mechanism is based on the fundamental mode velocity depending on the source wavelength λ. This kind of dispersion caused by the source wavelength range is usually called ***chromatic dispersion***, including material and waveguide dispersion, because both depend on $\Delta\lambda$. As a first approximation, these two dispersion effects can be simply added together, so that the overall dispersion per unit length is approximately given as:

$$\frac{\Delta\tau}{L} = |D_{\mathrm{m}} + D_{\mathrm{w}}|\Delta\lambda \tag{4-32}$$

This is shown in Fig. 4-17 where $D_m + D_w$, defined as the chromatic dispersion coefficient (D_{ch}) passes through zero at a certain wavelength (λ_0). For the example in Fig. 4-17, chromatic dispersion is zero at around 1300 nm.

Since D_w depends on the geometry of the waveguide, as shown in equation (4-31), the zero-dispersion wavelength λ_0 can be shifted by appropriately designing the waveguide. For example, by reducing the core radius and increasing the core doping, λ_0 can be moved to 1550 nm where the optical attenuation is the smallest in the fiber. Such fibers are called dispersion shifted fibers.

Although the chromatic dispersion $D_m + D_w$ passes zero, this does not mean that dispersion does not exist at all. First, we should note that $D_m + D_w$ can be made zero for only one wavelength (λ_0), not at every wavelength within the spectrum ($\Delta\lambda$) of the source. Further, other second order effects also contribute to dispersion.

4.4.4 Profile and Polarization Dispersion Effects

Although material and waveguide dispersion are the main factors in the spread of propagating light pulses, there are other dispersion effects. Since the group velocity $V_g(01)$ of the base model also depends on the refractive index difference Δ, there is an additional dispersion mechanism called ***profile dispersion***. If Δ changes with wavelength, then different wavelengths from the source would have different group velocities and experience different group delays, resulting in the spread of the pulse. It is part of chromatic dispersion because it depends on the input spectrum $\Delta\lambda$:

$$\frac{\Delta\tau}{L} = |D_p|\Delta\lambda \tag{4-33}$$

where D_p is the profile dispersion coefficient, which can be calculated. Typically, D_p is less than 1 ps • km^{-1} • nm^{-1}, so it is negligible compared to D_w. The overall chromatic dispersion coefficient then becomes $D_{ch} = D_m + D_w + D_p$. It should be mentioned that the reason Δ exhibits a wavelength dependence is due to material dispersion characteristics, i. e. , n_1 vs. λ and n_2 vs. λ behavior, so that, in reality, profile dispersion originates from material dispersion.

When the fiber is not completely symmetrical and uniform, that is, the refractive index is not isotropic, ***polarization dispersion*** will occur. When the refractive index depends on the direction of the electric field, the propagation constant of a given mode depends on its polarization. Due to various factors in the manufacturing process, such as glass composition, geometry, and induced local strain, the refractive indices n_1 and n_2 may not be isotropic. Assume that it is parallel to the x axis and y axis of the electric field, the values of n_1 are n_{1x} and n_{1y}, respectively, as shown in Fig. 4-18. Even if the light source wavelength is monochromatic, the propagation constants of the fields along the x and y directions would be different, $\beta_x(01)$ and $\beta_y(01)$, resulting in different group delays and dispersions. The situation in reality is more complicated because n_{1x} and n_{1y} would vary along the fiber length and there will be interchange of energy between these modes as well. However, the final dispersion depends on the degree of n_{1x}-n_{1y} anisotropy, which is kept to a minimum through various manufacturing processes. In fact, the polarization dispersion is less than a fraction of one picosecond per kilometer of fiber, and the dispersion does not scale linearly with the fiber length L.

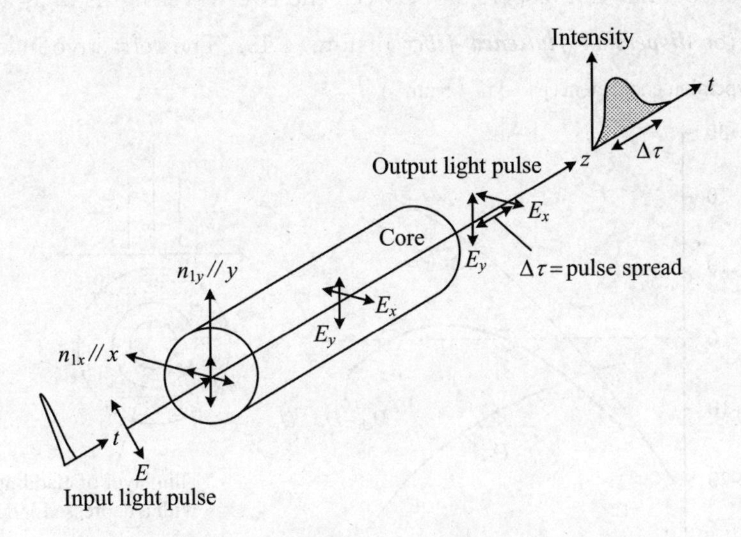

Fig. 4-18 Core refractive index has different values

Assume that the core refractive index has different values along two orthogonal directions corresponding to electric field oscillation directions (polarizations). Here, we define the x axis and y axis to be taken along these directions. The input light will propagate along the fiber at different group velocities of the E_x and E_y polarizations, thus reaching the output at different times.

4.4.5 Dispersion Flattened Fibers

The doping of the core material changes the dispersion of the material (represented by D_m), which causes the dispersion to extend to longer wavelengths and the attenuation of the signal increases. Further, it may be desirable to have minimal dispersion over a range of wavelengths not just at the zero-crossing wavelength λ_0 as in Fig. 4-17. The dispersion of the waveguide represented by D_w can be adjusted by changing the geometry of the waveguide. As mentioned earlier, the waveguide dispersion is caused by the dependence of the group velocity V_g on the wavelength λ. As the wavelength increases, the field penetrates more into the cladding, which changes the proportion of light energy carried by the core and the cladding and thereby changes V_g. Therefore, change the geometry of the waveguide, that is, the refractive index profile, and control D_w to produce a flat full dispersion between the two wavelengths of λ_1 and λ_2, as shown in the ***dispersion-flattened fiber*** of Fig. 4-19. The refractive index profile

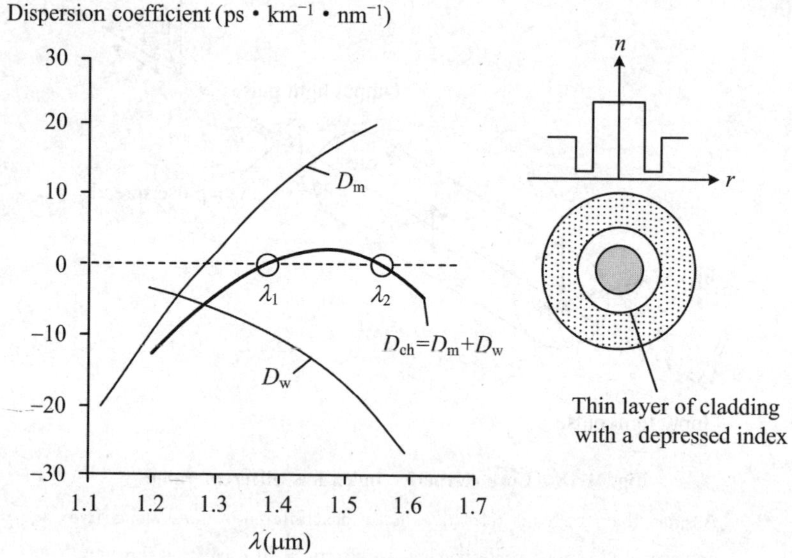

Fig. 4-19 Dispersion flattened fiber example

The material dispersion coefficient (D_m) for the core material and waveguide dispersion coefficient (D_w) for the doubly clad fiber result in a flattened small chromatic dispersion between λ_1 and λ_2.

of this fiber looks like a W, and the cladding is a thin layer with a reduced refractive index. This fiber is called a double cladding. The simple step index fiber is singly clad. Greater control on waveguide dispersion can be obtained by using multiply clad fibers. This type of fiber is difficult to manufacture, but can exhibit excellent dispersion (1～3 ps · km^{-1} · nm^{-1}) in the wavelength range of 1. 3～1. 6 μm. Of course, the low dispersion in a wavelength range allows wavelength multiplexing, for example, multiple wavelengths of 1. 3 μm and 1. 55 μm can be used as communication channels.

Example 4.4.1

Problem

Generally, the width $\Delta\lambda$ of the wavelength spectrum of the source and the dispersion $\Delta\tau$ refer to half-power widths and not widths from one extreme end to the other. Suppose that $\Delta r_{1/2}$ called the linewidth, is the width of intensity vs. wavelength spectrum between the half intensity points and $\Delta\tau_{1/2}$ is the width of the output light intensity vs. time signal between the half-intensity points.

Estimate the material dispersion effect per km of silica fiber operated from a light emitting diode (LED) emitting at 1. 55 μm with a linewidth of 100 nm. What is the material dispersion effect per km of silica fiber operated from a laser diode emitting at the same wavelength with a linewidth of 2 nm?

Solution

From Fig. 4. 19, at 1. 55 μm, the material dispersion coefficient $D_m = 22$ ps · km^{-1} · nm^{-1}. This is given as per km of length and per nm of spectral linewidth. For the LED, $\Delta\lambda_{1/2} = 100$ nm.

$$\Delta\tau_{1/2} \approx L \mid D_m \mid \Delta\lambda_{1/2} \approx (1\ \text{km})(22\ \text{ps} \cdot \text{km}^{-1} \cdot \text{nm}^{-1})(100\ \text{nm})$$
$$= 2200\ \text{ps} \quad \text{or} \quad 2.2\ \text{ns}$$

For the laser diode, $\Delta\lambda_{1/2} = 2$ nm and

$$\Delta\tau_{1/2} \approx L \mid D_m \mid \Delta\lambda_{1/2} \approx (1\ \text{km})(22\ \text{ps} \cdot \text{km}^{-1} \cdot \text{nm}^{-1})(2\ \text{nm})$$
$$= 44\ \text{ps} \quad \text{or} \quad 0.044\ \text{ns}$$

There is clearly a big difference between the dispersion effects of the two sources. The total dispersion, however, as indicated by Fig. 4-19, will be less. Indeed, if the fiber is properly dispersion shifted so that $D_m + D_w = 0$ at 1. 55 μm, then dispersion due to excitation from a typical laser diode will be a few picoseconds per km (but not zero)!

Example 4.4.2

Problem

Consider a single mode optical fiber with a core of SiO_2-13. 5% GeO_2 for which the material and waveguide dispersion coefficients are shown in Fig. 4-20. Suppose that the fiber is excited from a 1. 5 μm laser source with a linewidth $\Delta\lambda_{1/2}$ of 2 nm. What is the dispersion per km of fiber if the core diameter $2a$ is 8 μm? What should be the core diameter for zero chromatic dispersion at λ=1. 5 μm?

Solution

In Fig. 4-20, at λ=1. 5 μm, D_m=10 ps • km^{-1} • nm^{-1} and with a= 4 μm, D_w=−6 ps • km^{-1} • nm^{-1}, so that chromatic dispersion is

$$D_{ch} = D_m + D_w = 10 - 6 = 4 \text{ ps} \cdot \text{km}^{-1} \cdot \text{nm}^{-1}$$

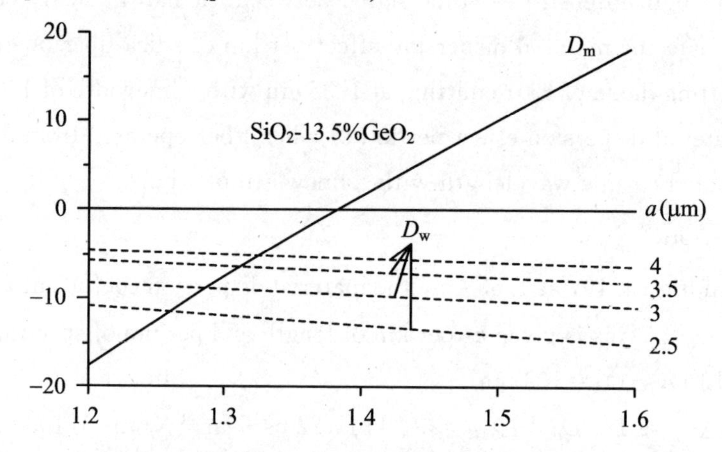

Fig. 4-20 Material and waveguide dispersion coefficients

In an optical fiber with a core SiO_2-13. 5% GeO_2, for a=2. 5 to 4 μm.

Total dispersion or chromatic dispersion per km of fiber length is then,

$$\Delta\tau_{1/2}/L = |D_{ch}|\Delta\lambda_{1/2} = (4 \text{ ps} \cdot \text{km}^{-1} \cdot \text{nm}^{-1})(2 \text{ nm}) = 8 \text{ ps} \cdot \text{km}^{-1}$$

Dispersion will be zero at 1. 5 μm when D_w=−D_m or when D_w=−10 ps • km^{-1} • nm^{-1}. Examination of D_w vs. λ for a=2. 5 to 4 μm shows that a should be about 3 μm. It should be emphasized that although D_m+D_w=0 at 1. 5 μm, this is only at one wavelength, whereas the input radiation is over a range $\Delta\lambda$ of wavelengths so that in practice chromatic dispersion is never actually zero. In this example, when the core radius $a=3\ \mu\text{m}$, it is the smallest at $1.5\ \mu\text{m}$.

New Words and Expressions

[1] **material dispersion**：材料色散。它是由材料自身特性造成的一种色散。

[2] **group velocity**：群速度。它是指包络波上任一恒定相位点的传播速度。

[3] **waveguide dispersion**：波导色散。它是对于光纤的某一传输模式，在不同的光波长下群速度不同而引起的脉冲展宽。

[4] **chromatic dispersion**：色散。它是指复色光分解为单色光而形成光谱的现象。

[5] **profile dispersion**：折射率分布色散。光纤的折射率分布随波长变化所引起的色散。

[6] **polarization dispersion**：偏振色散。光纤非对称和非均匀性引起的折射率的各向异性分布，从而导致不同方向分布的模式有不同的有效折射率，也即偏振相关特性引起的光脉冲的展宽。

[7] **dispersion-flattened fiber**：色散平坦光纤。在一定波段内，总色散接近于零、色散曲线较平坦的光纤，可用于高速率的密集波分复用、频分复用光纤通信。

4.5 Attenuation in Optical Fibers

When light propagates through an optical fiber, it is attenuated by many processes that depend on the wavelength of the light. Define the optical power of the optical fiber with the input length L as P_{in}, the optical power at the output end as P_{out}, and the intensity at the distance x distance from the input end in the optical fiber as P. The ***attenuation coefficient*** α is defined as the fractional decrease of the optical power per unit distance, which can be written as:

$$\alpha=-\frac{1}{p}\frac{\mathrm{d}P}{\mathrm{d}x} \tag{4-34}$$

In addition, we can also relate the fraction of the unit fiber length L to α and P_{out} and P_{in} by

$$\alpha = -\frac{1}{L}\ln\left(\frac{P_{in}}{P_{out}}\right) \tag{4-35}$$

Light intensity can also be used to define attenuation, because optical power is usually used to characterize fiber attenuation. If α is known, then the relationship of P_{out} can always be obtained from P_{in}, $P_{out}=P_{in}\exp(-\alpha L)$.

Generally, the optical power attenuation in an optical fiber is expressed in decibels per unit length of optical fiber, usually dB/km. The attenuation per unit length of the signal can be given as:

$$\alpha_{dB} = \frac{1}{L}10\log\left(\frac{P_{in}}{P_{out}}\right) \tag{4-36}$$

Substituting for P_{in}/P_{out} from above we obtain:

$$\alpha_{dB} = \frac{10}{\ln 10}\alpha = 4.34\alpha \tag{4-37}$$

Fig. 4-21 shows the relationship between the attenuation coefficient (dB/km) of a typical silica glass-based fiber and the wavelength. The absorption of energy by the "lattice vibration" of the constituent ions of the glass material leads to a sharp increase in the attenuation at wavelengths above 1.6 μm in the infrared region. Fundamentally, energy absorption in this region corresponds to the stretching of the Si-O bonds in ionic polarization induced by the electromagnetic wave. When the ***resonance wavelength*** of the Si-O bond is about 9 μm, the

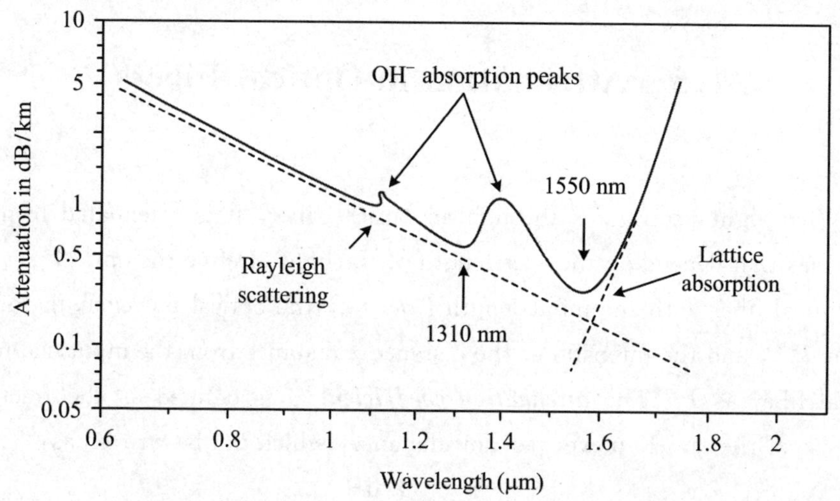

Fig. 4-21 A typical attenuation vs. wavelength characteristics of a silica-based optical fiber

There are two communications channels at 1310 nm and 1550 nm.

absorption increases as the wavelength increases. In the case of Ge-O glasses, this is further away, around 11 μm. In addition, because photons excite electrons from the valence band to the conduction band of glass, there is another inherent material absorption in the region below 500 nm, which is not shown in the figure.

There is an obvious attenuation peak centered at 1.4 μm, and a barely discernible minor peak at about 1.24 μm. Since it is difficult to remove all traces of hydroxyl products during the fiber production process, these attenuation areas are caused by the presence of hydroxyl ion impurities in the glass structure. Further, in the production process, hydrogen atoms easily diffuse into the glass structure at high temperatures, resulting in the formation of hydrogen bonds between the silicon structure and OH^- ions. Absorbed energy is mainly by the ***stretching vibrations*** of the OH^- bonds within the silica structure, which has a fundamental resonance in the infrared region (beyond 2.7 μm) but overtones or harmonics at lower wavelengths (or higher frequencies). It can be seen from the figure that the first overtone is most significant around 1.4 μm. The second overtone is about 1 μm, which can be ignored in high-quality fibers. When a small loss peak appears for SiO_2 around 1.24 μm, the first overtone of the OH^- vibration is combined with the fundamental frequency. There are two important windows in the attenuation and wavelength behavior, where the attenuation is the smallest. The window at around 1.3 μm is the region between two neighboring OH^- absorption peaks. This window is widely used in ***optical communications*** at 1310 nm. The window at around 1.55 μm is between the first harmonic absorption of OH^- and the infrared lattice absorption tail and represents the lowest attenuation. Current technological drive is to use this window for long-haul communications. Therefore, it is important to keep the hydroxyl content of the fiber at a tolerable level.

There is a background attenuation process that decreases with wavelength and is due to the Rayleigh scattering of light by the local variations in the refractive index. Glass has a noncrystalline or an amorphous structure, which means that there is no long range order to the arrangement of the atoms but only a short range order, typically a few bond lengths. The glass structure is as if the structure of the melt has been suddenly frozen. We can only define the number of bonds a given atom in the structure will have. Random variations in the bond angle from atom to atom lead to a disordered structure. There is therefore a random local variation in the density over a few atomic lengths. There random fluctuation in the refractive index give rise to light scattering and hence light attenuation along the fiber. It should be apparent that since a degree of structural randomness is an intrinsic property of the glass structure this scattering process is unavoidable and represents the lowest attenuation possible through a glass medium. As one may surmise, attenuation by scattering in a medium is minimum for light propagating through a "perfect" crystal. In this case the only scattering mechanisms will be due to thermodynamic defects (vacancies) and the random thermal vibrations of the lattice atoms.

As mentioned above, the Rayleigh scattering process decreases with wavelength and according to Rayleigh, it is inversely proportional to λ^4. The expression for the attenuation α_R in a single component glass due to Rayleigh scattering is approximately given by

$$\alpha_R \approx \frac{8\pi^3}{3\lambda^4}(n^2 - 1)\beta_T K_B T_f \tag{4-38}$$

where λ is the free space wavelength, n is the refractive index at the wavelength of interest, β_T is the isothermal compressibility (at T_f) of the glass, K_B is the Boltzmann constant, and T_f is a quantity called the fictive temperature (roughly the softening temperature of glass) at which the liquid structure during the cooling of the fiber is frozen to become the glass structure. Fiber is drawn at high temperatures and as the fiber cools eventually the temperature drops sufficiently for the atomic motions to be so sluggish that the structure become essentially "frozen-in" and remains like this even at room temperature. Thus, T_f marks the temperature below which the liquid structure is frozen and hence the density fluctuations are also frozen into the glass structure. It is apparent that Rayleigh scattering represents the lowest attenuation window at 1. 5 μm may be lowered to approach the Rayleigh scattering limit.

The most important attenuation caused by external factors is ***micro-bending loss*** and ***macro-bending loss***. The micro-bending loss is the "sharp" local bending of the optical fiber that changes the geometry and refractive index distribution of the waveguide, which causes some light energy to radiate out of the guiding direction. The sharp bend will change the local waveguide geometry, such a tortuous light forms an incident angle θ', resulting in transmitted waves (refracted waves entering the cladding) or greater penetration of the covering layer, which is visually shown in Fig. 4-22. If $\theta' < \theta_c$, the critical angle, the waveguide core does not have total reflection, and a large amount of optical power will be radiated into the cladding layer and eventually to the external medium. Since stronger penetration will cause the light field to reach the outer boundary of the cladding, part of the light will disappear in the outer coating. Attenuation increases sharply with the extent of bending; as θ', gets narrow and TIR is lost, substantially more energy is transferred into the cladding. In addition, the highest modes propagate when the incident angle θ is close to θ_c, which means that these modes are most affected. Therefore, multi-mode fibers suffer more bending losses than single-mode fibers.

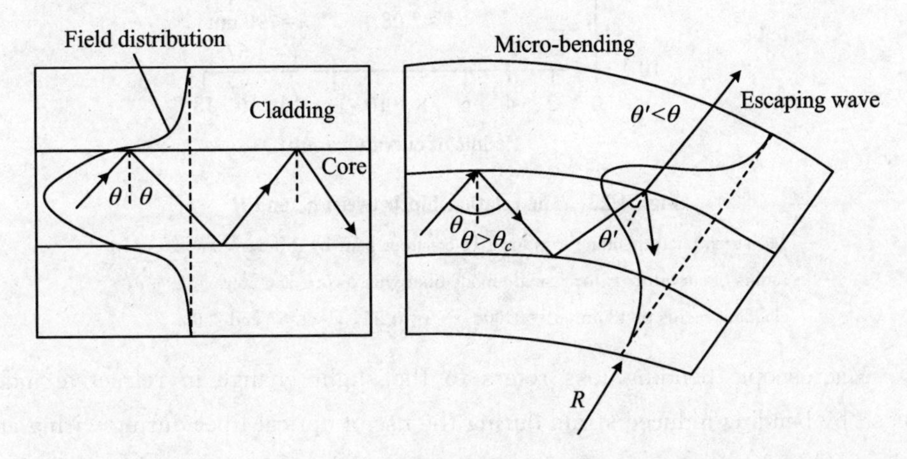

Fig. 4-22 Sharp bends change the local waveguide geometry that can lead to waves escaping

The zigzagging ray suddenly finds itself with an incidence angle θ' that gives rise to either a transmitted wave, or to greater cladding penetration; the field reaches the outside medium and some light energy is lost.

Micro-bending loss α_B increases rapidly with the increase of the bending "sharpness", that is, with the bending radius of curvature R (see Fig. 4-22 for definition). Fig. 4-23 shows the relationship between the typical micro-bending loss α_B and the radius of curvature R of a single-mode fiber at two different operating wavelengths. Obviously, α_B increases exponentially with R, which depends on the wavelength and fiber characteristics (such as the V-number). Generally, bending radii smaller than 10 mm will cause significant micro-bending losses.

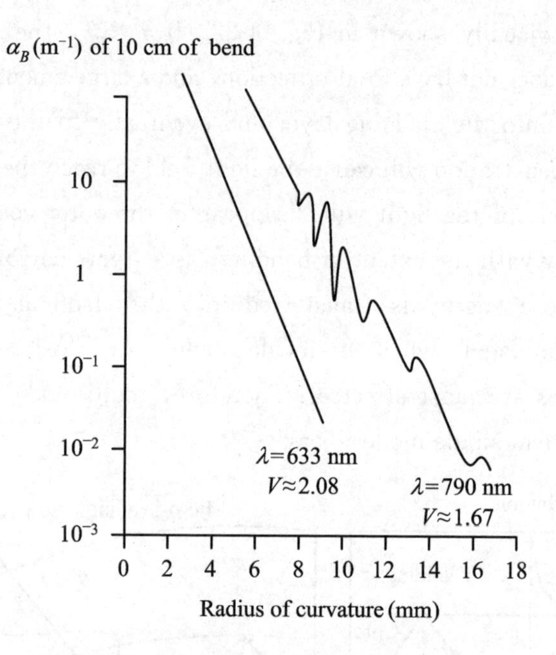

Fig. 4-23　The relationship between α_B and R

Measured micro-bending loss for a 10 cm fiber bent by different amounts of radius of curvature R for a single-mode fiber with a core diameter of 3. 9 μm, cladding radius of 48 μm, Δ=0. 004, NA=0. 11, $V\approx$1. 67 and 2. 08.

Macroscopic bending loss refers to the slight change in refractive index caused by bending induced strain during the use of optical fiber during wiring and laying. The induced strain changes n_1 and n_2, thereby affects the mode field diameter, that is, the field penetration into the cladding. Some of the increased cladding field will reach the cladding boundary and disappear in the external medium (radiation, absorption, etc.). Generally, when the radius of curvature is less than a few centimeters, the macro-bending loss will transit to the micro-bending loss.

Example　4.5.1

Problem

What is the attenuation due to Rayleigh scattering at around the λ=1. 55 μm window given that pure silica (SiO_2) has the following properties: T_f=1730 ℃ (softening temperature); $\beta_r = 7\times10^{-11}$ m^2 · N^{-1} (at high temperature); n= 1. 4446 at 1. 5 μm.

Solution

We simply calculate the Rayleigh scattering attenuation using:

$$\alpha_R \approx \frac{8\pi^3}{3\lambda^4}(n^2-1)^2\beta_r k_B T_f$$

so that

$$\alpha_R \approx \frac{8\pi^3}{3(1.55\times10^{-6})}(1.4446^2-1)^2(7\times10^{-11})(1.38\times10^{-23})(1730+273)$$

$$= 3.27\times10^{-5}\ \text{m}^{-1} \quad \text{or} \quad 3.27\times10^{-2}\ \text{km}^{-1}$$

Attenuation in dB per km is then:

$$\alpha_{\text{dB}} = 4.34\alpha_R = 4.34(3.27\times10^{-2}\ \text{km}^{-1}) = 0.142\ \text{dB/km}$$

This represents the lowest possible attenuation for a silica glass core fiber at 1.55 μm.

Example 4.5.2

Problem

The optical power emitted from the laser diode to the single-mode fiber is about 1 mW, while the photo-detector requires a minimum power of 10 nW to provide a clear signal (above noise). The fiber operates at 1.3 μm and has an attenuation coefficient of 0.4 dB/km. Without inserting a repeater (to regenerate the signal), what is the maximum length of the optical fiber?

Solution

We can use:

$$\alpha_{\text{dB}} = \frac{1}{L}10\log\left(\frac{P_{\text{in}}}{P_{\text{out}}}\right)$$

so that

$$L = \frac{1}{\alpha_{\text{dB}}}10\log\left(\frac{P_{\text{in}}}{P_{\text{out}}}\right) = \frac{1}{0.4}10\log\left(\frac{10^{-3}}{10\times10^{-9}}\right) = 125\ \text{km}$$

In addition, there will be additional losses, such as fiber bending loss, which is caused by fiber bending, which will reduce the length below this limit. For long-distance communication, the signal must be amplified by an optical amplifier, and finally regenerated by a repeater after a distance of about 50～100 km.

New Words and Expressions

［1］ **attenuation coefficient**：衰减系数。即光波在光纤中传播单位长度的光功率衰减量，用于表征光纤中的光波传输损耗。

［2］ **resonance wavelength**：共振波长。此处用于表征化学键相关的振动。

［3］ **stretching vibration**：伸缩振动。可分为对称和反对称伸缩振动。

［4］ **optical communication**：光通信。它是指以光波为信号载体的通信方式。

［5］ **micro-bending loss, macro-bending loss**：微观和宏观弯曲损耗。它是指由于微观或宏观的光纤弯曲导致的光波传输损耗。

4.6 Fiber Manufacture

4.6.1 Fiber Drawing

There are several ways to produce fiber optics for a variety of applications. The ***outside vapor deposition*** (***OVD***) technique is a commonly used method to process and produce low-loss fiber, which is also known as the outside vapor phase oxidation process.

Preform is a glass rod which has the right refractive index profile across its cross-section. It also has the right glass properties (e. g. , negligible impurity content). Preparing a preform is the first step of OVD. The diameter and length of the rod usually are 10～30 mm and 1～2 m, respectively. This rod is typically 10～30 mm in diameter and about one to two meters in length. There is a special ***fiber-drawing*** equipment to draw the optical fiber from this preform. This fabricate process is shown in Fig. 4-24.

As shown in the picture above, through the hot furnace, the preform rod flows like a viscous melt (resembling honey) around 1900～2000 ℃. The preform tip is pulled with the appropriate tension when the rod reaches the hot zone and its end begins to flow. It come out as a fiber and spooled on a rotating take-up drum. To achieve the required waveguide characteristics, the fiber diameter should be precisely controlled. The changes of the fiber diameter are monitored

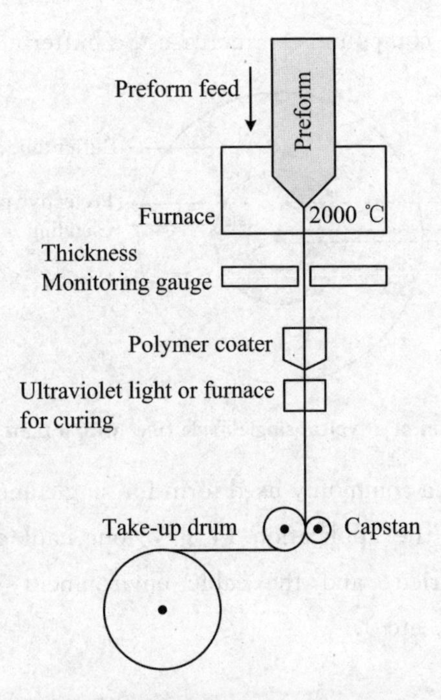

Fig. 4-24 Schematic illustration of a fiber drawing tower

by an optical thickness monitor which is used (in an automatic feedback control system) to adjust the speed of the preform feeder and the fiber-winding mechanism to maintain a constant fiber dimeter, typically better than 0.1%. The preform is hollow in some cases which simply collapses during the drawing and does not affect the final drawn fiber.

It is very important that when the fiber is stretched, it will be coated with a layer of polymer (e.g., urethane acrylate) to protect the surface of the fiber mechanically and chemically. The fiber will develop various microcracks on the surface which dramatically reduce the mechanical strength (fracture strength) when exposed to ambient conditions. The polymeric coating is initially a viscous liquid and is cured (hardened) when the coated optical fiber passes through the curing oven. It can also be cured by ultraviolet lamps to UV hardened. Sometimes a two-layer polymer coating is required. The cladding is usually 125~150 μm, and the total diameter of the polymer coating is 250~500 μm. Fig. 4-25 shows the schematic diagram of the cross-section of typical single-mode optical fiber. In this example, there is a thick polymeric buffer tube or a buffer jacket outside the fiber and coating to reduce the influence of mechanical stress and microbending (sharp bending) on the optical fiber. The optical fiber loose in the buffer tube filled with compound can increase the buffer ability.

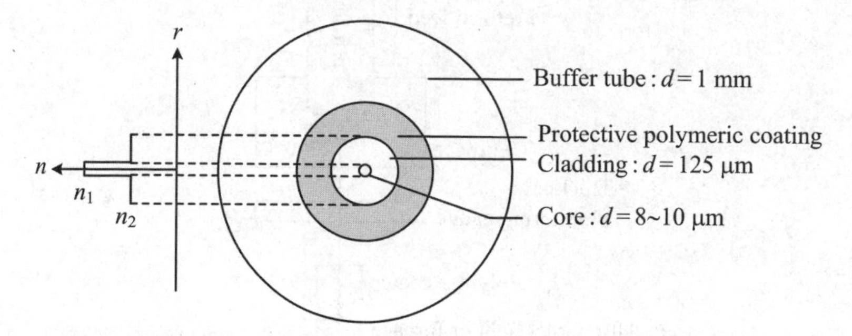

Fig. 4-25 The cross-section of a typical single-mode fiber with a tight buffer tube (d=diameter)

The cable form is a commonly used form for single and multiple fibers which structure depends on the application (c. g., long-haul communications), the number of fibers carried, and the cable environment (e. g., underground, underwater, overhead, etc.).

4.6.2 Outside Vapor Deposition (OVD)

Outside vapor deposition (OVD) is one of the vapor deposition techniques used to produce the rod preform used in fiber drawing. The OVD process has two stages and illustrated in Fig. 4-26. A fused silica glass rod (or a ceramic rod such

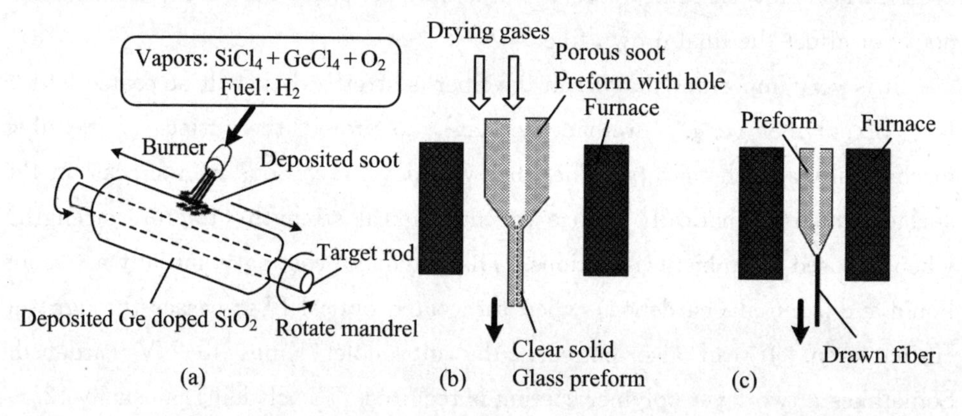

Fig. 4-26 Schematic illustration of OVD and the preform preparation for fiber drawing

(a) Reaction of gases in the burner flame produces glass soot that deposits on to the outside surface of the mandrel. (b) The mandrel is removed and the hollow porous soot preform is consolidated; the soot particles are sintered, fused together, to form a clear glass rod. (c) The consolidated glass rod is used as preform in fiber drawing.

as alumina) as a target rod is the first laydown stage, as shown in Fig. 4-26(a). This acts as a mandrel and is rotated. Through depositing glass soot particles, the required glass material for the preform with the right composition is grown on the outside surface of this target rod. Burning various gases in an oxyhydrogen burner(torch) flame can achieve the deposition and the glass soot is produced as reaction products.

Suppose that we need a higher refractive index preform with a core that has germania (GeO_2) in silica glass. The required gases, $SiCl_4$(silicon tetrachloride), $GeCl_4$(germanium tetrachloride), and fuel in the form of oxygen O_2 and hydrogen H_2, are burnt in a burner flame over the target rod surface as shown in Fig. 4-26 (a). The important reactions of the gases in the flame are

$$SiCl_4\,(gas) + O_2\,(gas) \rightarrow SiO_2\,(solid) + 2Cl_2\,(gas)$$

$$GeCl_4\,(gas) + O_2\,(gas) \rightarrow GeO_2\,(solid) + 2Cl_2\,(gas)$$

The "soot" deposit on the outside surface of the target rod and form a porous glass layer as the burner travels along the mandrel. It is the fine glass particles, silica and germania produced in these reactions. By actually moving the burner or moving the mandrel (the result is the same), the burner is slowly moved up and down along the length of the rotating mandrel to build the glass preform layer by layer. The layers for the core region are deposited firstly and then deposited the layers for then cladding region by adjusting the gas composition. The final preform may have about 200 layers commonly. By adjusting the relative amounts of $SiCl_4$ and $GeCl_4$ gases fed into the burner for chemical reaction, the composition and hence the refractive index of each layer can be controlled. Indeed, by controlling the layer compositions, any desired refractive index profile.

Once all the necessary glass layers have been deposited, the central target mandrel is removed, which leaves a hollow porous glass preform rod, i. e. , a porous opaque glass tube. Sintering this porous glass rod is the second consolidation stage as shown in Fig. 4-26(b). The porous preform is sent to a consolidation furnace (1400～1600 ℃), where the fine glass particles are sintered (fused) at high temperature into a dense, transparent solid, the glass preform. In order to avoid unacceptably high attenuation, drying gases (such as chlorine or thionyl chloride) are forced in to remove water vapors and hydroxyl impurities. As shown in Fig. 4-26(c) and described above, this clear glass preform is then fed into a draw furnace to draw the fiber. The central hollow simply collapses and fuses at the high temperatures of the draw process. Typically, making a preform and the subsequent drawing of a fiber from the preform take a few more hours, respectively. The manufacturing cost reflects in the fiber cost, which is in excess of $25 per km.

Example 4.6.1

Problem

Drawing a fiber usually uses a preform of length 110 cm and diameter 20 mm. The fiber drawing rate is supposed to $5\ \mathrm{m \cdot s^{-1}}$. If the last 10 cm of the preform is not drawn and the fiber diameter is 125 μm, what is the maximum length of fiber that can be drawn from this preform? And how long does it take to draw the fiber?

Solution

The length of the preform is assumed to L_p to draw as fiber. The volume must also be conserved because the density is the same and the mass is conserved. The fiber and preform diameters is d_f and d_p, and the lengths of the fiber and drawn preform is L_f and L_p, respectively, then:

$$L_f d_f^2 = L_p d_f^2$$

For example,

$$L_f = \frac{(1.1 - 0.1\ \mathrm{m})(20 \times 10^{-3}\ \mathrm{m})^2}{(125 \times 10^{-6}\ \mathrm{m})^2} = 25600\ \mathrm{m} \quad \text{or} \quad 25.6\ \mathrm{km}$$

Since the rate is 5 m/s, the time it takes to draw the fiber in minutes is

$$\text{Time (hrs)} = \frac{\text{Length(km)}}{\text{Rate(km/hr)}} = \frac{25600\ \mathrm{m}}{(5\ \mathrm{m/s})(60 \times 60\ \mathrm{s/hr})} = 1.4\ \text{hrs}$$

Typical drawing rates are in the range 5～20 m/s, so that 1.4 hrs would be on the long-side.

New Words and Expressions

[1] **outside vapor deposition (OVD)**：外部气相沉积法。它是指化学气体或蒸汽在基质表面反应合成涂层或纳米材料的方法。

[2] **fiber-drawing**：光纤拉丝。它是一种将预制棒拉制成符合标准的光纤的工艺。

Questions

1. There are three questions about dielectric slab waveguide.

(1) Consider the rays 1 and 2 in Fig. 4-4. Derive the waveguide condition.

(2) Consider the rays 1 and 2 in Fig. 4-5. Show that the phase difference when they meet at C at a distance y above the guide center is

$$\Phi_m = k_1 2(a-y)\cos\theta_m - \phi_m$$

(3) Using the waveguide condition, show that

$$\Phi_m = \Phi_m(y) = m\pi - \frac{y}{a}(m\pi + \phi_m)$$

2. This question is about TE field pattern in slab waveguide.

Consider two parallel rays 1 and 2 interfering in the guide as in Fig. 4-5. Given the phase difference (as in Question 1).

$$\Phi_m = \Phi_m(y) = m\pi - \frac{y}{a}(m\pi + \phi_m)$$

Between the waves at C, distance y above the guide center, find the electric field pattern $E(y)$ in the guide. With a core thickness of 20 μm, $n_1 = 1.455$, $n_2 = 1.44$, light wavelength of 1.3 μm, plot the field pattern for the first three modes taking a planar dielectric guide.

3. This question is about TE and TM modes in dielectric slab waveguide.

Consider a planar dielectric guide with a core thickness of 20 μm, $n_1 = 1.455$, $n_2 = 1.44$, light wavelength of 1.3 μm. Given the waveguide condition, equation (4-3) in Section 4.1, and the expressions for phase changes ϕ and ϕ' in TIR for the TE and TM modes, respectively,

$$\tan\left(\frac{1}{2}\phi_m\right) = \frac{\left[\sin^2\theta_m - \left(\frac{n_2}{n_1}\right)^2\right]^{1/2}}{\cos\theta_m}, \quad \tan\left(\frac{1}{2}\phi_m'\right) = \frac{\left[\sin^2\theta_m - \left(\frac{n_2}{n_1}\right)^2\right]^{1/2}}{\left(\frac{n_2}{n_1}\right)^2\cos\theta_m}$$

Find the angle θ for the fundamental TE and TM modes by using a graphical solution and compare their propagation constants along the guide.

4. This question is about group velocity.

We can use a convenient mathematical software package to calculate the group velocity of a given mode as a function of frequency ω. Assume that the mathematical software package can perform symbolic algebra such as partial differentiation (the author uses Maple's Mathview, but other ones can also be used). The propagation constant of a given mode is $\beta = k_1 \sin\theta$ in which β and θ imply β_m and θ_m. The objective is to express β and ω in terms of θ. Since $k_1 = n_1\omega/c$, the waveguide condition in equation (4-11) in Section 4.1 is

$$\tan\left(a\frac{\beta}{\sin\theta}\cos\theta - m\frac{\pi}{2}\right) = \frac{\left[\sin^2\theta - \left(\frac{n_2}{n_1}\right)^2\right]^{1/2}}{\cos\theta}$$

so that

$$\beta = \frac{\tan\theta}{a}\left[\arctan\left(\sec\theta\sqrt{\sin^2\theta - \left(\frac{n_2}{n_1}\right)^2}\right) + m\frac{\pi}{2}\right] = F_m(\theta) \tag{4-39}$$

The frequency is given by

$$\omega = \frac{c\beta}{n_1\sin\theta} = \frac{c}{n_1\sin\theta}F_m(\theta) \tag{4-40}$$

Both β and ω are now functions of θ. Then the group velocity is

$$V_g = \frac{d\omega}{d\beta} = \frac{d\omega}{d\theta}\cdot\frac{d\theta}{d\beta} = \frac{c}{n_1}\left[\frac{F'_m(\theta)}{\sin\theta} - \frac{\cos^2\theta}{\sin\theta}F_m(\theta)\right]\left[\frac{1}{F'_m(\theta)}\right]$$

For example:

$$V_g = \frac{c}{n_1\sin\theta}\left[1 - \cos^2\theta\frac{F_m(\theta)}{F_m^2(\theta)}\right] \tag{4-41}$$

For a given m value, equation (4-40) and equation (4-41) can be plotted parametrically, that for each θ value we can calculate ω and V_g and plot V_g vs. ω. Fig. 4-27 shows an example for a guide with the characteristics in the caption. Use easy-to-use math software packages, or through other means, to get the same V_g vs. ω behavior and discuss the intermodal dispersion and whether equation (4-40) in Section 4.2 is appropriate.

5. This question is about dielectric slab waveguide.

Consider a dielectric slab waveguide with a thin layer of GaAs with a thickness of 0.2 μm between two AlGaAs layers. The refractive index of GaAs and the AlGaAs layers are 3.66 μm and 3.4 μm, respectively. Assuming that the refractive index does not vary greatly with wavelength, what is the cut-off wavelength at which only a single mode can propagate in the waveguide? If radiation with a wavelength of 870 nm (corresponding to bandgap radiation) propagates in the GaAs layer, what is the penetration of the evanescent wave to

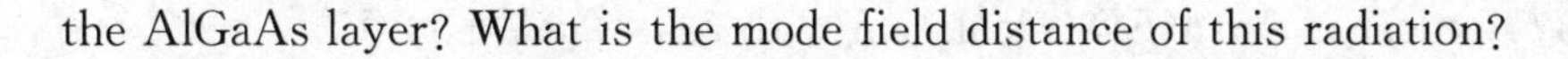

the AlGaAs layer? What is the mode field distance of this radiation?

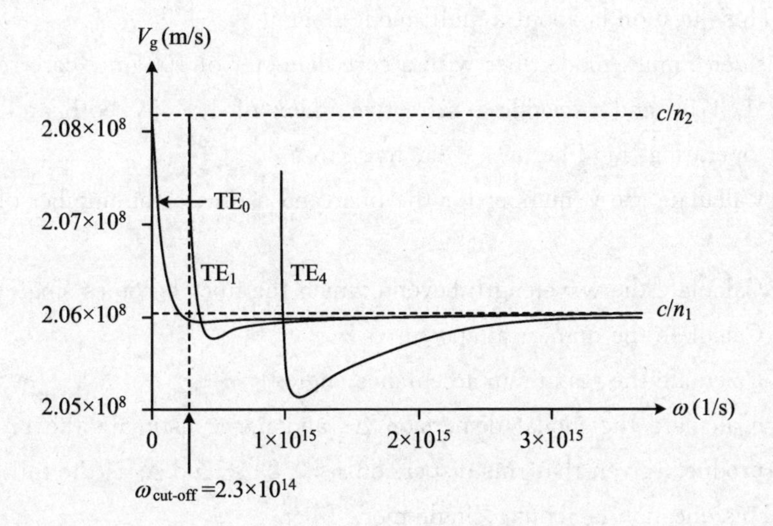

Fig. 4-27 Group velocity vs. angular frequency for three modes for a planar dielectric waveguide which has $n_1 = 1.455$, $n_2 = 1.440$, $a = 10$ μm

Results from Mathview, Waterloo Maple math—software application. TE_0 is for $m=0$, etc.

6. This question is about dielectric slab waveguide.

Consider a slab dielectric waveguide with a core thickness ($2a$) of 10 μm, $n_1=3$, $n_2=1.5$. Solution of the waveguide condition in equation (4-11) in Example 4.1.1 gives the mode angles θ_0 and θ_1 for the TE_0 and TE_1 modes for selected wavelengths as summarized in the Table 4-3. For each wavelength calculate ω and β_m and then plot ω vs. β_m. On the same plot shows the lines with slopes c/n_1 and c/n_2. Compare your plot with the dispersion diagram in Fig. 4-11.

Table 4-3 Question 6

λ(μm)	15	20	25	30	40	45	50	70	100	150	200
θ_0(°)	77.8	74.52	71.5	68.7	63.9	61.7	59.74	53.2	46.4	39.9	36.45
θ_1(°)	65.2	58.15	51.6	45.5	35.5	32.02	30.17	—	—	—	—

7. This question is about dielectric slab waveguide.

Consider a planar dielectric waveguide with a core thickness of 10 μm, $n_1=1.4446$, $n_2=1.4440$. Calculate the V-number, the mode angle θ_m for $m=0$ (use a graphical solution, if necessary), penetration depth, and mode field distance ($MFD = 2a + 2\delta$), for light wavelengths of 1 μm and 5 μm. What is your conclusion? Compare your *MFD* calculation with $2\omega_0 = 2a(V+1)/V$.

8. This question is about a multi-mode fiber.

Consider a multi-mode fiber with a core diameter of 100 μm, core refractive index of 1. 475, and a cladding refractive index of 1. 455, both at 850 nm. Consider operating this fiber at $\lambda = 850$ nm.

(1) Calculate the V-number for the fiber and estimate the number of guided modes.

(2) Calculate the wavelength beyond which the fiber becomes single-mode.

(3) Calculate the numerical aperture.

(4) Calculate the maximum acceptance angle.

(5) Calculate the modal dispersion $\Delta\tau$ and hence estimate the bit rate × distance product, given that rms dispersion $\sigma \approx 0.29\Delta\tau$ and $\Delta\tau$ is the full spread.

9. This question is about a single-mode fiber.

Consider a SiO_2-13. 5%GeO_2 fiber with a core diameter of 8 μm, a refractive index of 1. 468, and a cladding refractive index of 1. 464. The fiber is operated with a laser source with a half maximum width of 2 nm.

(1) Calculate the V-number for the fiber. Is this a single-mode fiber?

(2) Calculate the wavelength at which the fiber becomes multi-mode.

(3) Calculate the numerical aperture.

(4) Calculate the maximum acceptance angle.

(5) Obtain the material dispersion and waveguide dispersion, thereby estimating the bit rate × distance product of the fiber.

10. This is a question about a single-mode fiber design.

The pure SiO_2 and SiO_2-13. 5% GeO_2 n vs. λ is provided by the Sellmeier equation. The refractive index increases linearly with the addition of GeO_2 to SiO_2 from 0 to 13. 5%. A single-mode step index fiber needs to have the following properties: $NA = 0.1$, core diameter of 9 μm, and a core of SiO_2-13. 5%GeO_2. What should the cladding composition be?

11. This is a question about material dispersion.

If N_{g1} is the group refractive index of the core material of a step fiber, then the propagation time (group delay time) of the fundamental mode is

$$\tau = \frac{L}{V_g} = \frac{LN_{g1}}{c}$$

Since N_g will depend on the wavelength, it is proved that the material dispersion coefficient D_m is approximately given by

$$D_m = \frac{d\tau}{L d\lambda} \approx \frac{\lambda}{c} \frac{d^2 n}{d\lambda^2}$$

Using the Sellmeier equation, evaluate the material dispersion of pure silicon dioxide (SiO_2) and SiO_2-13.5%GeO_2 glass at $\lambda = 1.55\ \mu m$.

12. This is a question about waveguide dispersion.

Waveguide dispersion is the result of the correlation between the propagation constant and the V-number, and the V-number depends on the wavelength. It exists even if the refractive index is constant without material dispersion. Let us assume that n_1 and n_2 are independent of wavelength (or k). Assume that β is the propagation constant of mode lm and $k = 2\pi/\lambda$, where λ is the free-space wavelength. Then the relationship between the normalized propagation constant b and the propagation constant k is

$$\beta = n_2 k (1 + b\Delta) \tag{4-42}$$

The group velocity is defined and given by

$$V_g = \frac{d\omega}{d\beta} = c\frac{dk}{d\beta}$$

prove that the propagation time or the group delay time, τ of the mode is

$$\tau = \frac{L}{V_g} = \frac{L n_2}{c} + \frac{L n_2 \Delta}{c} \frac{d(kb)}{dk} \tag{4-43}$$

given the definition of V:

$$V = ka\,(n_1^2 - n_2^2)^{1/2} \approx k a n_2 (2\Delta)^{1/2} \tag{4-44}$$

and

$$\frac{d(Vb)}{dV} = \frac{d}{dV}[bkan_2(2\Delta)]^{1/2} = an_2(2\Delta)^{1/2}\frac{d}{dV}(bk) \tag{4-45}$$

Show that,

$$\frac{d\tau}{d\lambda} = -\frac{L n_2 \Delta}{c\lambda} V \frac{d^2(Vb)}{dV^2} \tag{4-46}$$

and that the waveguide dispersion coefficient is

$$D_w = \frac{d\tau}{L d\lambda} = -\frac{n_2 \Delta}{c\lambda} V \frac{d^2(Vb)}{dV^2} \tag{4-47}$$

Fig. 4-28 shows the dependence of $V[d^2(Vb)/dV^2]$ on the V-number. In the range $1.5 < V < 2.4$,

$$V\frac{d^2(Vb)}{dV^2} \approx \frac{1.984}{V^2}$$

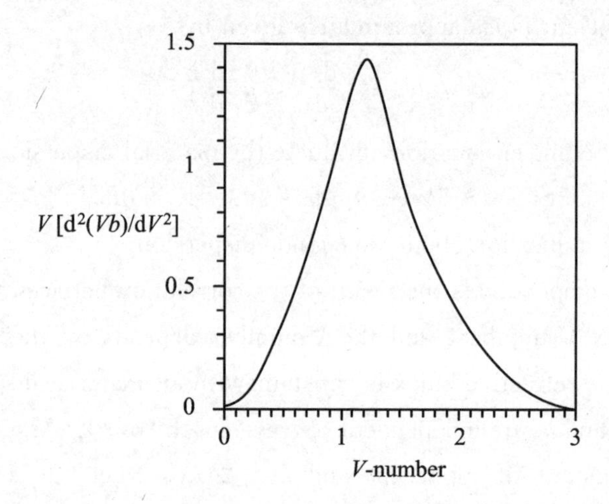

Fig. 4-28 $V[d^2(Vb)/dV^2]$ vs. V-number for a step index fiber

show that,

$$D_w \approx -\frac{n_2\Delta}{c\lambda}\frac{1.984}{V^2} = -\frac{(n_1 - n_2)}{c\lambda}\frac{1.984}{V^2} \tag{4-48}$$

which simplifies to

$$D_w \approx -\frac{1.984}{c(2\pi a)^2 2n_2}\lambda \tag{4-49}$$

Equation (4-43) should really have N_{g2} instead of n_2 in which case equation (4-50) would be

$$D_w \approx -\frac{1.984 N_{g2}}{c(2\pi a)^2 2n_2^2}\lambda \tag{4-50}$$

Consider a fiber with a core diameter of 8 μm, a refractive index of 1.468, and a cladding refractive index of 1.464. Both have a refractive index of 1300 nm. Assume that a 1.3 μm laser diode with a spectral linewidth of 2 nm is used to provide input light pulses. Use the equation (4-47) and equation (4-50) to estimate the waveguide dispersion per kilometer of the fiber.

13. This is a question about profile dispersion.

The material dispersion and waveguide dispersion are the primarily reasons for the total dispersion in a single-mode step index fiber. However, there is an additional dispersion mechanism called distributed dispersion, which is produced by the propagation constant β of the fundamental mode and also depends on the refractive index difference Δ. Consider a light source with a wavelength range of $\delta\lambda$ coupled into a step index fiber. We can view this as a change $\delta\lambda$ in the input wavelength λ. Assume that n_1, n_2, hence Δ depends on the wavelength λ. The propagation time, or the group delay time, τ_g per unit length is

$$\tau_g = \frac{1}{V_g} = \frac{1}{c}\left(\frac{d\beta}{dk}\right) \tag{4-51}$$

Since β depends on n_1, Δ, and V, let us consider τ_g as a function of n_1, Δ (thus n_2), and V. A change $\delta\lambda$ in λ will change each of these quantities. Using the partial differential chain rule:

$$\frac{\delta\tau_g}{\delta\lambda} = \frac{\partial\tau_g}{\partial n_1}\frac{\partial n_1}{\partial\lambda} + \frac{\partial\tau_g}{\partial V}\frac{\partial V}{\partial\lambda} + \frac{\partial\tau_g}{\partial\Delta}\frac{\partial\Delta}{\partial\lambda} \tag{4-52}$$

The mathematics tums out to be complicated, but the statement in equation (4-52) is equivalent to

Total dispersion= Material dispersion (due to $\partial n_1/\partial\lambda$)
+Waveguide dispersion (due to $\partial V/\partial\lambda$)
+Profile dispersion (due to $\partial\Delta/\partial\lambda$)

in which the last term is due to Δ depending on λ; although small, this is not zero. Even though the statement in equation (4-52) above is oversimplified, it still provides insight into the problem. The total intramode (chromatic) dispersion coefficient D_{ch} is then given by

$$D_{ch} = D_m + D_w + D_p \tag{4-53}$$

in which D_m, D_w, and D_p are material, waveguide, and profile dispersion coefficients, respectively. The waveguide dispersion is given by equation (4-52), and the profile dispersion coefficient is (very) approximately:

$$D_p \approx -\frac{N_{g1}}{c}\left(V\frac{d^2(Vb)}{dV^2}\right)\left(\frac{d\Delta}{d\lambda}\right) \tag{4-54}$$

in which b is the normalized propagation constant and $V[d^2(Vb)/dV^2]$ vs. V is shown in Fig. 4-28. The term $V[d^2(Vb)/dV^2]\approx 1.984/V^2$.

Consider a fiber with a core of diameter of 8 μm. The refractive and group indices of the core and cladding at $\lambda=1.55$ μm are $n_1=1.4504$, $n_2=1.445$, $N_{g1}=1.4676$, $N_{g2}=1.4625$, and $d\Delta/d\lambda=161$ m^{-1}. Estimate the waveguide and distributed dispersion of the fiber per kilometer of input light linewidth per nanometer at this wavelength.

14. This is a question about a graded index fiber.

(1) Consider an optimal graded index fiber with a core diameter of 30 μm, a refractive index in the center of the core of 1.474, and a cladding refractive index

of 1.453. Assuming fiber coupling to 1300 nm laser diode transmitter and 3 nm spectral linewidth. Suppose that the material dispersion coefficient at this wavelength is about -5 ps · km^{-1} · nm^{-1}. Calculate the total dispersion and estimate the bit rate×distance product of the fiber. How does this compare to the performance of a multi-mode fiber with the same core radius, n_1 and n_2? What would the total dispersion and maximum bit rate be if an LED source of spectral width $\Delta\lambda_{1/2}\approx 80$ nm is used?

(2) If $\sigma_{\text{intermode}}(\gamma)$ is the rms dispersion in a graded index fiber with a profile index γ, and if γ_0 is the optimal profile index, then

$$\frac{\sigma_{\text{intermode}}(\gamma)}{\sigma_{\text{intermode}}(\gamma_0)}=\frac{2(\gamma-\gamma_0)}{\Delta(\gamma+2)}$$

Calculate the new dispersion and bit rate × distance product if γ is 10% greater than the optimal value γ_0.

·15. This is a question about a planar waveguide with stratified medium (approximation to a graded index fiber).

Fig. 4-29 shows a planar dielectric waveguide in which the refractive index changes from n_1 to n_2 to n_3 and so on along y at $y=\delta/2$, $3\delta/2$, $5\delta/2$, …. Thus, the refractive index decreases one step at a time from $y=0$ along y as depicted in the figure.

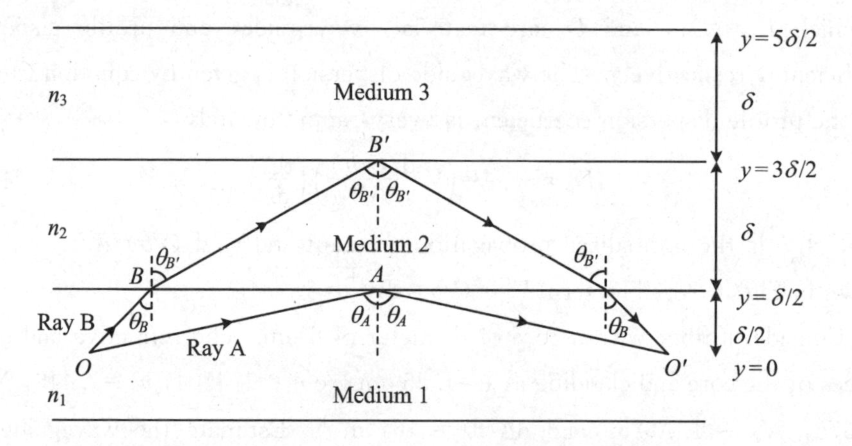

Fig. 4-29 Step-graded-index dielectric waveguide

Two rays are launched from the center of the waveguide at O at angle θ_A and θ_B such that ray A suffers TIR at A and ray B suffers TIR at B', both TIRs are at critical angles.

Consider the paths of two rays, ray A and ray B, starting from O. The first ray suffers total internal refraction at A and travels to O'. The launching angle for the first ray is such that θ_A is the critical angle for TIR. Ray B is launched at a small angle, penetrates medium 2 at point B, and then propagates toward point B' at the boundary between media 2 and 3. The emission of ray B makes the angle $\theta_{B'}$ at point B, the critical angle of TIR between media 2 and 3.

The question is what should be the relationship between n_1, n_2 and n_3, so that two rays of light can reach point O' at the same time, so that we can observe that there is no time dispersion. According to symmetry, if two rays reach O' at the same time, they must also reach their respective TIR points A and B' at the same time.

(1) Prove that the time taken for the first ray to travel from O to A is

$$t_{OA} = \frac{(1/2)\delta/\cos\theta_A}{\dfrac{c}{n_1}} = \frac{(1/2)\delta n_1}{c\left[1-\left(\dfrac{n_2}{n_1}\right)^2\right]^{1/2}} \tag{4-55}$$

(2) Prove that the time taken for the second ray to travel from O to B' is

$$t_{OB'} = \frac{(1/2)\delta n_1}{c\left[1-\left(\dfrac{n_3}{n_1}\right)^2\right]^{1/2}} + \frac{\delta n_2}{c\left[1-\left(\dfrac{n_3}{n_2}\right)^2\right]^{1/2}} \tag{4-56}$$

Let the step variations of n at $y=\delta/2, 3\delta/2, \cdots$ obey

$$n^2 = n_1^2\left[1-2\Delta\left(\frac{y}{a}\right)^{\gamma}\right] \tag{4-57}$$

in which Δ is some constant (less than unity), and γ is an index that describes the profile of the refractive index along y. Obviously, at $y=0$, $n=n_1$, as we expect. show that,

$$n_2^2 = n_1^2(1-\varepsilon)$$

in which

$$\varepsilon = 2\Delta\left(\frac{\delta}{2a}\right)\gamma \tag{4-58}$$

and

$$n_3^2 = n_1^2[1-\varepsilon(3^{\gamma})] \tag{4-59}$$

(3) Consider the condition $t_{OA} - t_{OB'} = 0$ so that the two rays arrive at the same time at O'. Using the results in equation (4-55) and equation (4-56) and the refractive indices in equation (4-58) and equation (4-59), it shows that the two rays arrive at the same time if

$$\frac{2(1-\varepsilon)}{(3^{\gamma}-1)^{1/2}} + \frac{1}{3^{\gamma/2}} - 1 = 0 \tag{4-60}$$

As the layer thickness δ becomes small we have $\varepsilon \to 0$. Show that as $\varepsilon \to 0$,

$\gamma=2.067$ is a solution of equation (4-60).

Do you draw conclusions from this exercise? Which type of graded index fiber would you recommend to achieve the smallest inter-mode dispersion? Schematically plot the refractive index distribution along the radial direction.

What is the main theoretical limitation of the treatment above? Is $\varepsilon\to 0$ a valid assumption? What will happen to γ if k was not zero but a small number?

16. This is a question about GRIN rod lenses.

Graded index (GRIN) rod lens is a glass rod whose refractive index changes parabolically from the central axis of its maximum refractive index. It is like a very thick, short graded index fiber, which may be 0.5～5 mm in diameter. This GRIN rod lens of different lengths can be used to focus or collimate light, as shown in Fig. 4-30. The working principle can be understood by considering the ray trajectory in the layered medium, as shown in Fig. 4-28 and Fig. 4-26(b), where the ray trajectory is a sinusoidal path. A pitch (P) is a complete cycle of the ray trajectory along the rod axis. Fig. 4-30(a), (b) and (c) show half pitch ($0.5P$), quarter pitch ($0.25P$), and $0.23P$ GRIN rod lenses. The point O in Fig. 4-30 (a) and (b) is located at the center of the face, while in Fig. 4-30 (c) it is slightly away from the face.

(1) How would you represent Fig. 4-30 (a) using two conventional converging lenses. What are O and O'?

(2) How would you represent Fig. 4-30(b) using a conventional converging lens What is O'?

(3) Sketch ray paths for a GRIN rod with a pitch between $0.25P$ and $0.5P$ starting from O at the face center. Where is O'?

(4) What use is $0.23P$ GRIN rod lens in Fig. 4-30(c)?

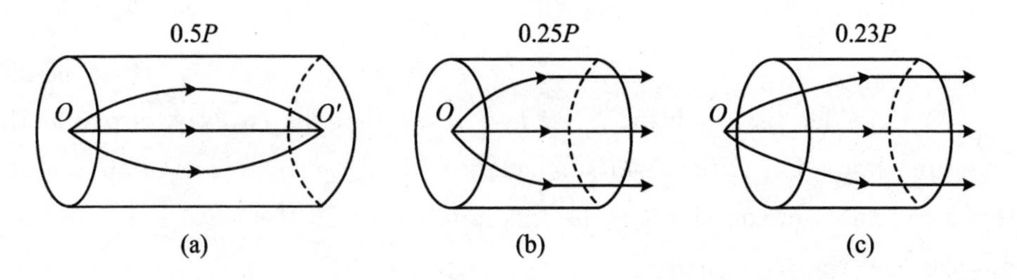

Fig. 4-30　Graded index (GRIN) rod lenses of different pitches

(a) Point O is on the rod face center and the lens focuses the rays onto O' on to the center of the opposite face. (b) The rays from O on the rod face center are collimated out. (c) O is slightly away from the rod face and the rays are collimated out.

17. This is a question about optical fibers.

Consider the manufacturing and materials used for the optical fiber.

(1) What factors would reduce dispersion?

(2) What factors would reduce attenuation?

18. This is a question about micro-bending loss.

It is found that for a single-mode fiber with a cut-off wavelength $\lambda_c = 1180$ nm, operating at 1300 nm, the micro-bending loss reaches 1 dB · m^{-1} when the radius of curvature of the bend is roughly 6 mm for $\Delta = 0.00825$, 12 mm for $\Delta = 0.0055$, and 35 mm for $\Delta = 0.00275$. Explain these findings.

19. This is a question about micro-bending loss.

The fiber characteristics and wavelength determine the micro-bending loss α_B. We will use the single-mode fiber micro-bending loss equation to approximate the α_B for a given fiber parameter:

$$\alpha_B = \frac{\pi^{1/2}\, k^2}{2\,\gamma^{3/2}\, V^2\, [K_1(\gamma a)]^2} R^{-1/2} \exp\left(-\frac{2\,\gamma^3}{3\beta^2} R\right)$$

where R is the bend radius of curvature, α is the fiber radius, β is the propagation constant, determined by b, normalized propagation constant, which is related to V, $\beta = n_2 k(1 + b\Delta)$; $k = 2\pi/\lambda$ is the free-space wave vector; $\gamma = \sqrt{\beta^2 - n_2^2 k^2}$; $k = \sqrt{n_1^2 k^2 - \beta}$, and $K_1(\chi)$ is a first-order modified Bessel function, readily available in math software packages. And $b = (1.1428 - 0.996V^{-1})^2$. Consider a single-mode fiber with $n_1 = 1.45$, $n_2 = 1.446$, $2a$ (diameter) $= 3.9\ \mu$m. Plot α_B vs. R for $\lambda = 633$ nm and 790 nm from $R = 2$ mm to 15 mm. What is your conclusion?

Chapter 5
Application of Optoelectronic Information Technology

The 20th century is the century of microelectronics; the 21st century is the century of photoelectron. At present, optoelectronic information technology has been widely used in all walks of life. This chapter introduces the typical application of optoelectric information technology.

5.1 Holography Technology

Holography is a technique which enables three-dimensional images (holograms) to be made. It involves the use of a laser, interference, diffraction, light intensity recording and suitable illumination of the recording. The holographic image, like Fig. 5-1, changes as the position and orientation of the

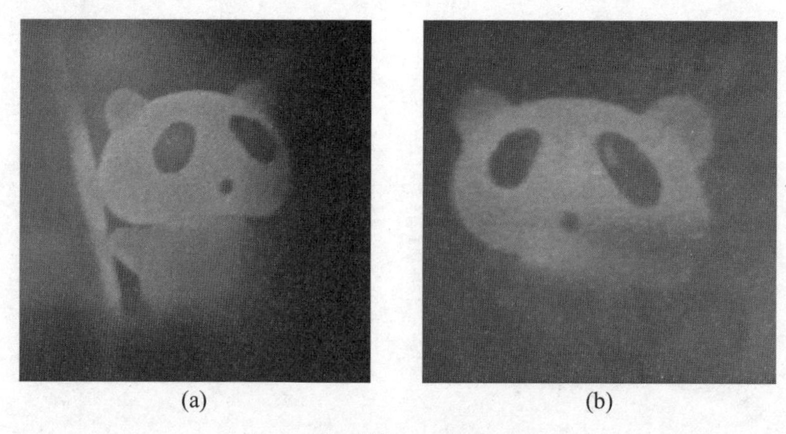

(a) (b)

Fig. 5-1 Holographic images

viewing system changes exactly in the same way as if the object were still present, thus making the image appear three-dimensional.

The holographic recording itself is not an image; it consists of an apparently random structure of either varying intensity, density or profile.

5.1.1 Overview and History of Holography

The Hungarian-British physicist, Dennis Gabor, was awarded the Nobel Prize in physics in 1971 "for his invention and development of the holographic method". His work, done in the late 1940s, built on pioneering work in the field of X-ray microscopy by other scientists including Mieczysław Wolfke in 1920 and W. L. Bragg in 1939. The discovery was an unexpected result of research into improving electron microscopes at the British Thomson-Houston (BTH) Company in Rugby, England, and the company filed a patent in December, 1947 (patent GB685286). The technique as originally invented is still used in electron microscopy, where it is known as electron holography, but optical holography did not really advance until the development of the laser in 1960.

The development of the laser enabled the first practical optical holograms that recorded 3D objects to be made in 1962 by Yuri Denisyuk in the Soviet Union and by Emmett Leith and Juris Upatnieks at the University of Michigan, USA. Early holograms used silver halide photographic emulsions as the ***recording medium***. They were not very efficient as the produced grating absorbed much of the incident light. Various methods of converting the variation in transmission to a variation in refractive index (known as "bleaching") were developed which enabled much more efficient holograms to be produced.

Several types of holograms can be made. ***Transmission holograms***, such as those produced by Leith and Upatnieks, are viewed by shining laser light through them and looking at the reconstructed image from the side of the hologram opposite the source. A later refinement, the "rainbow transmission" hologram, allows more convenient illumination by white light rather than by lasers. ***Rainbow holograms*** are commonly used for security and authentication, for example, on credit cards and product packaging.

Another kind of common hologram, the ***reflection hologram*** or Denisyuk hologram, can also be viewed using a white-light illumination source on the same side of the hologram as the viewer and is the type of hologram normally seen in holographic displays. They are also capable of multicolour-image reproduction.

Specular holography is a related technique for making three-dimensional images by controlling the motion of specularities on a two-dimensional surface. It works by reflectively or refractively manipulating bundles of light rays, whereas Gabor-style holography works by diffractively ***reconstructing wavefronts***.

In its early days, holography required high-power expensive lasers, but nowadays, mass-produced low-cost semi-conductor or diode lasers, such as those found in millions of DVD recorders and used in other common applications, can be used to make holograms and have made holography much more accessible to low-budget researchers, artists and dedicated hobbyists.

5.1.2 How Holography Works

Holography is a technique that enables a light field, which is generally the product of a light source scattered off objects, to be recorded and later reconstructed when the original light field is no longer present, due to the absence of the original objects. Holography can be thought of as somewhat similar to sound recording, whereby a sound field created by vibrating matter like musical instruments or vocal cords, is encoded in such a way that it can be reproduced later, without the presence of the original vibrating matter.

Holograms are recorded using a flash of light that illuminates a scene and then imprints on a recording medium, much in the way a photograph is recorded. In addition, however, part of the light beam must be shone directly onto the recording medium-this second light beam is known as thc ***reference beam***. A hologram requires a laser as the sole light source. Lasers can be precisely controlled and have a fixed wavelength, unlike sunlight or light from conventional sources, which contain many different wavelengths. To prevent external light from interfering, holograms are usually taken in darkness, or in low level light of a different color from the laser light used in making the hologram. Holography requires a specific ***exposure time*** (just like photography), which can be controlled using a shutter, or by electronically timing the laser.

As is shown in Fig. 5-2, when the two laser beams reach the recording medium, their light waves intersect and interfere with each other. It is this interference pattern that is imprinted on the recording medium. The pattern itself is seemingly random, as it represents the way in which the scene's light interfered with the original light source — but not the original light source itself. The interference pattern can be considered an encoded version of the scene, requiring a particular key — the original light source — in order to view its contents.

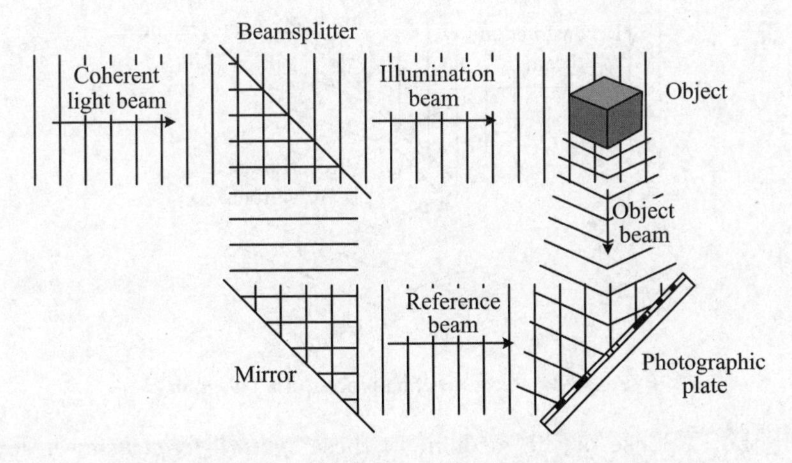

Fig. 5-2 Recording process of a hologram

This missing key is provided later by shining a laser, identical to the one used to record the hologram, onto the developed film. When this beam illuminates the hologram, it is diffracted by the hologram's surface pattern, as is shown in Fig. 5-3. This produces a light field identical to the one originally produced by the scene and scattered onto the hologram. The image this effect produces in a person's retina is known as a virtual image.

5.1.3 Holographic Recording Media

The recording medium has to convert the original interference pattern into an optical element that modifies either the amplitude or the phase of an ***incident light beam*** in proportion to the intensity of the original light field.

The recording medium should be able to resolve fully all the fringes arising from interference between object and reference beam. These fringe spacings can range from tens of micrometers to less than one micrometer, i. e., spatial frequencies ranging from a few hundred to several thousand cycles/mm, and ideally, the recording medium should have a response which is flat over this

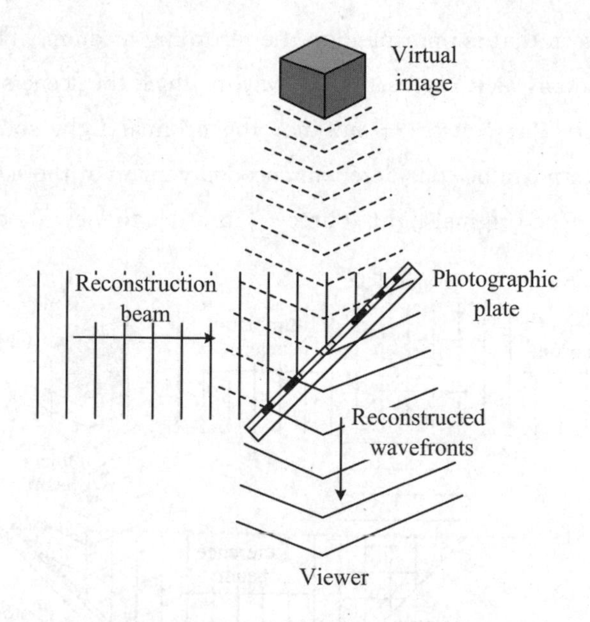

Fig. 5-3 Reconstructing process of a hologram

range. If the response of the medium to these ***spatial frequencies*** is low, the ***diffraction efficiency*** of the hologram will be poor, and a dim image will be obtained. Standard photographic film has a very low or even zero response at the frequencies involved and cannot be used to make a hologram-see, for example, Kodak's professional black and white film whose resolution starts falling off at 20 lines/mm — it is unlikely that any reconstructed beam could be obtained using this film.

If the response is not flat over the range of spatial frequencies in the interference pattern, then the resolution of the reconstructed image may also be degraded.

Table 5-1 shows the principal materials used for ***holographic recording***. Note that these do not include the materials used in the mass replication of an existing hologram, which are discussed in the next section. The resolution limit given in the table indicates the maximal number of interference lines/mm of the gratings. The required exposure, expressed as millijoules (mJ) of photon energy impacting the surface area, is for a long exposure time. Short exposure times (less than 1/1000 of a second, such as with a pulsed laser) require much higher exposure energies, due to reciprocity failure.

Table 5-1 General properties of recording materials for holography

Material	Reusable	Processing	Type of hologram	Theoretical maximum efficiency	Required exposure (mJ/cm^2)	Resolution limit (mm^{-1})
Photographic emulsions	No	Wet	Amplitude	6%	1.5	5000
			Phase (bleached)	60%		
Dichromated gelatin	No	Wet	Phase	100%	100	10000
Photoresists	No	Wet	Phase	30%	100	3000
Photothermoplastics	Yes	Charge and heat	Phase	33%	0.1	5001200
Photopolymers	No	Post exposure	Phase	100%	10000	5000
Photorefractives	Yes	None	Phase	100%	10	10000

5.1.4 Application of Holography

Early on, artists saw the potential of holography as a medium and gained access to science laboratories to create their work. Holographic art is often the result of collaborations between scientists and artists, although some holographers would regard themselves as both an artist and a scientist.

Holography can be put to a variety of uses other than recording images. ***Holographic data storage*** is a technique that can store information at high density inside crystals or photopolymers. The ability to store large amounts of information in some kind of media is of great importance, as many electronic products incorporate storage devices. As current storage techniques such as Blu-ray Disc reach the limit of possible data density (due to the diffraction-limited size of the writing beams), holographic storage has the potential to become the next generation of popular storage media. The advantage of this type of data storage is that the volume of the recording media is used instead of just the surface.

Holographic interferometry (HI) is a technique that enables static and dynamic displacements of objects with optically rough surfaces to be measured to optical interferometric precision (i. e., to fractions of a wavelength of light). It can also be used to detect optical-path-length variations in transparent media, which enables, for example, fluid flow to be visualized and analyzed. It can also be used to generate contours representing the form of the surface.

Holographic scanners are in use in post offices, larger shipping firms, and automated conveyor systems to determine the three-dimensional size of a package. They are often used in tandem with checkweighers to allow automated pre-packing of given volumes, such as a truck or pallet for bulk shipment of goods. Holograms produced in elastomers can be used as stress-strain reporters due to its elasticity and compressibility, the pressure and force applied are correlated to the reflected wavelength, therefore its color.

New Words and Expressions

[1] **holography**：全息。它特指一种技术，可以让从物体发射的衍射光能够被重现，其位置和大小同之前一模一样。从不同的位置观测此物体，其显示的像也会发生变化。因此，这种技术拍下来的照片是三维的。

[2] **recording medium**：记录介质。

[3] **transmission holograms**：透射全息。

[4] **rainbow hologram**：彩虹全息。它是利用记录时在光路的适当位置加狭缝，再现时同时再现狭缝像，观察再现像时将受到狭缝再现像的限制。当用白光照明再现时，对不同颜色的光，狭缝和物体的再现像位置都不同，在不同的位置将看到不同颜色的像，颜色的排列顺序与波长顺序相同，犹如彩虹一样，因此这种全息技术称为彩虹全息。

[5] **reflection hologram**：反射全息。

[6] **specular holography**：镜面反射全息。

[7] **reconstructing wavefronts**：波前重建。

[8] **reference beam**：参考光。

[9] **exposure time**：曝光时间。

[10] **incident light beam**：入射光束。

[11] **spatial frequency**：空间频率。它是指每度视角内图像或刺激图形的亮暗作正弦调制的栅条周数，单位是周/度。

[12] **diffraction efficiency**：衍射效率。它是指在某一个衍射方向上的光强与入射光强的比值。

[13] **holographic recording**：全息记录。

[14] **dichromated gelatin**：重铬酸盐明胶。

[15] **holographic data storage**：全息数据存储。

[16] **holographic interferometry**：全息干涉法。它是利用全息照相获得物体变形前后的光波波阵面相互干涉所产生的干涉条纹图，以分析物体变形的一种干涉量度方法，是实验应力分析方法的一种。

[17] **holographic scanners**：全息扫描仪。

5.2 3D Display Technology

A 3D display (also ***stereo display***), as shown in Fig. 5-4, is a display device capable of conveying depth perception to the viewer by means of ***stereopsis for binocular vision***.

(a) (b)

Fig. 5-4 3D displays

The basic technique of stereo displays is to present offset images that are displayed separately to the left and the right eyes. Both of these 2D offset images are then combined in the brain to give the perception of 3D depth. Although the term "3D" is ubiquitously used, it is important to note that the presentation of dual 2D images is distinctly different from displaying an image in three full dimensions. The most notable difference to real 3D displays is that the observer's head and eyes movements will not increase information about the 3-dimensional objects being displayed. For example holographic displays do not have such limitations. Similar to how in sound reproduction, it is not possible to recreate a full 3-dimensional sound field merely with two stereophonic speakers, it is likewise an overstatement of capability to refer to dual 2D images as being "3D". The accurate term "stereoscopic" is more cumbersome than the common misnomer "3D", which has been entrenched after many decades of unquestioned misuse. It is note that although most stereoscopic displays do not qualify as real 3D displays, all of the real 3D displays are also stereoscopic displays because they meet the lower criteria as well.

5.2.1 Types of 3D Displays

Based on the principles of stereopsis, described by Sir Charles Wheatstone in the 1830s, stereoscopic technology provides a different image to the viewer's left and right eyes. The following are some of the technical details and methodologies employed in some of the more notable stereoscopic systems that have been developed.

5.2.1.1 Side-by-side Images

Traditional stereoscopic photography consists of creating a 3D illusion starting from a pair of 2D images, a stereogram. The easiest way to enhance depth perception in the brain is to provide the eyes of the viewer with two different images, representing two perspectives of the same object, with a minor deviation exactly equal to the perspectives that both eyes naturally receive in binocular vision.

If ***eyestrain*** and ***distortion*** are to be avoided, each of the two 2D images preferably should be presented to each eye of the viewer so that any object at infinite distance seen by the viewer should be perceived by that eye while it is oriented straight ahead, the viewer's eyes being neither crossed nor diverging. When the picture contains no object at infinite distance, such as a horizon or a cloud, the picture should be spaced correspondingly closer together.

The side-by-side method is extremely simple to create, but it can be difficult or uncomfortable to view without optical aids.

5.2.1.2 Transparency Viewers

Pairs of stereo views printed on a transparent base are viewed by transmitted light. One advantage of transparency viewing is the opportunity for a wider, more realistic dynamic range than is practical with prints on an opaque base; another is that a wider ***field of view*** may be presented since the images, being illuminated from the rear, may be placed much closer to the lenses.

The practice of viewing film-based stereoscopic transparencies dates back to at least as early as 1931, when Tru-Vue began to market sets of stereo views on strips of 35 mm film that were fed through a hand-held Bakelite viewer. In 1939, a modified and miniaturized variation of this technology, employing cardboard disks containing seven pairs of small Kodachrome color film transparencies, was introduced as the View-Master.

5.2.1.3 ***Head-mounted Displays***

The user typically wears a helmet or glasses with two small LCD or OLED displays with magnifying lenses, one for each eye. The technology can be used to show stereo films, images or games. Head-mounted displays as shown in Fig. 5-5 may also be coupled with head-tracking devices, allowing the user to "look around" the virtual world by moving their head, eliminating the need for a separate controller.

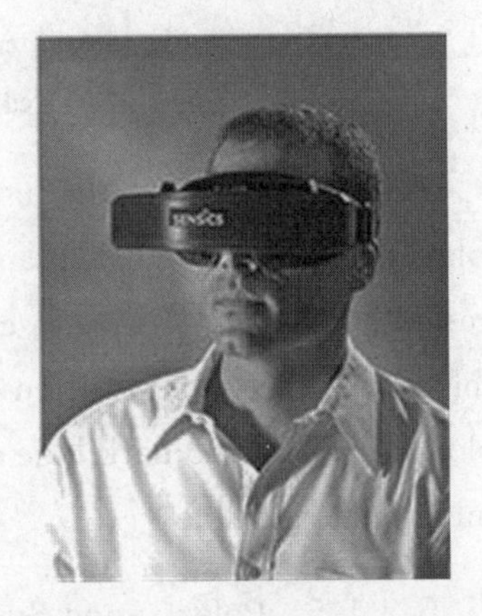

Fig. 5-5 A professional head-mounted display (HMD)

Owing to rapid advancements in computer graphics and the continuing miniaturization of video and other equipment these devices are beginning to become available at more reasonable cost. Head-mounted or wearable glasses may be used to view a see-through image imposed upon the real world view, creating what is called augmented reality. This is done by reflecting the video images through partially reflective mirrors. The real world view is seen through the mirrors' reflective surface.

5.2.1.4 Anaglyph

In an anaglyph, the two images are superimposed in an additive light setting through two filters, one red and one cyan (see Fig. 5-6). In a subtractive light setting, the two images are printed in the same complementary colors on white paper. Glasses with ***colored filters*** in each eye separate the appropriate images by canceling the filter color out and rendering the complementary color black. A compensating technique, commonly known as Anachrome, uses a slightly more transparent cyan filter in the patented glasses associated with the technique. Process reconfigures the typical anaglyph image to have less parallax.

Fig. 5-6 Stereo monochrome image anaglyphed for red (left eye) and cyan (right eye) filters

An alternative to the usual red and cyan filter system of anaglyph is ColorCode 3D, a patented anaglyph system which was invented in order to present an anaglyph image in conjunction with the NTSC television standard, in which the red channel is often compromised. ColorCode uses the complementary colors of yellow and dark blue on-screen, and the colors of the glasses' lenses are amber and dark blue.

5.2.1.5 Polarization Systems

To present a stereoscopic picture, two images are projected superimposed onto the same screen through different ***polarizing filters***. The viewer wears eyeglasses which also contain a pair of polarizing filters oriented differently (clockwise/counter-clockwise with circular polarization or at 90 degree angles, usually 45 and 135 degrees, with linear polarization). As each filter passes only that light which is similarly polarized and blocks the light polarized differently, each eye sees a different image. This is used to produce a three-dimensional effect by projecting the same scene into both eyes, but depicted from slightly different perspectives. Additionally, since both lenses have the same color, people with one dominant eye (amblyopia), where one eye is used more, are able to see the 3D effect, previously negated by the separation of the two colors.

Circular polarization has an advantage over linear polarization, in that the viewer does not need to have their head upright and aligned with the screen for the polarization to work properly. With linear polarization, turning the glasses sideways causes the filters to go out of alignment with the screen filters causing the image to fade and for each eye to see the opposite frame more easily. For circular polarization, the polarizing effect works regardless of how the viewer's head is aligned with the screen, such as tilted sideways, or even upside down. The left eye will still only see the image intended for it, and vice versa, without fading or crosstalk.

Polarized light reflected from an ordinary motion picture screen typically loses most of its polarization. So an expensive silver screen or aluminized screen with negligible polarization loss has to be used. All types of polarization will result in a darkening of the displayed image and poorer contrast compared to non-3D images. Light from lamps is normally emitted as a random collection of polarizations, while a polarization filter only passes a fraction of the light. As a result, the screen image is darker. This darkening can be compensated by increasing the brightness of the projector light source. If the initial polarization filter is inserted between the lamp and the image generation element, the light intensity striking the image element is not higher than the normal without the polarizing filter, and overall image contrast transmitted to the screen is not affected.

5.2.1.6 Autostereoscopy

In this method, glasses are not necessary to see the stereoscopic image. ***Lenticular lens*** and ***parallax barrier technologies*** involve imposing two (or more) images on the same sheet, in narrow, alternating strips, and using a screen that either blocks one of the two images' strips (in the case of parallax barriers) or uses equally narrow lenses to bend the strips of image and make it appear to fill the entire image (in the case of lenticular prints). To produce the stereoscopic effect, the person must be positioned so that one eye sees one of the two images and the other eye sees the other image. The optical principles of ***multiview*** autostereoscopy have been known for over a century.

Both images are projected onto a high-gain, corrugated screen which reflects light at acute angles. In order to see the stereoscopic image, the viewer must sit within a very narrow angle that is nearly perpendicular to the screen, limiting the size of the audience. Lenticular was used for theatrical presentation of numerous shorts in Russia from 1940 to 1948 and in 1946 for the feature length film Robinzon Kruzo.

Though its use in theatrical presentations has been rather limited, lenticular has been widely used for a variety of novelty items and has even been used in amateur 3D photography. Recent use includes the Fujifilm FinePix Real 3D with an autostereoscopic display that was released in 2009. Other examples for this technology include autostereoscopic LCD displays on monitors, notebooks, TVs, mobile phones and gaming devices, such as the Nintendo 3DS.

5.2.2 Applications of 3D Display

True 3D representations can enable faster and more adequate visualization, simulation, and collaboration, e. g. , entertainment including TV, computer monitor and mobile screen, film; fighter combat training, situational awareness, telepresence, battlefield visualization, undersea navigation, and medical visualization; and other areas such as multispectral LIDAR/LADAR data, the 3D structure of molecular docking simulations, and the very multidimensional data that is analyzed in combinatorial chemistry. 3D display device will be a very important tool for modern advanced technologies such as space and nuclear power and etc. , even the education and advertisement need 3D display with vivid images. The realization of 3D displays is a long fostered dream for mankind, because the world is three dimensions. Above 85% information is related the space position and 3D display is the best presentation of the nature. It is predicted that 3D stereo TVs will follow HDTVs and become the trend of image broadcast. If we regard black and white TVs as the first generation production, and regard color TV as the second generation production, then 3D stereo TV will be the third generation production. 3D display is called one of the greatest revolutionary technologies in the 21st century.

New Words and Expressions

[1] **stereo display**：立体显示。它是虚拟现实的一种实现方式。立体显示主要有以下几种方式：双色眼镜、主动立体显示、被动同步的立体投影设备、立体显示器、真三维立体显示、其他更高级的设备。

[2] **stereopsis for binocular vision**：立体双目视觉。

[3] **eyestrain**：眼睛疲劳。

[4] **distortion**：失真。

[5] **field of view**：视场角。

[6] **head-mounted displays**：头戴式显示器。

[7] **colored filters**：彩色滤光片。

[8] **polarizing filters**：偏振片。它是一种可以使天然光变成偏振光的光学元件。

[9] **lenticular lens**：柱透镜。

[10] **parallax barrier technology**：视差挡板技术。

[11] **multiview**：多视点。

5.3 Photolithography Technology

Photolithography, also termed optical lithography or ***UV lithography***, is a process used in ***microfabrication*** to pattern parts of a thin film or the bulk of a substrate. It uses light to transfer a geometric pattern from a ***photomask*** to a light-sensitive chemical "***photoresist***", or simply "resist", on the substrate. A series of chemical treatments then either engraves the exposure pattern into, or enables deposition of a new material in the desired pattern upon, the material underneath the photo resist. For example, in complex integrated circuits, a modern CMOS wafer will go through the photolithographic cycle up to 50 times.

Photolithography shares some fundamental principles with photography in that the pattern in the etching resist is created by exposing it to light, either without using a mask (directly) or with a projected image using an optical mask. This procedure is comparable to a ***high precision*** version of the method used to make printed circuit boards. Subsequent stages in the process have more in common with etching than with ***lithographic printing***. It is used because it can create extremely small patterns (down to a few tens of nanometers in size), it affords exact control over the shape and size of the objects it creates, and because it can create patterns over an entire surface cost-effectively. Its main disadvantages are that it requires a flat substrate to start with, it is not very effective at creating shapes that are not flat, and it can require extremely clean operating conditions.

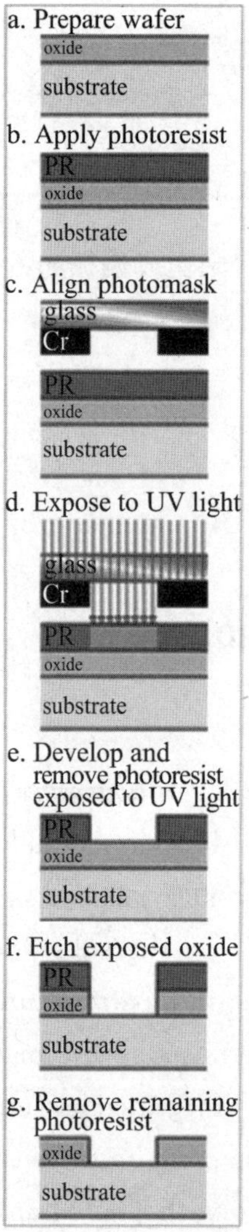

Fig. 5-7　Basic procedure of photolithography

5.3.1　Basic Procedure of Photolithography

A single iteration of photolithography combines several steps in sequence (see Fig. 5-7). Modern cleanrooms use automated, robotic wafer track systems to coordinate the process. The procedure described here omits some advanced treatments, such as thinning agents or edge-bead removal.

5.3.1.1　Cleaning

If organic or inorganic contaminations are present on the wafer surface, they are usually removed by wet chemical treatment, e. g. , the RCA clean procedure based on solutions containing hydrogen peroxide.

5.3.1.2　Preparation

The wafer is initially heated to a temperature sufficient to drive off any moisture that may be present on the wafer surface. Wafers that have been in storage must be chemically cleaned to remove contamination. A liquid or gaseous "adhesion promoter", such as Bis (trimethylsilyl) amine ("hexamethyldisilazane", HMDS), is applied to promote adhesion of the photoresist to the wafer. The surface layer of silicon dioxide on the wafer reacts with HMDS to form tri-methylated silicon-dioxide, a highly water repellent layer not unlike the layer of wax on a car's paint. This water repellent layer prevents the aqueous developer from penetrating between the photoresist layer and the wafer's surface, thus preventing so-called lifting of small photoresist structures in the (developing) pattern.

5.3.1.3 Photoresist Application

The wafer is covered with photoresist by ***spin coating***. A viscous, liquid solution of photoresist is dispensed onto the wafer, and the wafer is spun rapidly to produce a uniformly thick layer. The spin coating typically runs at 1200 to 4800 rpm for 30 ~ 60 s, and produces a layer between 0.5 and 2.5 micrometres thick. The spin coating process results in a uniform thin layer, usually with uniformity of within 5 ~ 10 nm. This uniformity can be explained by detailed fluid-mechanical modelling, which shows that the resist moves much faster at the top of the layer than at the bottom of the layer, where viscous forces bind the resist to the wafer surface. Thus, the top layer of resist is quickly ejected from the wafer's edge while the bottom layer still creeps slowly radially along the wafer. In this way, any "bump" or "ridge" of resist is removed, leaving a very flat layer. Final thickness is also determined by the evaporation of liquid solvents from the resist. For very small, dense features ($<$125 nm or so), lower resist thicknesses ($<$0.5 m) are needed to overcome collapse effects at high aspect ratios; typical aspect ratios are $<$ 4 : 1.

The photo resist-coated wafer is then prebaked to drive off excess photoresist solvent, typically at 90 ~ 100 ℃ for 30 ~ 60 s on a hotplate.

5.3.1.4 Exposure and Developing

After prebaking, the photoresist is exposed to a pattern of intense light. The exposure to light causes a chemical change that allows some of the photoresist to be removed by a special solution, called "developer" by analogy with photographic developer. Positive photoresist, the most common type, becomes soluble in the developer when exposed; with negative photoresist, unexposed regions are soluble in the developer.

A post-exposure bake (PEB) is performed before developing, typically to help reduce ***standing wave*** phenomena caused by the destructive and constructive interference patterns of the incident light. In deep ultraviolet lithography, chemically amplified resist (CAR) chemistry is used. This process is much more sensitive to PEB time, temperature, and delay, as most of the "exposure" reaction (creating acid, making the ***polymer*** soluble in the basic developer) actually occurs in the PEB.

The develop chemistry is delivered on a spinner, much like photoresist. Developers originally often contained sodium hydroxide (NaOH). However, sodium is considered as an extremely undesirable contaminant in MOSFET fabrication because it degrades the insulating properties of gate oxides (specifically, sodium ions can migrate in and out of the gate, changing the threshold voltage of the transistor and making it harder or easier to turn the transistor on over time). Metal-ion-free developers such as tetramethylammonium hydroxide (TMAH) are now used.

The resulting wafer is then "hard-baked" if a non-chemically amplified resist was used, typically at 120 ~ 180 ℃ for 20 ~ 30 min. The hard bake solidifies the remaining photoresist, to make a more durable protecting layer in future ion implantation, wet chemical etching, or ***plasma etching***.

5.3.1.5 Etching

In etching, a liquid ("wet") or plasma ("dry") chemical agent removes the uppermost layer of the substrate in the areas that are not protected by photoresist. In semiconductor fabrication, dry etching techniques are generally used, as they can be made ***anisotropic***, in order to avoid significant undercutting of the photoresist pattern. This is essential when the width of the features to be defined is similar to or less than the thickness of the material being etched (i. e., when the ***aspect ratio*** approaches unity). Wet etch processes are generally isotropic in nature, which is often indispensable for microelectromechanical systems, where suspended structures must be "released" from the underlying layer.

The development of low-defectivity anisotropic dry-etch process has enabled the ever-smaller features defined photolithographically in the resist to be transferred to the substrate material.

5.3.1.6 Photoresist Removal

After a photoresist is no longer needed, it must be removed from the substrate. This usually requires a liquid "resist stripper", which chemically alters the resist so that it no longer adheres to the substrate. Alternatively, photoresist may be removed by a plasma containing oxygen, which oxidizes it. This process is called ashing, and resembles dry etching.

5.3.2 Light Sources of Photolithography

Historically, photolithography has used ultraviolet light from gas-discharge lamps using mercury, sometimes in combination with noble gases such as xenon. These lamps produce light across a broad spectrum with several strong peaks in the ultraviolet range. This spectrum is filtered to select a single spectral line. From the early 1960s through the mid-1980s, Hg lamps had been used in lithography for their spectral lines at 436 nm ("g-line"), 405 nm ("h-line") and 365 nm ("i-line"). However, with the semiconductor industry's need for both higher resolution (to produce denser and faster chips) and higher throughput (for lower costs), the lamp-based lithography tools were no longer able to meet the industry's requirements.

This challenge was overcome when in a pioneering development in 1982, ***excimer*** laser lithography was proposed and demonstrated at IBM by Kanti Jain, and now excimer laser lithography machines (steppers and scanners) are the primary tools used worldwide in microelectronics production. With phenomenal advances made in tool technology in the last two decades, it is the semiconductor industry view that excimer laser lithography has been a crucial factor in the continued advance of Moore's Law, enabling minimum features sizes in chip manufacturing to shrink from 0. 5 m in 1990 to 45 nm and below in 2010. This trend is expected to continue into this decade for even denser chips, with minimum features approaching 10 nm. From an even broader scientific and technological perspective, in the 50-year history of the laser since its first demonstration in 1960, the invention and development of excimer laser lithography has been highlighted as one of the major milestones.

The commonly used deep ultraviolet excimer lasers in lithography systems are the krypton fluoride laser at 248 nm wavelength and the argon fluoride laser at 193 nm wavelength. The primary manufacturers of excimer laser light sources in the 1980s were Lambda Physik (now part of Coherent, Inc.) and Lumonics. Since the mid-1990s Cymer Inc. has become the dominant supplier of excimer laser sources to the lithography equipment manufacturers, with Gigaphoton Inc. as their closest rival. Generally, an excimer laser is designed to operate with a specific gas mixture. Therefore, changing wavelength is not a trivial matter, as the method of generating the new wavelength is completely different, and the absorption characteristics of materials change. For example, air begins to absorb significantly around the 193 nm wavelength; moving to sub-193 nm wavelengths would require installing vacuum pump and purge equipment on the lithography tools (a significant challenge). Furthermore, insulating materials, such as silicon dioxide, when exposed to photons with energy greater than the bandgap, release free electrons and holes which subsequently cause adverse charging.

Optical lithography has been extended to feature sizes below 50 nm using the 193 nm ArF excimer laser and liquid immersion techniques. Also termed immersion lithography, this enables the use of optics with ***numerical apertures*** exceeding 1. The liquid used is typically ultra-pure, deionised water, which provides for a refractive index above that of the usual ***air gap*** between the lens and the wafer surface. The water is continually circulated to eliminate thermally-induced distortions. Water will only allow NA's of up to 1.4, but materials with higher refractive indices will allow the effective NA to be increased further.

Experimental tools using the 157 nm wavelength from the F2 excimer laser in a manner similar to current exposure systems have been built. These were once targeted to succeed 193 nm lithography at the 65 nm feature size node but now all have been eliminated by the introduction of immersion lithography. This was due to persistent technical problems with the 157 nm technology and economic considerations that provided strong incentives for the continued use of 193 nm excimer laser lithography technology. High-index immersion lithography is the newest extension of 193 nm lithography to be considered. In 2006, features less than 30 nm were demonstrated by IBM using this technique.

An option, especially if and when wavelengths continue to decrease to extreme UV or X-ray, is the free-electron laser (or one might say xaser for an X-ray device). These can produce high quality beams at arbitrary wavelengths.

New Words and Expressions

[1] **photolithography**：光刻技术。它是指在集成电路制造中，利用光学-化学反应原理和化学、物理刻蚀方法，将电路图形传递到单晶表面或介质层上，形成有效图形窗口或功能图形的工艺技术。

[2] **UV lithography**：紫外光刻技术。

[3] **microfabrication**：精密加工。

[4] **photomask**：光学掩模板。它是在薄膜、塑料或玻璃基体材料上制作各种功能图形并精确定位，以便用于光致抗蚀剂涂层选择性曝光的一种结构。

[5] **photoresist**：光刻胶，光阻材料，又称光致抗蚀剂。它是由感光树脂、增感剂(见光谱增感染料)和溶剂三种主要成分组成的对光敏感的混合液体。

[6] **high precision**：高精度。它是指测量值与真值的接近程度很高。

[7] **lithographic printing**：平版印刷。

[8] **spin coating**：旋涂。它是指在电子工业中，基片垂直于自身表面的轴旋转，同时把液态涂覆材料涂覆在基片上的工艺。

[9] **standing wave**：驻波。它是指频率和振幅均相同、振动方向一致、传播方向相反的两列波叠加后形成的波。

[10] **polymer**：聚合物，即高分子化合物。它是指那些由众多原子或原子团主要以共价键结合而成的相对分子量在一万以上的化合物。

[11] **plasma etching**：等离子体刻蚀。

[12] **anisotropic**：各向异性的。晶体的各向异性即沿晶格的不同方向，原子排列的周期性和疏密程度不尽相同，由此导致晶体在不同方向的物理化学特性也不同，这就是晶体的各向异性。

[13] **aspect ratio**：纵横比。它是一个图像的宽度与它的高度之比。

[14] **excimer**：激态原子。

[15] **numerical apertures**：数值孔径，又叫镜口率，简写为 NA。

[16] **air gap**：气隙。它是电机定转子之间的空隙。

5.4 *Photoelectric Detection*

Modern information technology includes ***information acquisition technology***, information transmission technology and ***information processing technology***. Information acquisition is the use of different sensitive devices to convert all kinds of original information into electrical signal, and these are sensors. Photoelectric detection is the use of photoelectric sensor to realize all kinds of tests, which converts the measurand into ***luminous flux measurement***, and then into electricity. Finally online and automatic detection of various physical quantities will be realized by comprehensive utilization of information transmission technology (modern communication technology) and information processing technology (electronic and computer technology).

5.4.1 The Basic Method of Photoelectric Detection

Photoelectric sensor is composed of three parts, which are light source, optical system and ***photoelectric conversion device***. According to the different positions of the above three parts, photoelectric sensor can be divided into the following three kinds:

(1) ***Direct-type***: Photoelectric converter is placed to face the light source. Their optical axis coincide with each other, which makes the direction of optical axis be the maximum of light flux for the light source, and the maximum of sensitivity for photoelectric conversion device.

For avoiding the influence of stray light on the measurement for practicality, some methods generally can be adopted like the use of ***camera obscura***, the enhancement of light source intensity, the ***modulation of light flux*** and the appropriate placement of light source and ***photoelectric receiver***, etc. For example, when measuring count of large objects on ***conveyor belt***, the stray light direction should be in the same direction with the light sources for large objects, and should be in the opposite direction for small objects.

(2) ***Reflection-type***: This type can be divided into specular reflection and diffuse reflection. Measured quantities for the former type have smooth surface, or are affixed reflectors. Photoelectric receiver receives one-way reflected light of measured quantities. However, measured quantities for the latter type have rough surface, and photoelectric receiver receives diffuse light.

(3) ***Radiation-type***: The measured quantities itself is a radiation source. Measurement of photoelectric receiver can be realized by receiving the radiation of

light energy.

According to the detection principle, the basic method of the photoelectric detection is the direct method, differential method, compensation method and pulse measurement method, etc.

5.4.1.1 Direct Method

Luminous flux controlled by the measured quantities is converted into electric signal through photo can be obtained directly by the testing institution after being converted into electricity by photoelectric receiver. Measuring block diagram is shown in Fig. 5-8, calibration means measuring by the ***benchmark***, adjust the system's magnification or ratio, and make the output value is the same as the reference.

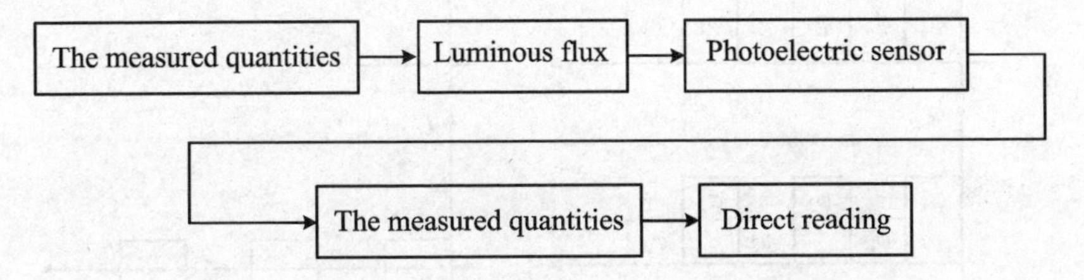

Fig. 5-8 The measuring block diagram of direct action

5.4.1.2 Differential Method

By comparing the measured quantities with the standard amount, the income difference reflects the size of the measurement. For example, measure the length of the object with double optical path differential method, as shown in Fig. 5-9,

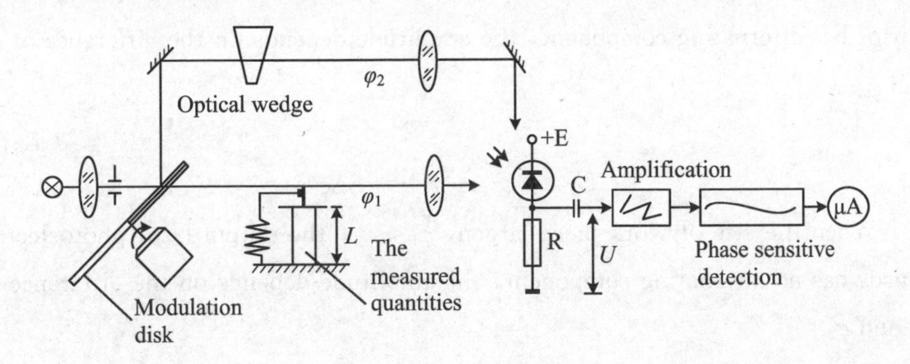

Fig. 5-9 The differential method of measuring the length of the object by double light path

the half of modulation disk is open, and installs a mirror in the other half. A beam of light is turned into two beams of light by the modulation disk's rotation, φ_1 and φ_2 are alternately.

1. Adjustment

Put in the work piece which has the standard size, and adjust the optical wedge, make $\varphi_1 = \varphi_2$ and the reading of μA is "0".

2. Measurement

When the size of work piece is without error, $\varphi_1 = \varphi_2$, the output U of photoelectric sensor has no alternating component, as shown in Fig. 5-10.

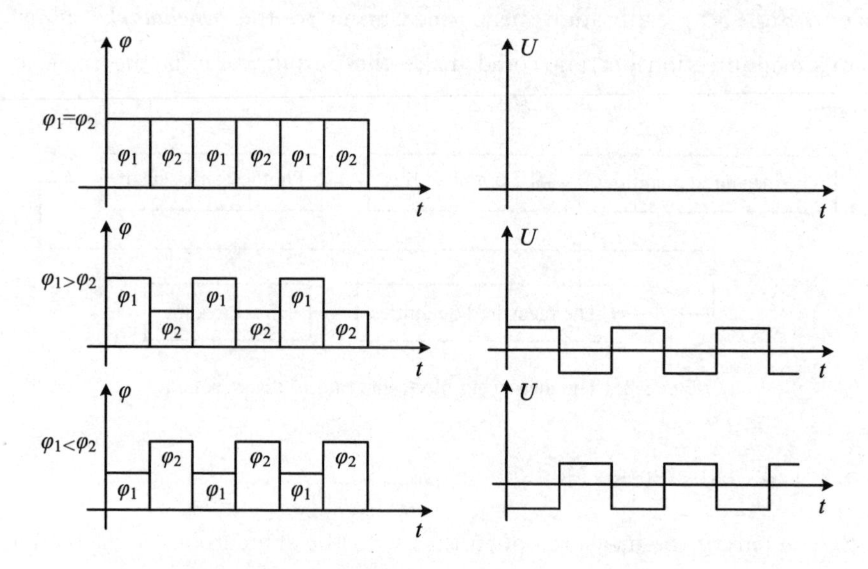

Fig. 5-10 The output of photoelectric sensor

When the size of work piece lessen, $\varphi_1 > \varphi_2$, the output U of photoelectric sensor has alternating component, the amplitude depends on the difference of φ_1 and φ_2.

$$U = S(\varphi_1 - \varphi_2) = S\Delta\varphi \tag{5-1}$$

When the size of work piece largen, $\varphi_1 < \varphi_2$, the output U of photoelectric sensor has no alternating component, the amplitude depends on the difference of φ_1 and φ_2.

$$U = S(\varphi_1 - \varphi_2) = - S\Delta\varphi \tag{5-2}$$

3. Conclusion

(1) The size of the measured value is decided by the amplitude of U, the positive and negative of measured values is decided by the phase of the U, can be obtained by ***phase-sensitive detector***.

(2) Measurement by double optical path can eliminate the stray light, the measuring error which caused by light source fluctuation, temperature variation, and mains voltage fluctuation, measuring precision and sensitivity is greatly increased.

4. The Theory of Phase Sensitive Detector (PSD)

The core of PSD is a multiplier and a filter, as shown in Fig. 5-11. One is useful signal U_S, the other is the reference signal U_R, now assume:

$$U_i = U_S = E_i \sin(\omega_1 t + \theta_1) \tag{5-3}$$

$$U_R = E_R \sin(\omega_2 t + \theta_2) \tag{5-4}$$

so

$$U_O = U_i \cdot U_R = 1/2 \cdot E_i E_R \cos[2\pi(f_1 - f_2)t + \theta_1 - \theta_2] - 1/2 \cdot E_i E_R \cos[2\pi(f_1 + f_2)t + \theta_1 + \theta_2] \tag{5-5}$$

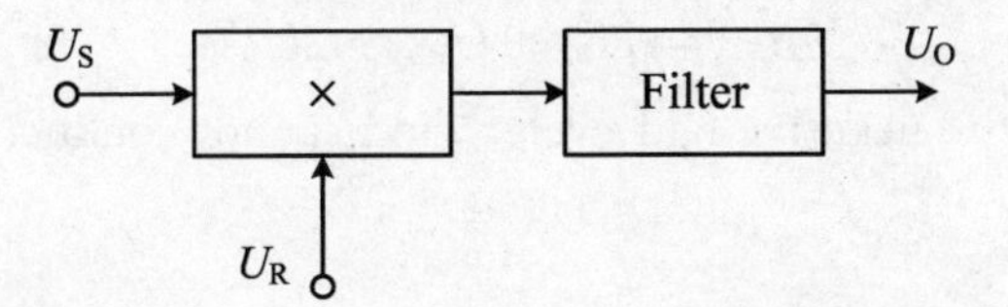

Fig. 5-11 The principle of phase sensitive detector

After the high frequency signals are filtered by the filter, we can get:

$$U_O = 1/2 \cdot E_i E_R \cos[2\pi(f_1 - f_2)t + \theta_1 - \theta_2] \tag{5-6}$$

If $f_1 = f_2$, then

$$U_O = 1/2 \cdot E_i E_R \cos(\theta_1 - \theta_2) = 1/2 \cdot E_i E_R \cos \Delta\theta \tag{5-7}$$

The positive and negative of output is designed by the phase difference of two signal, if $\Delta\theta = \theta_1 - \theta_2 = 0$, then the output of U_O is positive, else $\Delta\theta = \theta_1 - \theta_2 = \pi$, then the output of U_O is negative.

So, phase-sensitive detector can be used to measure amplitude, phase, also can put high frequency signal into frequency signal (only if $\theta_1 = \theta_2$, $f_1 > f_2$, then can get the difference frequency signal).

A kind of phase sensitive detector circuit as shown in Fig. 5-12, in the figure, U_S' is induced voltage of the signal coil to U_S, U_R' is induced voltage of reference coil to U_R, U_R and U_S need to satisfy the following three conditions:

$f_1 = f_2$, $U_R \gg U_S$, and U_R and U_S in-phase or reverse phase.

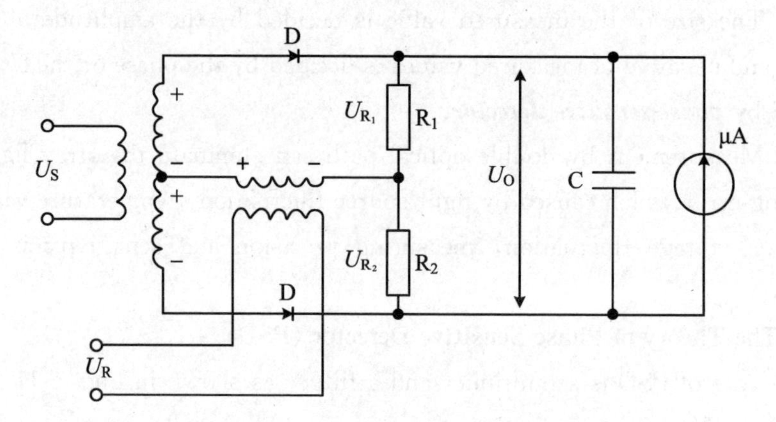

Fig. 5-12 The circuit of phase-sensitive detector

In-phase:

When U_R for the positive half cycle,

$$U_S' + U_R' = U_{R1} \tag{5-8}$$

$$U_S' - U_R' = U_{R2} \tag{5-9}$$

$$U_O = U_{R1} + U_{R2} = 2U_S' \tag{5-10}$$

When U_R for the negative half cycle, diode is not conducting.

$$U_O = 0$$

Reverse phase:

When U_R for the positive half cycle,

$$-U_S' + U_R' = U_{R1} \tag{5-11}$$

$$-U_S' - U_R' = U_{R2} \tag{5-12}$$

$$U_O = U_{R1} + U_{R2} = -2U_S' \tag{5-13}$$

When U_R for the negative half cycle, diode is not conducting.

$$U_O = 0$$

So, as long as to judge the positive and negative of U_O, you will know that the positive and negative deviation of measured work piece, and only you measure the size of the U_O, you will know the deviation of the work piece.

In practical, replace transformer coil with PSD, as shown in Fig. 5-13. In the figure, the ***field effect tube*** as electronic switch, 9 V is broken. The circuit is actually to multiplied signal voltage by the unit square wave which was used as the reference signal voltage, that is using the positive half cycle of unit square wave to control signal open, correspond to $1 \times U_S = U_S$, and use the negative half cycle of unit square wave to control signal inverting open, correspond to $-1\times(-U_S)=U_S$. Similarly, when the reference signal and the signal voltage are inverting, use the positive half cycle of unit square wave to control signal inverting open, correspond to $1\times(-U_S)=-U_S$, and use the negative half cycle of unit square wave to control signal open, correspond to $-1\times U_S=-U_S$. In the figure, capacitance C is a filter.

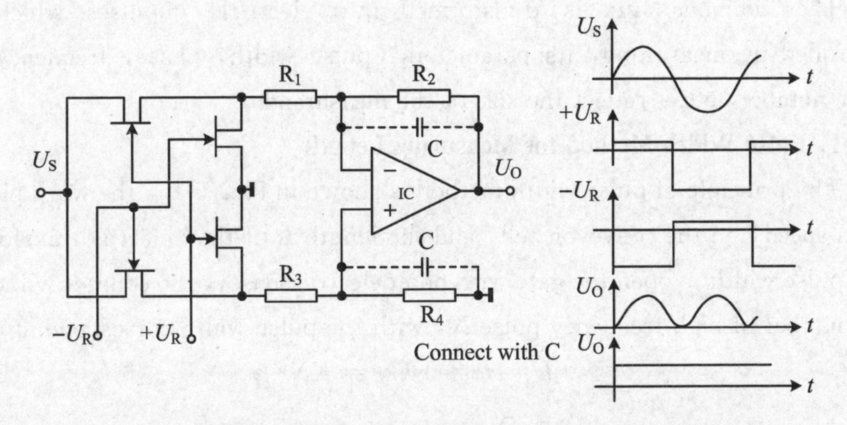

Fig. 5-13 The circuit and output waveform of phase-sensitive detector

In the phase sensitive detector, the reference signal should have the same frequency of measurement signal, we can install a photoelectric receiver in modulating disk as shown in Fig. 5-14.

5.4.1.3 Compensation Method

The change of luminous flux cased by the change measurand is compensated in the method of electricity and light, the read station indicates the offset value which connects with the moving element of the compensator. The sizes of the compensation values reflect the size of measurand's change.

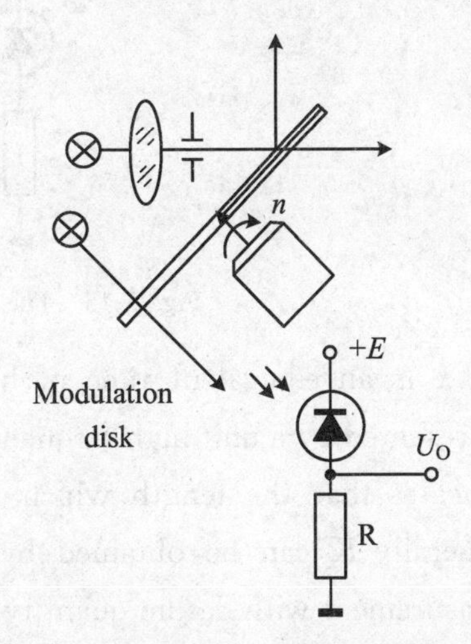

Fig. 5-14 Obtain the reference signal

In the measurement of double light path difference method, to direct the moved up and down of the optical wedge by using phase-sensitive detector, until $\varphi_1 - \varphi_2 = 0$, that is to say, the output of PSD is 0, and optical wedge moved up and down connected to a reading device, then get readings from a reading device just reflect the change of the luminous flux, that is, the value of the measurand.

5.4.1.4 Pulse Measurement Method

The luminous flux is transformed into electrical impulses which are controlled by measurand, its parameters (pulse width, phase, frequency and pulse number, etc.) reflect the size of the measurand.

1. Pulse Width Method for Measuring Length

The principle of pulse width method is shown in Fig. 5-15, the work piece L has a speed v in the conveyor belt, and the length L of the object is transformed into pulse width to open the gate with photoelectric sensor, the counter will count the number of high frequency pulse N, with the pulse width corresponding, so

$$L = vt = vkN = KN \tag{5-14}$$

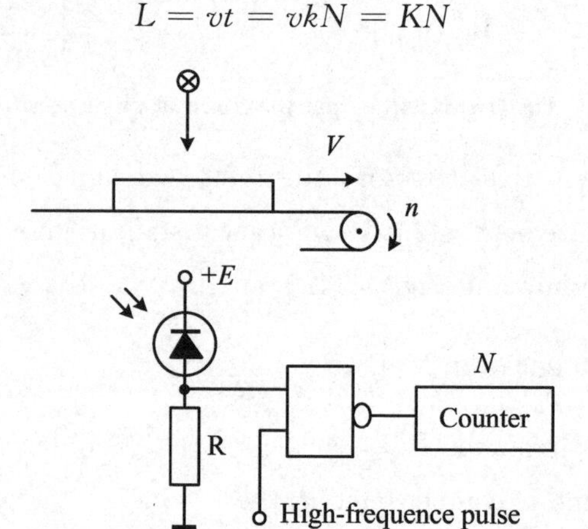

Fig. 5-15　The principle of pulse width method

and k is an equivalent time of high frequency, indicates that the time which represented by a unit high frequency pulse. But K is the length of the equivalent, indicates that the length which represented by a unit high frequency pulse. Generally K can be obtained by scaling, can be obtained by making actual measurement with datum quantity, and then K value is obtained by calculation.

In the above measurements, the assumption is that v is constant. However, due to the vibration of the object, v is not uniform because of the fluctuations of the motor voltage frequency, so that it can bring the measurement error. In order to eliminate the error, a ***photoelectric rotary converter*** installed on the conveyor belt to produce optical pulses in the method of electricity and light.

2. Phase Method for Measuring Distance

The telescope with ranging function launches a pulsed modulated laser beam, and then receives emitted light from object. The phase difference between transmitting and receiving signals reflects the distance of object. If V_o and V_i represent transmitting and receiving pulsed signals respectively, the phase difference between these two signals will be measured using high frequency fill method, and then transmitted to the corresponding distance, as shown in Fig. 5-16.

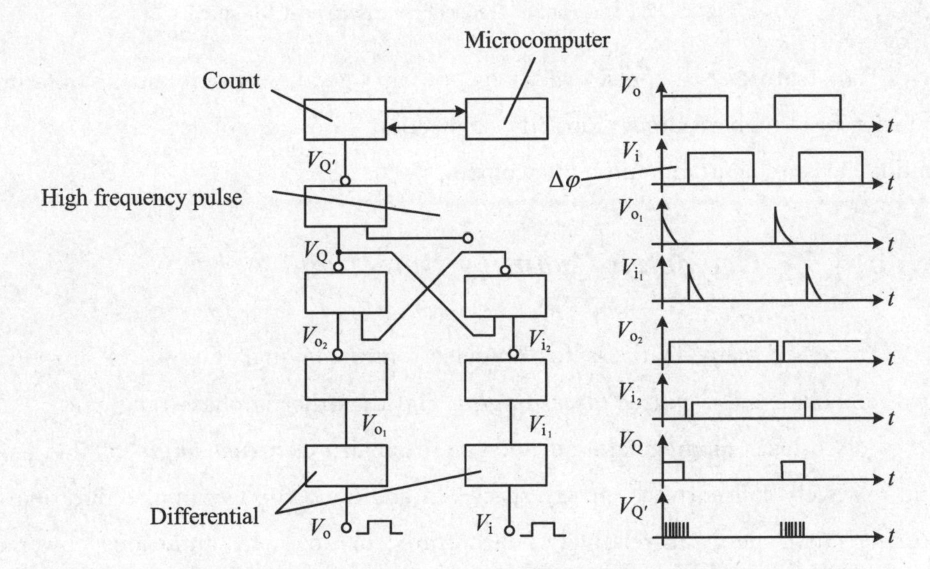

Fig. 5-16 Diagram and waveform of phase ranging method

3. Frequency Method for Measuring Speed

The basic principle of frequency measurement for speed is shown in Fig. 5-17. Reflective sheets are evenly pasted on turning wheel, and ***optoelectronics sensors*** can receive optical pulse corresponding to turning speed. If m and n represent the number of reflective sheets and round per minute respectively, then the following equations will be obtained:

$$f = nm/60 = N/t$$

$$n = 60N/(mt) \qquad (4\text{-}15)$$

As long as the counter N is controlled in a certain time t, the turning speed of wheel can be calculated.

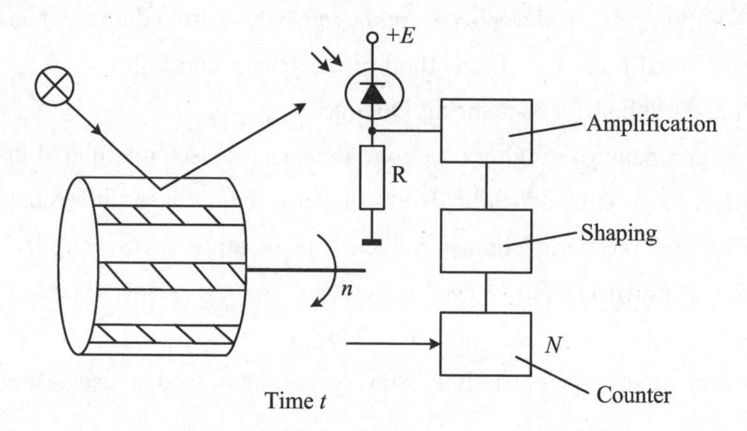

Fig. 5-17 Diagram of frequency measurement for speed

The features of pulse method contain good characteristics of anti-interference, high accuracy, directly connection with computer, easy to achieve online measurement and automatic control.

5.4.2 *Geometric Quantity Detection*

There are many methods for optoelectronics ranging. Here, we introduce two methods, one is ***pulsed laser ranging*** and the other is phase ranging.

Pulsed laser ranging take advantages of small ***divergence angle*** of the laser and relatively concentrated energy space. In addition, the laser pulse duration is so short that energy is relatively concentrated in time. Instantaneous power is very big and can reach up to the level of MW. Due to above two features, pulsed laser ranging have very far ranging distance with the existence of reflector. The reflector is no need for close distance measurement (a few kilometers) and low accuracy, and ranging can be completed through reflective signals from object to pulsed laser. At present, pulsed laser ranging has widely application including topographic survey, tactical frontier ranging, engineering survey, clouds and the height measurement of airplane, orbit tracking of missile, artificial earth satellite ranging, distance measurement between the earth and the moon and so on. Currently, there are many models of pulsed laser ranging instrument that have the same basic working principle.

The working principle of pulsed laser ranging is shown in Fig. 5-18. Laser generated from place 1 shoots to place 2 after the measured distance. There is reflective equipment at the place 2 and D is the measured distance. If there is one kind of equipment that can measure the time required for the pulsed laser from place 1 to place 2 and return 1, the following equation should be met:

$$D = ct/2$$

where c is speed of light.

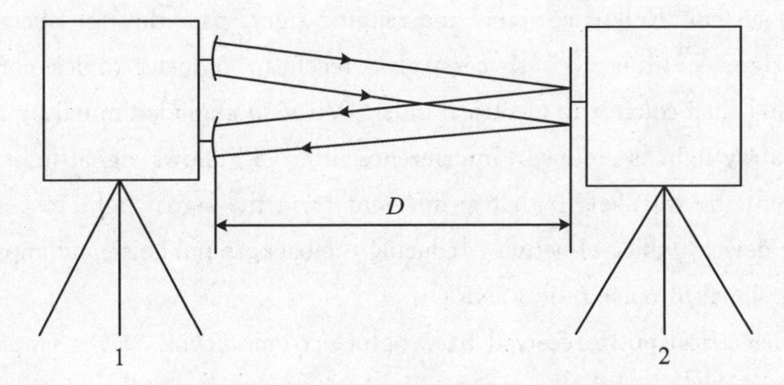

Fig. 5-18 Working principle of pulsed laser ranging

Block diagram of pulsed laser ranging is shown in Fig. 5-19. It consists of launch system of pulsed laser, receiving system, control circuit, ***clock pulse oscillator***, and count display circuit.

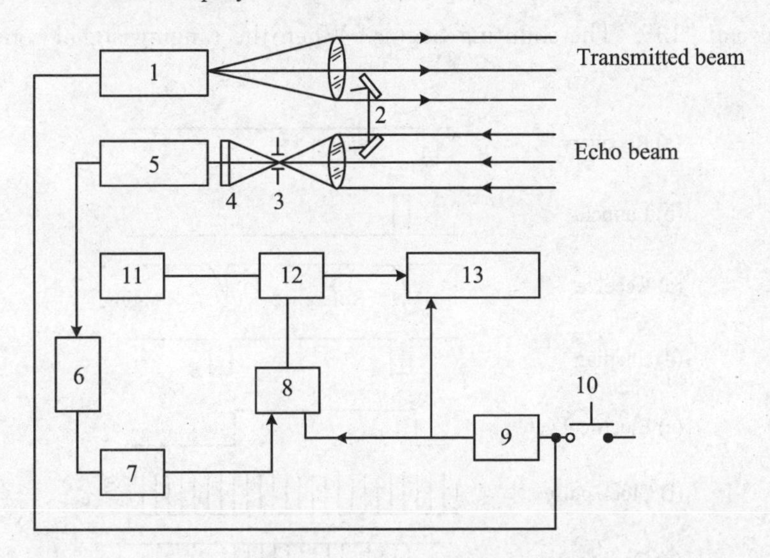

Fig. 5-19 Diagram of pulsed laser ranging

The working process is as follows: When pressing the start button "10", recovery signal from recovery circuit "9" make the whole system recover and the system is prepared for measurement. And at the same time, pulse laser generator "1" is triggered and launches laser pulse. The most of this laser pulse shoot to the ranging target except for a small part of energy that is directly transmitted to receiving system by reference signal sampler "2" as starting point of time. Part of energy of laser pulse is reflected to the ranging instrument and received by the receiving system. Reference signal and ranging signal pass through aperture "3" and interference filter "4" successively, reach to optoelectronics conversion device, and then convert to electrical pulse. Viewing angle is limited by aperture "3" and stray light is stopped. Interference filter "4" allows signal light to pass and prevent the wavelength that is different from the signal light to get to the receiving device, which effectively reducing the background noise and improve the receiving signal to noise ratio (SNR).

The electrical pulse received from optoelectronic device "5" is amplified by amplification "7", and then negative pulse with a certain waveform is transmitted to control circuit "8". The negative pulse "A" generated from reference signal, as shown in Fig. 5-20 (d), open electrical gate "12" after passing through control circuit "8". Clock oscillator "11" with a certain ***oscillation frequency*** produces clock pulse that can pass through electrical gate "12" and enter the counting display circuit "13". The counting begins. When the ranging signal comes, the

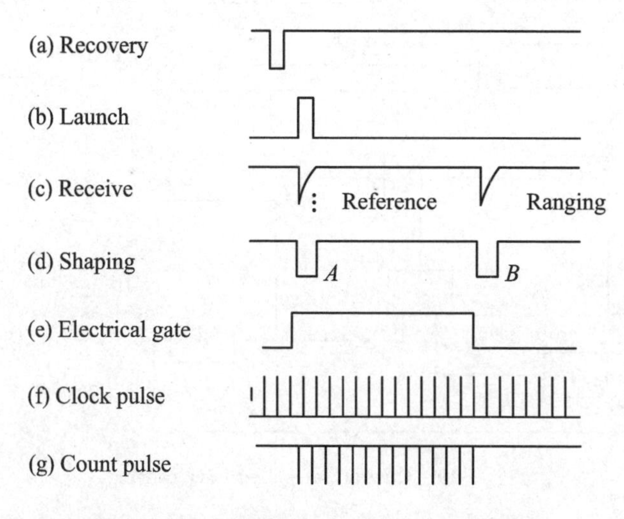

Fig. 5-20 Waveform of pulsed laser ranging

electrical "12" is closed. The counting stops. The counting and display pulse are shown in Fig. 5-20(g). The time from counting start to stop is proportional to time between the reference signal and the ranging signal.

Pulsed laser ranging equipment has simple principle and structure, and far ranging distance. Its main shortcoming is lower absolute distance accuracy.

Phase ranging method can be applied in higher precision.

New Words and Expressions

[1] **photoelectric detection**：光电检测。

[2] **information acquisition technology**：信息采集技术。

[3] **luminous flux measurement**：光通量测量。光通量指人眼所能感觉到的辐射功率，它等于单位时间内某一波段的辐射能量和该波段的相对视见率的乘积。由于人眼对不同波长光的相对视见率不同，所以当不同波长光的辐射功率相等时，其光通量并不相等。

[4] **information processing technology**：信息处理技术。

[5] **photoelectric conversion device**：光电转换装置。光电转换过程的原理是光子将能量传递给电子，使其运动，从而形成电流。这一过程有两种解决途径：最常见的一种是使用以硅为主要材料的固体装置，另一种则是使用光敏染料分子来捕获光子的能量。染料分子吸收光子能量后将使半导体中的带负电的电子和带正电的空穴分离。

[6] **camera obscura**：暗箱。

[7] **modulation of light flux**：光通量调制。

[8] **photoelectric receiver**：光电接收器。

[9] **conveyor belt**：输送带。

[10] **direct-type**：直射型光电传感器。光电转换器对着光源放置，并使它们的光轴重合，即对于光源为发射光通量最大的方向，对于光电转换器件为灵敏度最高的方向。使用时，一般可通过使用暗箱、提高光源强度、调制光通量和适当放置光源和光电接收器等办法来避免杂散光对测量的影响。如在传送带上对物体计数测量，测量大物体时，应使杂散光方向与光源方向一致；如测量小物体时，应使杂散光方向与光源方向相反。

[11] **reflection-type**：反射型光电传感器。它可分为单向反射和漫反射两种。前者被测物表面光滑或贴上反射镜，光电接收器接收被测物的单向反射光；后者被测物表面粗糙，光电接收器接收被测物的漫反射光。

[12] **radiation-type**：辐射型光电传感器。被测物本身就是一个辐射源，光电接收器通过接收被测物的辐射光能量实现测量。

[13] **benchmark**：标杆。

[14] **phase-sensitive detector**：相敏检波器。

[15] **field effect tube**：场效应管。由多数载流子参与导电，也称为单极型晶体管。它属于电压控制型半导体器件。具有输入电阻高（$10^7 \sim 10^{12}\ \Omega$）、噪声小、功耗低、动态范围大、易于集成、没有二次击穿现象、安全工作区域宽等优点，现已成为双极型晶体管和功率晶体管的强大竞争者。

[16] **photoelectric rotary converter**：光电旋转式变流器。

[17] **optoelectronics sensors**：光电传感器。它是采用光电元件作为检测元件的传感器。它首先把被测量的变化转换成光信号的变化，然后借助光电元件进一步将光信号转换成电信号。光电传感器一般由光源、光学通路和光电元件三部分组成。

[18] **geometric quantity detection**：几何量检测。

[19] **pulsed laser ranging**：脉冲激光测距。

[20] **divergence angle**：发散角。

[21] **clock pulse oscillator**：时钟脉冲振荡器。

[22] **oscillation frequency**：振荡频率。

5.5 Optical Disc Storage

Information storage is defined as a technique combination which includes recording such useful data as characters, documents, sounds and images into device temporarily or permanently by writing station and utilizing readout station to reappear the information from the storage medium. As the development of science and technology, the storage technique has gone through the process of paper writing, micro photos, mechanical records, tapes and discs. Nowadays, the storage technique comes into a new phase of optical recording ways. In recent years, ***optical storage*** technique has been making great progress not only in scanning ***hologram*** in pure optics, but also in optical disc using light-thermal deformation and in optical disk using the photo magnetic effect and the thermo magnetic effect. All of them have vast potential for future development. Optical storage technique gathers the essence and technical know-how of modern photo electronic techniques. Photo electricity methods about measurement, modulation, tracking and control have been used adequately. What introduced below emphasis on the optical disc ***storage system*** using light-thermal deformation.

Optical disc is a round-shape information storage device. It utilizes modulated narrow laser beam to change the optical properties of different positions in disc medium, and records the data prepared. When the laser beam illuminates the surface of medium, storage information will be extracted due to the difference of optical properties in each information point. The device we use to write information into disc is called disc record system, for example, the document disc recorder. The device we use to read data from the disc is called disc replay system, such as video disc player. Most disc devices have both functions of recording and replaying.

1. Read-only Type

Fig. 5-21 shows the principle of writing in and reading out of optical disc. Fig. 5-22 shows ser beam carrying sound, video or document information will be focused into 1 μm-diameter-spot by focus lens. Narrow-beam laser with high energy density will heat the surface of disc's record medium, makes part position deform permanently, which causes the binarization of optical properties on medium surface. Since optical disc rotates, and writing head translates, tiny spiral pit or other types of information points will be formed on disc. Different coding schemes of these information points represent the information data stored.

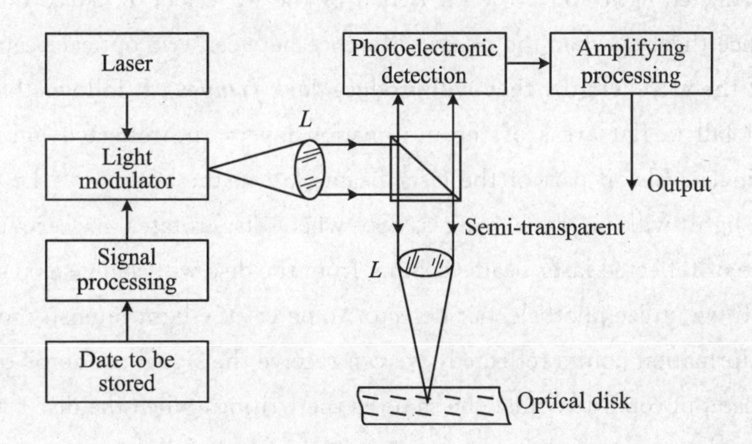

Fig. 5-21 The principle of writing in and reading out of the optical disk

Width of pits in disc are generally 0. 4 μm, depth are 0. 11 μm, one fourths of the wave length of read-out beam. Separation distance between spiral traces is 1. 67 μm.

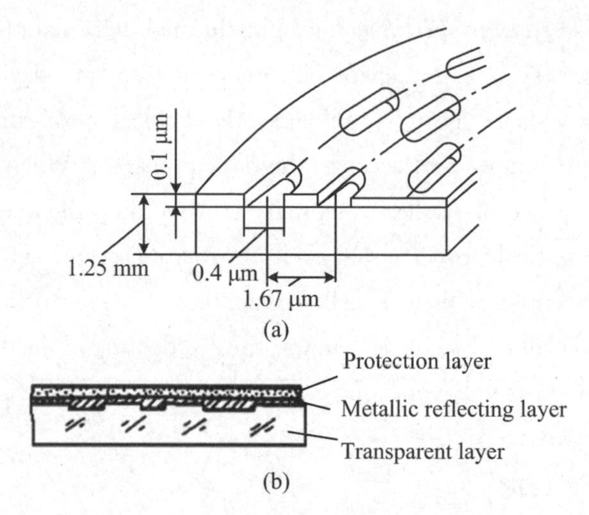

Fig. 5-22 Cross-section shape of disk

In reading out condition, illuminating laser beam will be focused on the information layer of the disc. When laser beam illuminates the flat area of the information layer, most light will be reflected to the objective. Reflected light falling on the pit fringe will spread to side because of the ***diffraction effect***, only a little reflected light can return the objective. Since depth of the pit is one fourths of the wave length, the phase difference between reflected light at bottom of the pit and reflected light above the pit is half of the wave length. According to the interference theory, when the phase difference between two optical beams comes to half of the wave length, they will produce ***dark fringes***. It follows that if laser beams all fall in flat areas of the information layer, the reflected light will be bright fringes. Else if part of the laser beams fall in the bottom of the pits, the reflected light will be dark fringes. So when disc rotates in certain speed, luminance of reflected laser beams coming from the disc will change as pits on disc change. If we utilize photoelectric detector to detect the light intensity modulated by the information points reflected, we can receive the signal of "zero" or "one".

In read-out condition, just the same as the writing, when the disc rotates and writing head translates, they combine into ***helical motion***.

2. Erasable Type

Similar to the read-only type disc, this type of disc is only different when laser beams illuminate the disc; they magnetize the magnetic films on disc surface or change the ***crystallization*** state of medium surface.

Another kind of optical record method adopts the magnetic disc device based on the photo magnetic effect and thermo magnetic effect. Its basic device is similar to the optical disc device. Their difference mainly in the record medium, erasable discs use magnetic medium. Under the modulation effect of the narrow laser beam, we accomplish the information storage by changing the direction of magnetization of the magnetic medium. When reading out the information, we detect the polarization of light instead of detecting the reflectivity of light to achieve the magnetization direction of the information point.

New Words and Expressions

[1] **optical storage**：光存储。它是指受光盘表面的介质影响，光盘上有凹凸不平的小坑，光照射到上面有不同的反射，再转化为0、1的数字信号的过程。

[2] **hologram**：全息图。它是以激光为光源，用全景照相机将被摄体记录在高分辨率的全息胶片上构成的图。

[3] **storage system**：存储系统。它是指计算机中由存放程序和数据的各种存储设备、控制部件及管理信息调度的设备（硬件）和算法（软件）所组成的系统。

[4] **diffraction effect**：衍射效应。波在传播时，若被一个大小接近于或小于波长的物体阻挡，就绕过这个物体，继续进行；若通过一个大小接近于或小于波长的孔，则以孔为中心，形成环形波向前传播。

[5] **dark fringes**：暗条纹。

[6] **helical motion**：螺旋运动。如果带电粒子进入均匀磁场B时，其速度 v 与B之间成 θ 角，则粒子将做螺旋运动。

[7] **crystallization**：结晶化、具体化。

5.6 Light Guide Illumination System Based on Automatic Acquisition of Natural Light

Since the decrease of energy causes the increase utilization of clean energy and natural light, the utilization of natural light will be the inevitable trend in future development.

The objective of this project is to effectively collect the natural light, to save energy and reduce pollution, and to achieve green lighting in the end. The electricity saving ratio in one day (which mean the ratio of the electricity this system saved to the electricity ***artificial lighting*** system saved in 24 hours) is greater than 40% in same area by measuring.

Internationally, there is a ***natural light*** collection and guide system in Japan while its utilization of optical fiber to guide the light leads to the reduction of ***luminous flux***, so it can only use in small area lighting. Otherwise, its high cost makes large-scale use impossible. Although there are applications of ***light-pipe*** in America and Canada, they utilize ***artificial light source*** instead of natural light. Also there is report of natural light collection in Soviet Union, but the technology is still in its infancy.

This project utilizes the latest technology to guide the natural light into the room to achieve green, environmental and safe indoor lighting, and achieve the purpose of saving energy. The project is practical and has extended value, and it has a remarkable social and economic worth. Daylight lasts about 14 hours in summer and about 10 hours in winter. In spring and autumn, daylight lasts about 12 hours in average. With this system, we can use green natural light over 10 hours daily, which means that we don't need to turn on the light in 41.7% ($10 \div 24 = 41.7\%$) time per day. It shows great effects. Take one hundred T5 straight tube ***fluorescent lamp*** (28 W) as an example, the power of ***ballast*** is 4 W, we assume the electricity saving ratio of the system is 40%, after using this system we can saving 11212.8 kW • h [$100 \times (28+4) \div 1000 \times 365 \times 24 \times 40\% = 11212.8$ (kW • h)] per year. At the same time, it is energy-saving and green, also is "contribution in the contemporary era and benefit on the future generations". It conforms to China's policy and has significant social efficiency.

5.6.1 Technical Indicators

(1) The output of the device is about 160 thousand ***lumens*** when there is sunlight. The output diameter is less than 500 mm and output angle is less than 36 degree so that the device can match the light-tube well.

(2) It has the function of automatic scan, position and tracking in order that the optical axis of the system is able to direct at the sun.

(3) It has auto-protection device to prevent strong wind, heavy rain and snow, hailstone and sandstorm.

(4) It has light-control system and communication system. We set admin console in the duty room so that operator on duty can follow the situation in time and control it manually when it is needed. When the intensity of incident light is less than certain value, it will control the ***light bulbs*** automatically.

(5) It has cold light filter to prevent infrared light in case that the light-tube will heat up because of the high intensity of light. So only visible light can be collected.

5.6.2 Implementation Plan for the Project

The principle of automatic sunlight collection device is shown in Fig. 5-23.

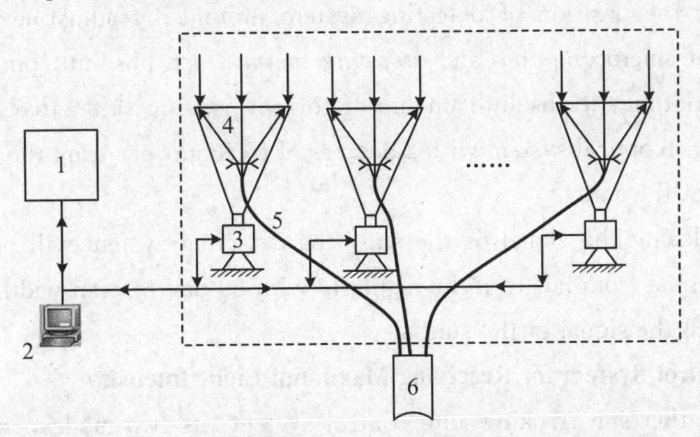

Fig. 5-23 Principle diagram of automatic sunlight collection device

1. Scan and measure system of sun position and light intensity; 2. Computer; 3. Max light intensity control system; 4. Optical system; 5. Flexible light guide system; 6. Fixed light guide system.

1. Optical and *Light Guide* System

Optical system can be implemented with reflective system or transmission system. In reflective system, we can utilize ***parabola*** which is made of aluminum to receive sunlight. But take the cost after batch production and convenience of installing and maintaining into consideration, we approach to use transmission system. We utilize telephoto system with large aperture ***Fresnel lens*** to guide direct sunlight into the transit light-guide system through Fresnel lens with proper exit aperture. The PC material is high heat-resistant and shock-resistant and has good physical and chemical properties, which can be used for natural light collection. Basic principle is shown in Fig. 5-23.

Since we press the PC material by molds to make Fresnel lens, most expense will be cost on the design and manufacture of the mold. So putting molds into batch production will lead to the fall in costs. At present, the diameter of lens made by domestic technique is only about 1500 mm, so if we expect to collect more sunlight, we should utilize Fresnel lens groups (more than five as required), then use several transit light pipe to guide the light into the following fixed light guide system.

Light-tube will be made of flexible glass cable for future sunlight transmission.

2. Measurement System of Sun Position and Light Intensity Scan

The system utilize ***PSD*** photosensitive sensor to locate the sun accurately, then adjust the position of detecting system in time by adjusting mechanism consisted of microcomputer and ***stepping motor***, use absolute photoelectrical ***encoder*** to output the absolute amount of current position, after that system will send it to each optical system with 2 degrees of freedom, ensuring the optical axis will direct at the sun.

When losing the signal of the sun, the detecting system will perform 180 degree scanning from left to right regularly with 60 degree scan width, until the system catch the signal of the sun.

3. Control System for Receiving Maximum Light Intensity

All weather sun tracking auto control system has two modes, one is using PSD device as mentioned above, and the other is astronomical tracking mode. The system can switch between two modes according to the weather condition. Both two modes can perform the accurate sun tracking with protecting function to deal with the strong wind to maintain stability.

4. Mechanical and Auto-protection System

Mechanical system ensures the stability and longevity of the device. This system has the function to prevent from strong wind, heavy rain and snow, hailstones and sandstorm.

5. Computer Communication and Control System

This system is installed with light control system and communication system. We set admin ***console*** in the duty room so that operator on duty can follow the situation in time and control it manually when it is needed. Otherwise, there is additional light control system; the system will control the light bulbs automatically when the intensity of incident light is less than the certain value.

5.6.3 Technical Innovations

Innovation 1: Utilizing light pipe to guide the natural light into the room.

Innovation 2: Having auto sun tracking and light intensity scan and measurement to collect the natural light. Tracking the sun whenever is cloudy or clear.

Innovation 3: Auto control of the incident angle of the light pipe without the influence of the sun position.

Innovation 4: Efficient transmission of natural light reducing the loss of light and implementing the long-range lighting.

Innovation 5: Filtering the IR and UV, reserving the visible light does good to health and avoid overheat in transmission.

Innovation 6: Illumination auto-control system reduces the influence of weather (clear or cloudy) to the indoor lighting.

Innovation 7: Protection function to resist the strong wind and hailstone.

Innovation 8: Communication function enables the mutual transmission of relevant data and auto control of artificial lighting.

This system has several advantages: safe, no existence of power risk; no ***stroboscopic effect*** or dazzle that hurts eyes; uniform illumination leads to great lighting effect; good color rendering and good color temperature; restore the true color of the object; easy maintaining and longevous.

New Words and Expressions

［1］ **artificial lighting**：人工照明。人工照明是为创造夜间建筑物内外不同场所的光照环境，补充白昼因时间、气候、地点不同而造成的采光不足，以满足工作、学习和生活的需求而采取的人为措施。

［2］ **natural light**：自然光，又称天然光。它是指不直接显示偏振现象的光。

［3］ **luminous flux**：光通量。它是指人眼所能感觉到的辐射功率，等于单位时间内某一波段的辐射能量和该波段的相对视见率的乘积。

［4］ **light-pipe**：光导管，光导照明，又叫管道式日光照明设备或者日光照明系统。

［5］ **artificial light source**：人造光源。它是随着人类的文明、科学技术的发展而逐渐制造出来的光源，按其出现的先后顺序，分别有了火把、油灯、蜡烛、电灯（白炽灯、日光灯、高压氙灯）等。

［6］ **fluorescent lamp**：荧光灯、日光灯。它是利用低气压的汞蒸气在放电过程中辐射紫外线，从而使荧光粉发出可见光的原理发光，因此它属于低气压弧光放电光源。

［7］ **ballast**：镇流器。它是日光灯上起限流作用和产生瞬间高压的设备，是在硅钢制作的铁芯上缠漆包线制作而成的，这样的带铁芯的线圈，在瞬间开/关上电时，就会自感产生高压，加在日光灯管两端的电极（灯丝）上。

［8］ **lumens**：流明。它是光通量的单位。发光强度为 1 坎德拉（cd）的点光源，在单位立体角（1 球面度）内发出的光通量为 1 流明。

［9］ **light bulbs**：电灯泡。照明用品，是一种经过通电，利用电阻把幼细丝线（现代通常为钨丝）加热至白炽，用来发光的灯。

［10］ **light guide**：光波导、光导纤维。它是由光透明介质（如石英玻璃）构成的传输光频电磁波的导行结构。

［11］ **parabola**：抛物面。

［12］ **Fresnel lens**：菲涅尔透镜。多是由聚烯烃材料注压而成的薄片，镜片表面一面为光面，另一面刻录了由小到大的同心圆。

［13］ **PSD**：相位灵敏探测器。属于半导体器件，一般做成 P+I+N 结构，具有高灵敏度、高分辨率、响应速度快和配置电路简单等优点，弱点主要是非线性。它的工作原理是基于横向光电效应。

［14］ **stepping motor**：步进电机、步进马达。它是将电脉冲信号转变为角位移或线位移的开环控制元步进电机件。

［15］ **encoder**：编码器、译码器。它是将信号（如比特流）或数据进行编制、转换为可用以通信、传输和存储的信号形式的设备。

[16] **console**：控制台、操纵台。

[17] **stroboscopic effect**：频闪效应。它是指在以一定频率变化的光线照射下，观察到的物体运动呈现出静止或不同于其实际运动状态的现象。

5.7 New Infrared Safety Light Curtain

5.7.1 Overview

Infrared safety ***light curtain*** is a modern newly developed photoelectric safety protection technology. Its main working principle is that several infrared beams in a plane become light wall by parallel or cross. When there is an object of a certain detection size through the alert plane, photosensitive element generates a switching signal to achieve the alarm effect. Infrared safety light curtain has a very wide application in many areas, such as the elevators, security, protection of machining centers and pipeline, etc.

Security level of curtain can be divided into 4, 3, 2, 1, B, of five levels from high to low. The division of the level mainly consults ***security standard*** IEC 61496-1 and IEC 61496-2s proposed by the International Electro Technical Commission. Some products consult standard of safety equipment in Europe or North America. Currently, more use and better performance of products mainly come from foreign manufacturers, such as Germany Leuze, Sic, American Banner, Honeywell, Japanese OMRON, and SUNX. Detection distance of their products is usually up to 15 m, detection height is up to 2.5 m, minimum diameter of object is 14 mm, and the security meets forth level requirements. Due to the late start of domestic technologies, security level of products mainly reach Level 2 or 3, but some such as Shandong Jinan Optoelectronics Technology is also designed to reach 4 light curtain safety standards, and to achieve commercialization, market.

Here we present an infrared safety light curtain sync scan mode, which can eliminate the problem of ***mutual interference*** between ***adjacent channels*** from the fundamental principle. Meanwhile, the design of the new ***cylindrical lens*** will also improve the detection accuracy of existing curtains.

5.7.2 Structure and Principle

Safety light curtains, as the name suggests, is the optical curtain used to ensure the safety. The structure is shown in Fig. 5-24. Each unit of infrared emission side emits a beam of rays, correspondingly, ***infrared receiver*** unit receives optical signal. The intensity of the beam depending on the interval of emission unit and beam size, so the detection accuracy of the safety light curtain consists of two parts: the diameter of the optical lens d_1 and the distance of adjacent lens center d_2. Safety light curtains can accurately detect the smallest object diameter $D = d_1 + d_2$, which is to ensure that at least one beam is completely blocked. Only part of the object which is smaller than the diameter will be obscured. Although the light curtain may detect at a certain time, the detection result is not accurate, which there is a certain probability of random errors. In order to ensure the accuracy of the results and to avoid ***malfunctions***, the actual detection of the object size must be greater than the detection accuracy D.

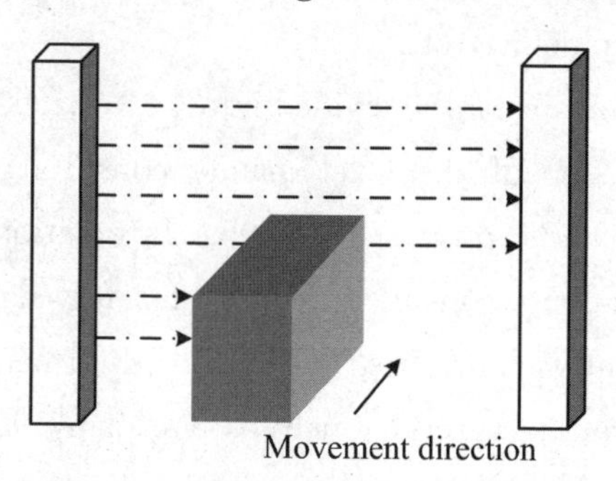

Fig. 5-24 Security light curtain structure

When there is no blocking object between the emitter and the receiver under normal conditions, the beam emitted by the emitting unit is received by the receiving unit completely, and the receiver unit outputs a normal signal. Once the abnormality occurs that there is blocking in detection region, at least one or several beam light will be blocked, part of the receiver unit will not receive the infrared beam, then the receiver outputs the alarm signal immediately.

The whole system of security screens can be divided into three parts: the infrared emitter module, the infrared receiver module, and the safety control module. Emitter module outputs the modulated infrared light signal, and the selection of the modulation frequency is closely related to the response time of the light curtain system, so the actual application needs to be considered. Receiver module receives the modulated optical signal according to the ***synchronization*** scan mode, and the corresponding signal amplification, filtering and ***demodulation***. Security control module determines receiver module output signal, controlling security according to the verdict, and controlling peripheral relay devices to take appropriate action. Due to the complexity of the security control module and functional diversity, selection is based on single chip microcomputer platform to achieve. In addition, for the control module, system protection, the line fault detection, and the controlling interface matching for monitored equipment, such as machine tools, should also be considered.

5.7.3 Circuit Systems Design and Analysis

5.7.3.1 Design Overview

Security light curtain takes synchronous scan mode, that is, from the first emitter-receiver stage, and one by one, there is only one emitter-receiver channel at work in each scanning unit. According to synchronous scan mode, design of a system structure is shown in Fig. 5-25. There is one single chip microcomputer control system in the emitter and the receiver respectively. In order to achieve the emitting and the receiving synchronization, setting a synchronization signal between two single chip microcomputer ensure that they are consistent with the work of the clock. SCM emitter outputs ***clock signal*** CP and ***gating signal*** S_x. When S_x takes a high level into the ***shift unit***, the shift unit starts to control multi-channel switch, and the infrared emission tubes in the emission circuits take one successive work. When the emitter works, the receiver takes the corresponding action because of the synchronization signal SN. The receiver circuit takes S_x' to the shift unit. The shift unit starts to control multi-channel switch, and reads light signals one by one. Signal is input into single chip microcomputer after appropriate shaping and amplified, and using software determines the detection results.

The key to realize the synchronous scan mode exists a synchronization signal

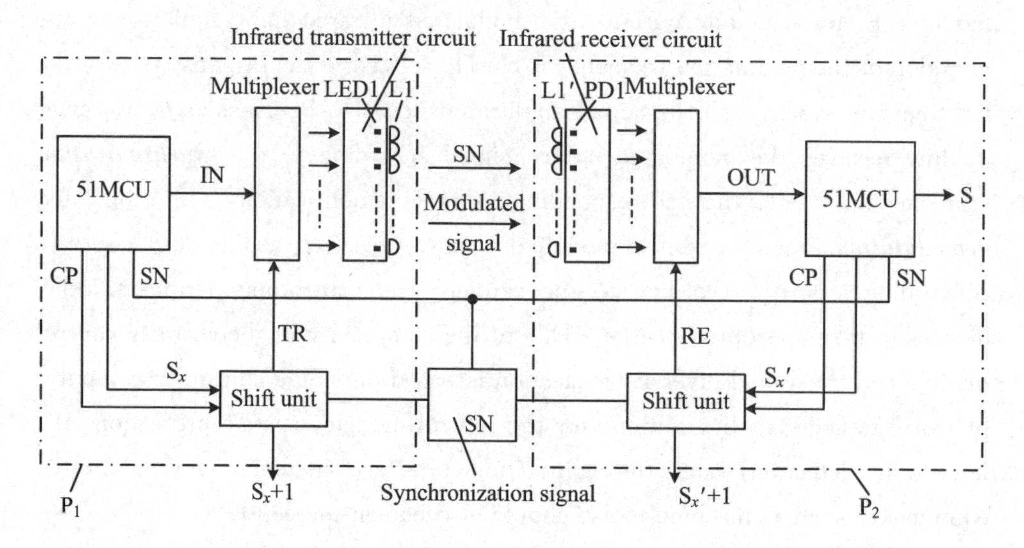

Fig. 5-25　Synchronous scanning system structure

SN between the emitter mode and the receiver mode, that is to say, they will start working at the same time. Time cannot appear deviation, otherwise, it will cause the error that i stage emits and the $i+1$ receives. Before each big scan cycle starts, the first emitter gives a synchronous coding module signal through synchronous communication port to the receiver mode, after the receiver mode receives the synchronization signal. The delay for a fixed period of time and the receiver outputs valid clock signal at the same time with the emitter mode, and then starting scanning. Since each big scan cycle requires a synchronous signal ***calibration***, so no cumulative timing errors appear. Scanning signal waveform is shown in Fig. 5-26, S_x is the gating signal from the emitter, presenting the start of the whole emission scan cycle; S_x' is the gating signal from the receiver, and the high level means the start of the receive scan cycle; Q_1, Q_1' presents the first emission and ***reception level***. High level presents that this stage is in working condition. Because the calibration can be synchronized by a synchronization signal SN between the emitter and the receiver, it can be substantially consistent with the emitting and the receiving clock signals, thus Q_1 and Q_1' can be completely synchronized, and there is no phase error. $Q_1(Q_1')$ signal period is equal to a full scan of the whole big cycle. Referring to $Q_2(Q_2')$ and $Q_3(Q_3')$ waveform, it can be analogized that the $Q_i(Q_i')$ and $Q_1(Q_1)$ waveform is consistent, but the phase shifts $i-1$ units.

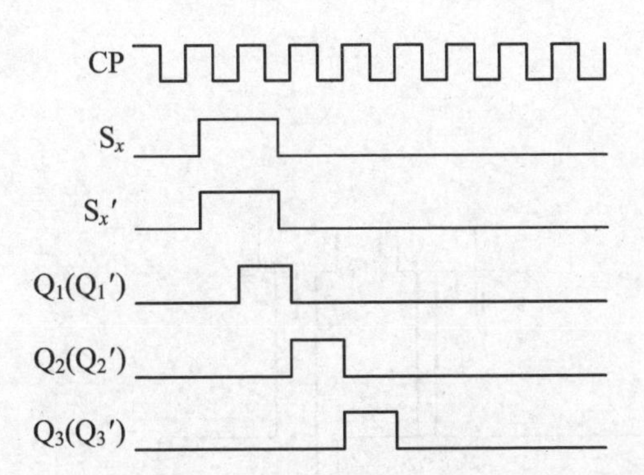

Fig. 5-26 Synchronous scanning system structure

5.7.3.2 Emitter Circuit Design

Fig. 5-27 is an emitter circuit diagram of the signal light which needs to enter five single chip microcomputer signals, and then CD4015BMS controls four SN74HC08 and emitting diodes to achieve the purpose of the emitted light signal. Because light signal needs to be emitted one by one, so using CD4015BMS and SN74HC08 is considered. The combination of these two devices can play a simple and practical role. The circuit in the lower left of the map is to achieve a voltage conversion, the first input is voltage of 24 V, the final output is 5 V regulator, 5 V regulated supply will have a significant role in the receiving circuit.

Fig. 5-28 is a receiver circuit diagram of the signal light. The first is a 5 V regulator supply which requires stability. If instability will lead to ***fluctuations*** in the output waveform, and affect the experimental results, then a photodiode receives light signal, and then the signal is amplified by two ***preamplifiers*** consisted of two transistors and one reverse amplifier consisted of operational amplifier (main amplifier). There will be 1-4 the same circuits because of a total of four light curtains. The signal light will be scattered, so 1-4 road photodiode will receive light signals simultaneously, and it is difficult to determine which way the light is in effect. For this, add an ***analog switch*** in every way which model is HCC4066B, and then chip CD4015BMS controls these four switches on or not. Finally, it needs to add a ***comparator*** (play a role in shaping) and a ***reference potential*** constructed by the transistor. D and C signals will get from the ***single chip microcomputer*** and 5 V regulated supply is provided by the emitter. This will form a relatively complete receiver circuit diagram of the signal light.

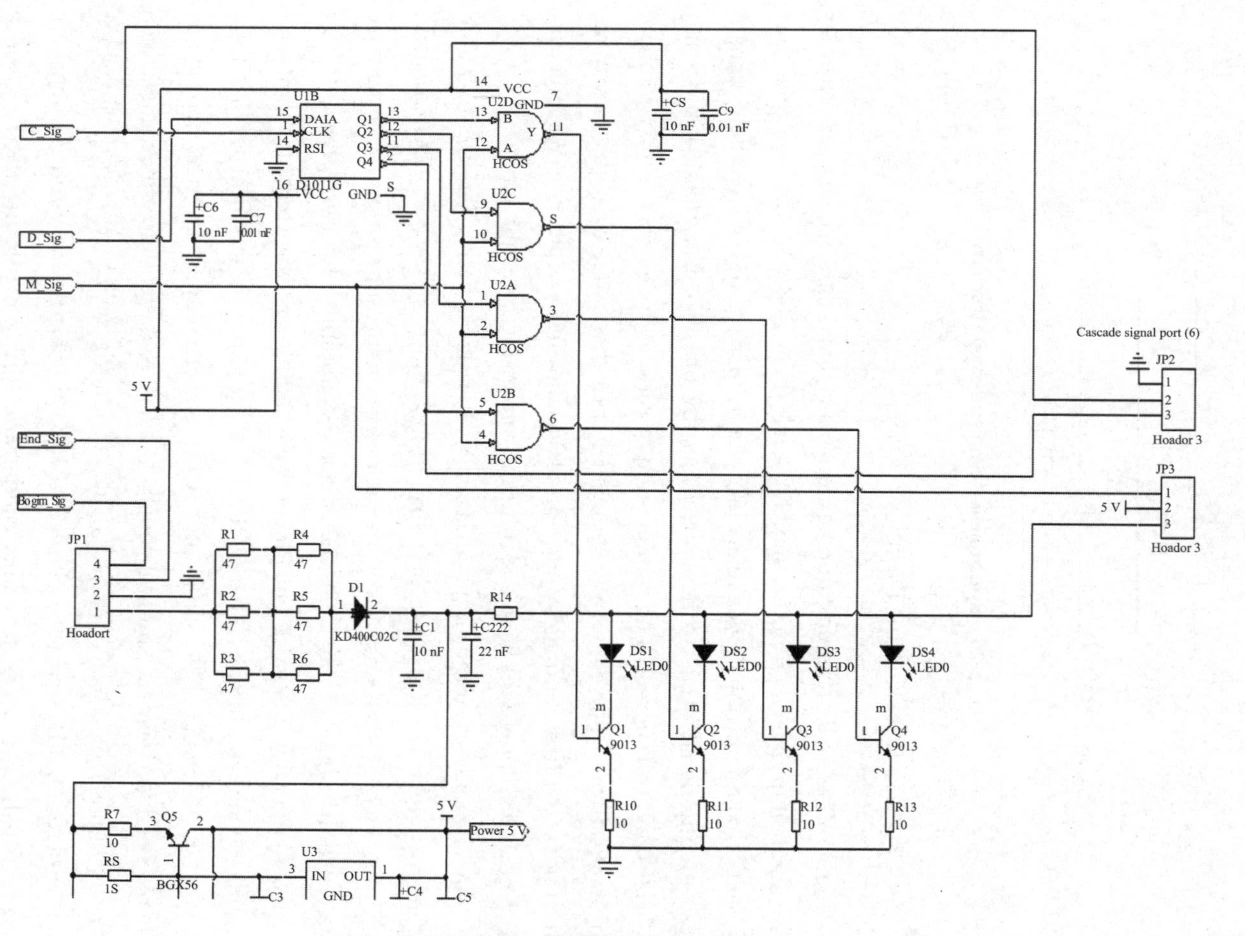

Fig. 5-27　Emitting circuit diagram

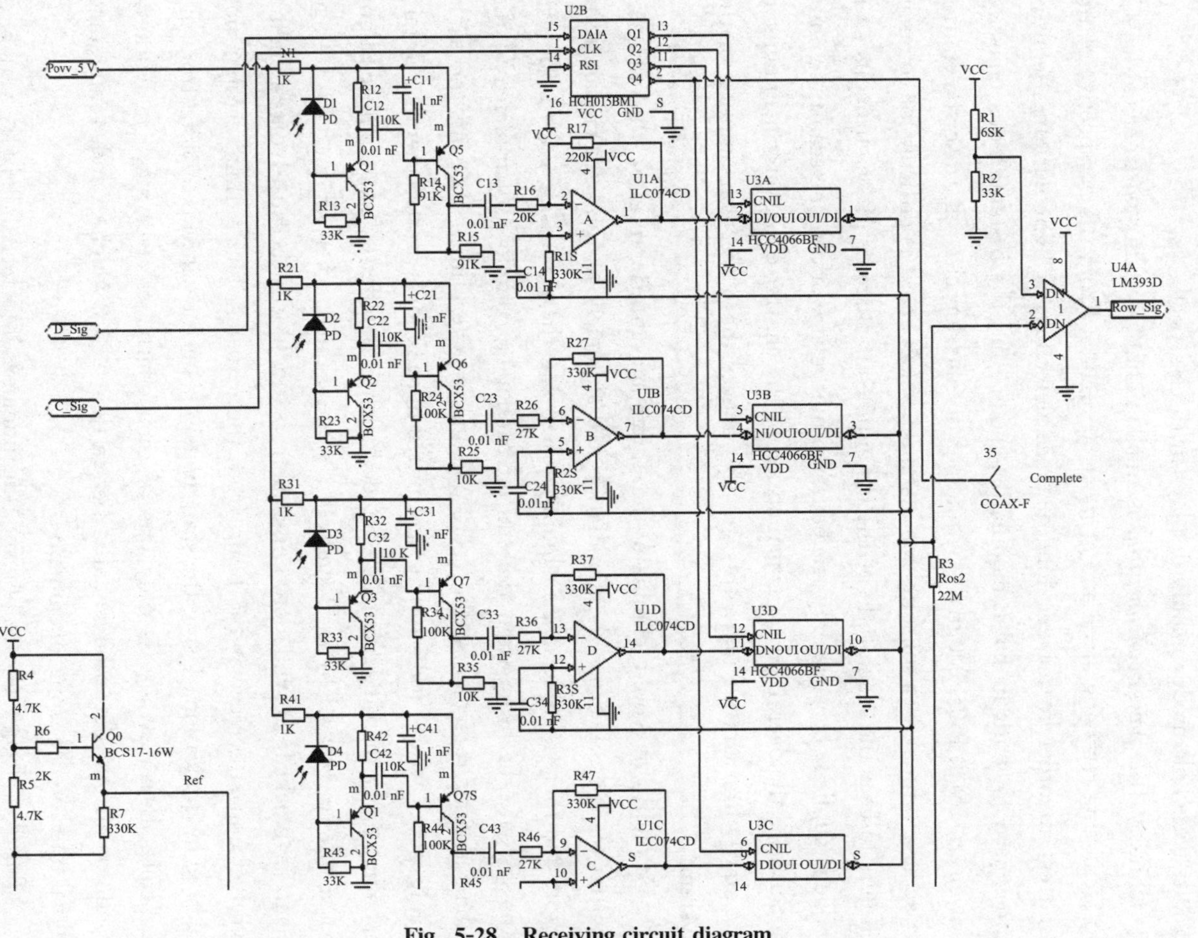

Fig. 5-28 Receiving circuit diagram

5.7.3.3 Analysis of the Advantages and Disadvantages

Synchronous scan mode has very obvious advantages compared to the traditional whole mode, mainly displays in three aspects:

Firstly, the ***power consumption*** is small. A light curtain system has at least 4 to 8 emitting-receiving stage and at most 80 to 100. From the perspective of the emitter, in order to achieve long-distance detection of light curtains, emitter tubes need to work in pulse modulation mode, and the instantaneous emission power can reach 150～200 mW. A number of emission circuits cause the emitter to achieve large power consumption instantaneously and it is difficult for common circuits to bear the magnitude of the instantaneous current. If synchronous scan mode is applied, because only one emission unit is at work in one scan time unit, in this case the overall power consumption of the emitter essentially maintains at the stage of an emitter unit, this greatly reduces the power consumption of the system.

Secondly, it could eliminate the interstage interference of the adjacent emitter-receiver. Because the installation and alignment of the light curtain have errors and there is allowance including design and the processing craft of optical lens system, the beam from the emitter cannot be completely paralleled to the tube and only received by the corresponding receiver tube. If using the same whole mode, we can foresee that the emitted light beam is scattered with a certain opening angle, then it cannot be excluded such a situation: i stage emitted light beam will be received by $i-1$ and $i+1$ stage. Thus, when $i-1$ or $i+1$ stage beam is blocked, it will determine there is no block, because corresponding receiver unit still receives the beam from the i stage. This serious error is not allowed in application. Using the scan mode, this error willed is avoided from the principle. Because there is only one of each emitter-receiver stage in working condition and the detection signal is also in this stage, which excluded interstage signal interference from fundamental.

Thirdly, design is flexible and the number of detected light is variable. Depending on the height of the applications, light curtain detection are uncertain. According to the overall design of system working mode, it is almost impossible to change the number of detection beam. But in the synchronous scan mode, it is very easy to achieve. For example, in a 20-channel light curtain, each 4-channel as a separate section and adjacent sections share or pass some information through the communication port. Structures of the final section and the first section are specific and each middle section is completely consistent, which can interchange with each other or remove some sections. So it can easily get 8, 12, 16, 20 channel four different detection height of the light curtain.

There are some drawbacks for light curtain system to use synchronous scan mode. With the emitter-receiver stage increasing in the number of scan, the response time will be extended for detection of the entire safety light curtain. Generally speaking, don't consider the external ***relay device***, the response time of the light curtain is 2 to 3 times as large scan cycle, and the values is related to the device circuits and software design. The solution is mainly from two aspects: first, improve the frequency of scanning signal and shorten scan cycle; second, choose the appropriate high-speed devices matched with signal frequency.

5.7.4 Optical System Design

The design of optical lens for the security light curtain has different requirements according to the applied occasions. General elevator light curtains usually do not need strict optical lens; they only need to select tubes that have smaller launch angle. For example, half-wave loss angle in $10^\circ \sim 20^\circ$ is OK. Moreover, for the security light curtains which are used for regional protection in industrial site, lens design will directly affect the stability and accuracy of the detection of the light curtain.

At present, the security light curtain in industrial site is generally composed of the infrared emission unit and the infrared receiver unit which are placed face to face. The emission unit is infrared LED array, and the receiver unit is infrared receiving tube array. Every infrared LED and an infrared receiving tube on the other side correspond to each other, and meanwhile, they are mounted on the same straight line. As a result of the existence of lens fixed outside the box, there are gaps between the beams through the lens to adjust the output of the diode emission. It makes security light curtain actually have undetectable ***blind spots***, and these blind spots will influence the detection accuracy of the light curtain. Furthermore, when the distance between the components is far, the received optical energy will be greatly reduced, and this is very easily leading to wrong judgments. Therefore, we must take measures to control the beam divergence angle within the allowable range.

We choose cylindrical lens to replace traditional curtain lens for reducing ***crosstalk*** between the unit levels by this optical design and eliminating the blind spot detection. This method is suitable for long-distance use. Specific approach is as follows (Fig. 5-29): Place a cylindrical lens in front of the launch tube to make a point source spread along the direction of the launch tube array into parallel light. Meanwhile, we placed a cylindrical mirror in front of the receiver tube as well to make the parallel light focus onto the photosensitive surface of the receiving tube. We place cylindrical lens closely to make the beam be together.

The security light curtain uses the way scanning one by one, however, the optical design of cylindrical lens increases, the light emitted by the LED array go through the cylindrical lens array spaced closely, and diffuse into closely contact parallel light longitudinally along the tube array, thereby a seamless curtain covering the entire detection space forms. At the receiving end, according to the intensity of the received light, it can be determined the presence or absence of the ***light shielding*** case, and the size of the shading object through the calculation. It is also useful for measuring the system light curtain.

5.7.5 Software Design

The ***program flowchart*** of this system is showed as Fig. 5-30.

5.7.6 Conclusion

The security light curtain's requirements of security and stability are particularly high because of the particularity of the environment and objects. All designs are carried around to reduce false positives, eliminate the interference, eliminate unstable factors and self-testing. Therefore, the principles and process of this system require careful consideration. Transmitting-receiving synchronous scanning mode can completely eliminate the interference between the adjacent optical signal channels in principle. Moreover, phase error of the clock can be calibrated by synchronous communication signal of software design. This basically can do transmitter and receiver synchronized scanning. According to the

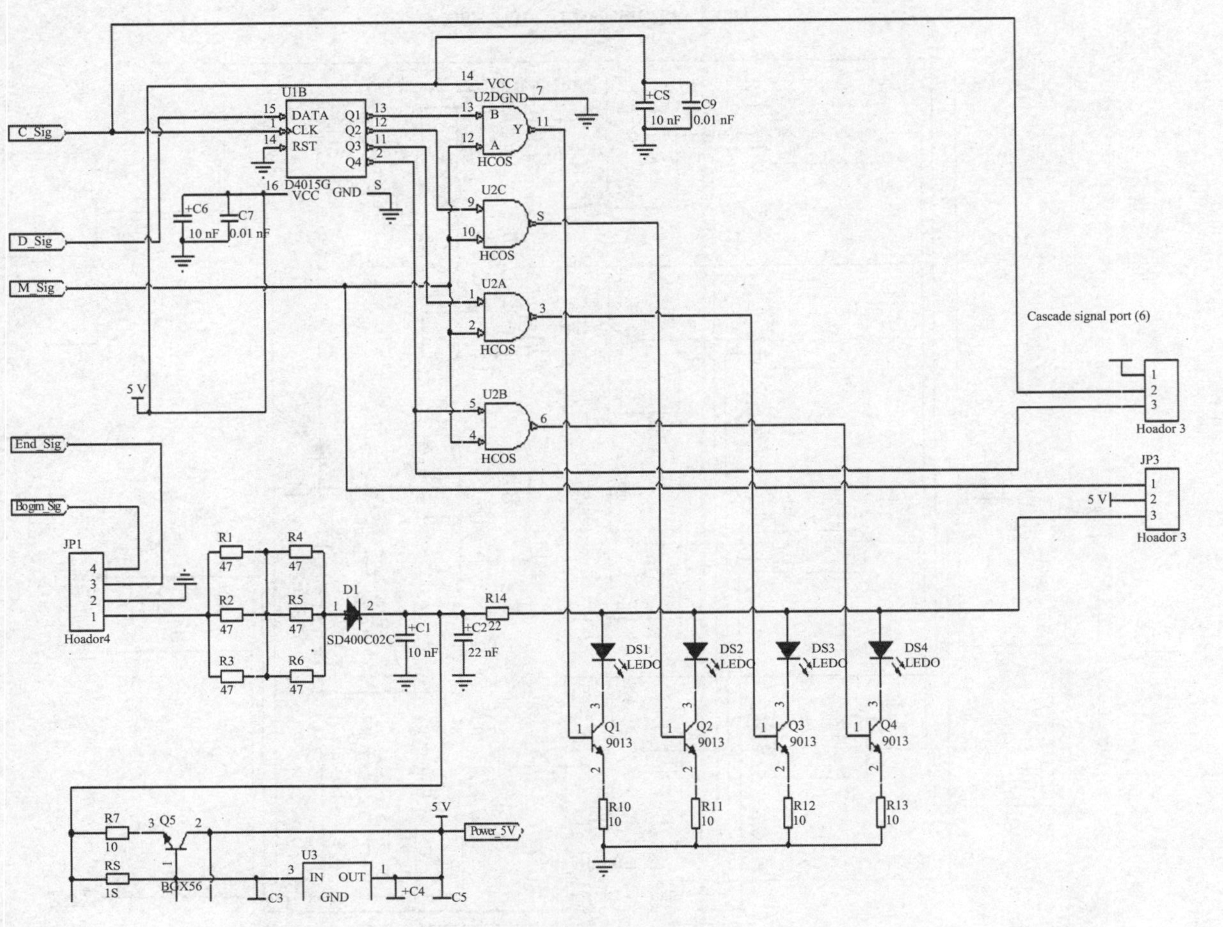

Fig. 5-29 Cylindrical lens system architecture

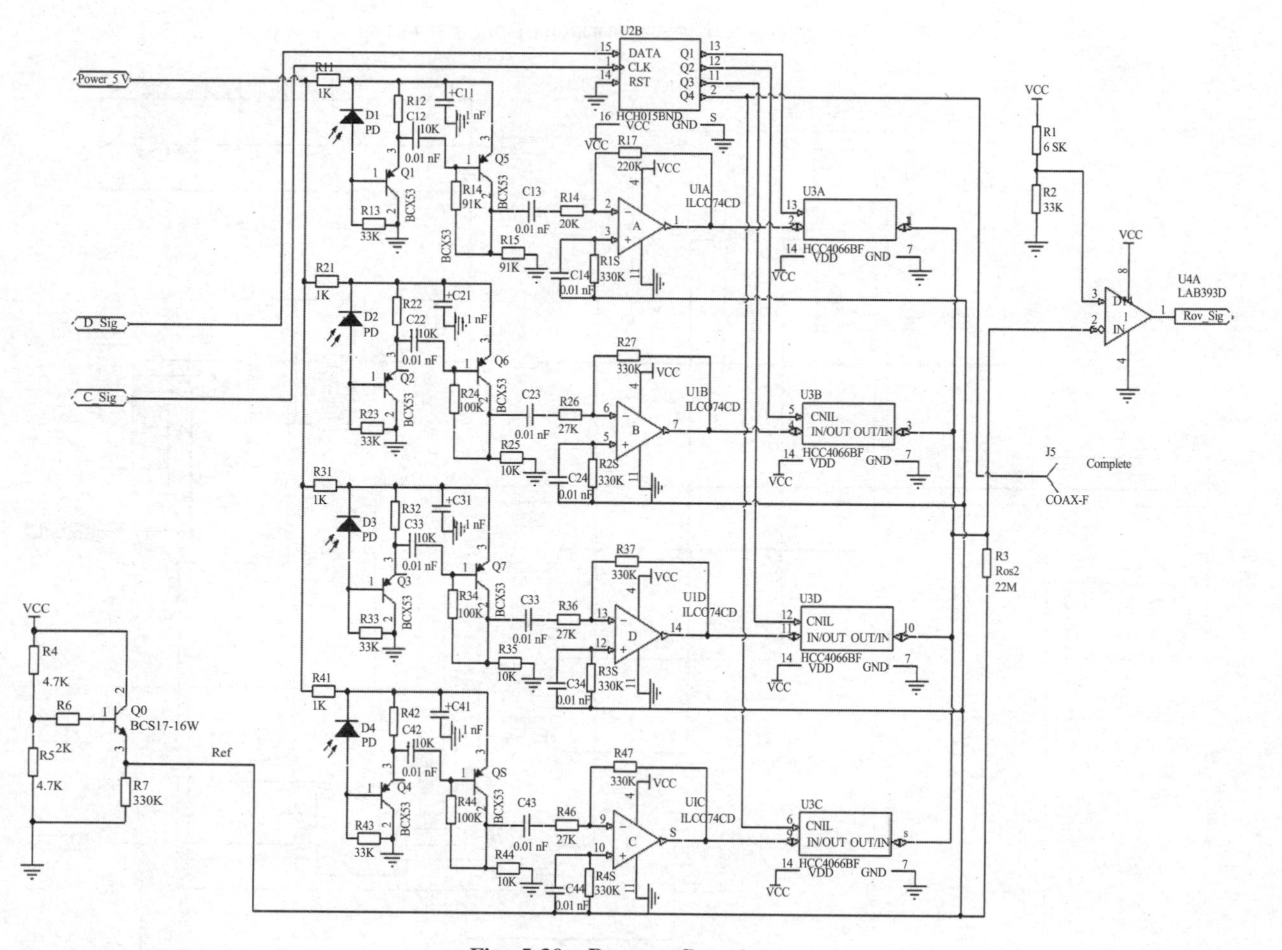

Fig. 5-30　Program flow chart

model system designed after a detailed experimental demonstration and testing, there has already a relatively mature product. The new cylindrical lens system design also made great breakthrough in practical applications, and has applied for national patent.

New Words and Expressions

[1] **light curtain**：光幕。它是利用光电感应原理制成的安全保护装置。此外，光幕也指钱背无文，又称素背、光背、素幕。

[2] **security standard**：安全标准。它是指为保护人体健康、生命和财产安全而制定的标准，是强制性标准，即必须执行的标准。

[3] **mutual interference**：互相干扰、互相干涉。

[4] **adjacent channels**：相邻信道。

[5] **cylindrical lens**：柱面镜。它是非球面透镜，可以有效减小球差和色差。它可分为平凸柱面透镜、平凹柱面透镜、双凸柱面透镜和双凹柱面透镜。它具有一维放大功能。

[6] **infrared receiver**：红外探测器。它是将入射的红外辐射信号转变成电信号输出的器件。

[7] **malfunctions**：失灵。它是指变得不灵敏或完全不起应有的作用。

[8] **synchronization**：同步。它是指两个或两个以上随时间变化的量在变化的过程中保持一定的相对关系。

[9] **demodulation**：检波、反调制、解调制。它是从携带消息的已调信号中恢复消息的过程。

[10] **clock signal**：时钟信号。它是时序逻辑的基础，它用于决定逻辑单元中的状态何时更新。

[11] **gating signal**：选通信号、门信号。

[12] **shift unit**：移位器装置、移位器移相器。它是能够对波的相位进行调整的一种装置。

[13] **calibration**：校准。它是指校对机器、仪器等使其准确。

[14] **reception level**：接收电平。

[15] **fluctuations**：波动、变动、起伏现象。在物理上，振动在空间的传播称为波动。

[16] **preamplifiers**：前置放大器。它是指把音频(AUX、MIC)信号放大至功率放大器所能接受的输入范围。

［17］ **analog switch**：模拟开关。它主要是完成信号链路中的信号切换功能。采用MOS管的开关方式关断或者打开信号链路；由于其功能类似于开关，但用模拟器件的特性实现，因此称为模拟开关。

［18］ **comparator**：比较仪。它是利用相对法进行测量的长度测量工具，主要由测微仪和比较仪座组成。

［19］ **reference potential**：基准电压、基准势、参考电位。它是指传感器置于0 ℃的温场（冰水混合物），在通以工作电流（100 μA）的条件下，传感器上的电压值。

［20］ **single chip microcomputer**：单片微型计算机。它是制作在一块集成电路芯片上的计算机，简称单片机。

［21］ **power consumption**：能量功耗。它是指设备、器件等输入功率和输出功率的差额。

［22］ **relay device**：中继装置、中继器。

［23］ **blind spots**：盲点。医学上，视网膜上无感光细胞的部位称为盲点。

［24］ **crosstalk**：串扰。它是两条信号线之间的耦合、信号线之间的互感和互容引起线上的噪声。

［25］ **light shielding**：光屏蔽。

［26］ **program flowchart**：程序流程图。它是程序分析中最基本、最重要的分析技术，是进行流程程序分析过程中最基本的工具。

References

［1］ 王爱红，王琼华，李大海，等. 三维立体显示技术［J］. 电子器件，2008，31(1)：28.

［2］ 杨永才，何国兴，马军山. 光电信息技术［M］. 上海：东华大学出版社，2009.

Instruction for Experiment

Detecting eye's frequency response by LD and measuring the relation between laser intensity and drive current (see Fig. 5-31).

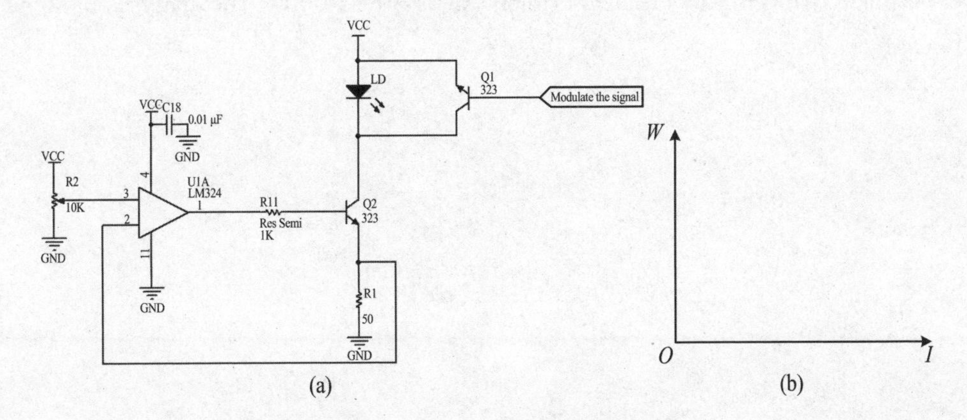

Fig. 5-31 Experimental circuit diagram

(a) Circuit diagram of detecting eye's frequency response. (b) Relationship between laser intensity and drive current.

1. Purpose: To understand the characteristics and the operation circuit of LD, detect eye's frequency response by LD and measure the relation between laser intensity and driver circuit.

2. Elements and device: LD, operational amplifier, transistor, resistance, capacitance, electrical wire, power supply, signal generator, bread board, optical power meter.

3. Steps:

(1) To design and construct the operation circuit of LD and optical detector for detecting the frequency response of your eyes.

(2) To gradually increase the frequency of the input digital signal from 10 Hz, observe the light emission from the LD, until you cannot feel the flash of the laser. Record the frequency of the input digital signal when you stop. This is your eye's frequency response, the value of which is ___________.

(3) To change the circuit's input voltage by adjusting the sliding resistance,

measure and record the voltage of feedback resistor and the corresponding power of the LD. Calculate the current of the driving circuit according to the formula $I=\frac{V}{R}$.

(4) To repeat the step (3) for five times and record the data.

(5) To draw the curve of the laser intensity versus the drive current. Writing down the report of experiment in English, including the principle of experiment, the circuits, the experiment steps, the results, the analyses of some phenomena and etc.